INTERNATIONAL

REVIEW OF CYTOLOGY

SUPPLEMENT 13

Biology of the Rhizobiaceae

ADVISORY EDITORS

INTERNATIONAL

Review of Cytology

EDITED BY

G. H. BOURNE
St. George's University School of Medicine
St. George's, Grenada
West Indies

J. F. DANIELLI
Worcester Polytechnic Institute
Worcester, Massachusetts

ASSISTANT EDITOR

K. W. JEON
Department of Zoology
University of Tennessee
Knoxville, Tennessee

SUPPLEMENT 13

Biology of the Rhizobiaceae

EDITED BY

KENNETH L. GILES
Department of Life Sciences
Worcester Polytechnic Institute
Worcester, Massachusetts

ALAN G. ATHERLY
Department of Genetics
Iowa State University
Ames, Iowa

ACADEMIC PRESS 1981
A Subsidiary of Harcourt Brace Jovanovich, Publishers
New York   London   Toronto   Sydney   San Francisco

ACADEMIC PRESS, INC.
111 Fifth Avenue, New York, New York 10003

United Kingdom Edition published by
ACADEMIC PRESS, INC. (LONDON) LTD.
24/28 Oval Road, London NW1 7DX

LIBRARY OF CONGRESS CATALOG CARD NUMBER: 74–17773

ISBN 0–12–364374–0

PRINTED IN THE UNITED STATES OF AMERICA

81 82 83 84   9 8 7 6 5 4 3 2 1

Contents

LIST OF CONTRIBUTORS xi

PREFACE.. xiii

SUPPLEMENTS IN SERIES xv

The Taxonomy of the Rhizobiaceae

GERALD H. ELKAN

I. Introduction 1

II. The Genus *Agrobacterium* 2

III. The Genus *Rhizobium* 6

IV. Relationship between *Agrobacterium* and *Rhizobium* 11

   References 12

Biology of *Agrobacterium tumefaciens:* Plant Interactions

L. W. MOORE AND D. A. COOKSEY

I. Introduction 15

II. Disease Cycle 16

III. Control of Crown Gall Disease 35

IV. Conclusions 41

   References 42

Agrobacterium tumefaciens in Agriculture and Research

FAWZI EL-FIKI AND KENNETH L. GILES

I. Introduction 47

II. Crown Gall Disease around the World 48

III. Host Range 50

IV. Factors Affecting Development of the Disease 50

V. Virulence and Avirulence of *Agrobacterium tumefaciens* 52

VI. Agricultural Control of *Agrobacterium tumefaciens* 54

VII. Agrocin and Its Mode of Action on *Agrobacterium tumefaciens* 54

VIII. Chemical Treatments 55

IX. Evaluation of Present Methods of Control 56

   References 56

Suppression of, and Recovery from, the Neoplastic State

ROBERT TURGEON

  I. Introduction 59
 II. The Potential of the Tumor Cell 60
III. Crown Gall Teratoma 62
 IV. Recovery from Crown Gall Tumor Disease 71
  V. Suppression and Recovery in Other Neoplastic Diseases 73
      References 78
      Note Added in Proof 81

Plasmid Studies in Crown Gall Tumorigenesis

STEPHEN L. DELLAPORTA AND RICK L. PESANO

  I. Introduction 83
 II. General 84
III. Ti Plasmid-Determined Traits 85
 IV. Genetic Analysis of Ti Plasmid DNA 90
  V. T-DNA 95
      References 101
      Note Added in Proof 104

The Position of *Agrobacterium rhizogenes*

JESSE M. JAYNES AND GARY A. STROBEL

  I. Introduction 105
 II. Biology of the Organism 106
III. Molecular Biology of the Hairy Root Plasmid 111
 IV. Future Prospects 122
      References 124

Recognition in *Rhizobium*–Legume Symbioses

TERRENCE L. GRAHAM

  I. Introduction 127
 II. Recognition in the Soybean–*Rhizobium japonicum* System 129
III. Recognition in the Clover–*Rhizobium trifolii* System 136
 IV. The *Rhizobium* Cell Surface 142
  V. Concluding Remarks 145
      References 145

CONTENTS

The *Rhizobium* Bacteroid State

W. D. SUTTON, C. E. PANKHURST, AND A. S. CRAIG

I.	Introduction	149
II.	Structural Aspects of the Bacteroid State	150
III.	Genetic Aspects of the Bacteroid State	156
IV.	Bacteroid Biochemistry and Physiology	159
	References	171

Exchange of Metabolites and Energy between Legume and *Rhizobium*

JOHN IMSANDE

I.	Introduction	179
II.	Root–Shoot Interactions	180
III.	Basic Reactions of Dinitrogen Fixation	181
IV.	Ammonia Incorporation and Translocation	182
V.	Cost of Dinitrogen Fixation	183
VI.	Mass and Composition of Nodules	185
VII.	Summary and Prospects	187
	References	188

The Genetics of *Rhizobium*

ADAM KONDOROSI AND ANDREW W. B. JOHNSTON

I.	Introduction	191
II.	Mutant Isolation	192
III.	Gene Transfer Systems	196
IV.	Chromosomal Mapping	208
V.	Arrangement of Genes in *Rhizobium*	212
VI.	Genetics of Rhizobiophages	214
VII.	Strain Construction in *Rhizobium*	217
VIII.	Conclusions	218
	References	219

Indigenous Plasmids of *Rhizobium*

J. DÉNARIÉ, P. BOISTARD, FRANCINE CASSE-DELBART,
A. G. ATHERLY, J. O. BERRY, AND P. RUSSELL

I.	Introduction: Early Genetic Evidence for Plasmid Control of Symbiotic Properties (1967–1976)	225

  II. Physical Evidence for the Presence of Large Plasmids 227
 III. Genetic Methods for Plasmid Studies 235
  IV. Plasmid Control of Early Functions in Symbiosis 238
   V. Plasmid Control of Late Functions in Symbiosis 240
  VI. Concluding Remarks.. 243
      References.. 245
      Note Added in Proof.. 246

Nodule Morphogenesis and Differentiation

WILLIAM NEWCOMB

    I. General Introduction 247
   II. Invasion of the Root Hair.. 248
  III. Role of the Root Cortex 256
   IV. Ineffective Root Nodules 284
    V. Cells Associated with Metabolite Transport 286
   VI. *Rhizobium* Nodules on Nonlegumes 289
  VII. General Conclusions 293
       References 295

Mutants of *Rhizobium* That Are Altered in Legume Interaction and Nitrogen Fixation

L. D. KUYKENDALL

    I. Introduction 299
   II. Spontaneous Derivatives... 300
  III. Symbiotic Mutants Obtained by Direct Isolation 301
   IV. Contaminants Mistaken for Mutants 302
    V. Drug-Resistant Mutants 302
   VI. Auxotrophs 304
  VII. Glutamine Synthetase Mutants... 306
 VIII. Polysaccharide-Deficient Mutants 307
   IX. Pleiotrophic Carbohydrate-Negative Mutants 307
    X. Interesting Mutants Retaining Symbiotic Nitrogen-Fixing Ability.......... 307
   XI. Overview 308
       References 308

The Significance and Application of *Rhizobium* in Agriculture

HAROLD L. PETERSON AND THOMAS E. LOYNACHAN

    I. Introduction 311

CONTENTS

II. Application of *Rhizobium* in Agriculture 312
III. Significance of Inoculation in Agriculture 324
IV. Factors Limiting Inoculation in Agriculture 327
References 328

INDEX 333

List of Contributors

Numbers in parentheses indicate the pages on which the authors' contributions begin.

A. G. ATHERLY (225), *Department of Genetics, Iowa State University, Ames, Iowa 50011*

J. O. BERRY (225), *Department of Genetics, Iowa State University, Ames, Iowa 50011*

P. BOISTARD (225), *Laboratoire de Biologie Moléculaire des Relations Plantes-Microorganismes, I.N.R.A., Castanet-Tolosan B. P. 12, France*

FRANCINE CASSE-DELBART (225), *Laboratoire de Biologie Cellulaire, I.N.R.A., 78000 Versailles, France*

D. A. COOKSEY (15), *Department of Botany and Plant Pathology, Oregon State University, Corvallis, Oregon 97331*

A. S. CRAIG (149), *Plant Physiology and Applied Biochemistry Divisions, D.S.I.R., Palmerston North, New Zealand*

STEPHEN L. DELLAPORTA (83), *Department of Life Sciences, Worcester Polytechnic Institute, Worcester, Massachusetts 01609*

J. DÉNARIÉ (225), *Laboratoire de Biologie Moléculaire des Relations Plantes-Microorganismes, I.N.R.A., Castanet-Tolosan B. P. 12, France*

FAWZI EL-FIKI (47), *Department of Life Sciences, Worcester Polytechnic Institute, Worcester, Massachusetts 01609*

GERALD H. ELKAN (1), *Department of Microbiology, North Carolina State University, Raleigh, North Carolina 27650*

KENNETH L. GILES (47), *Department of Life Sciences, Worcester Polytechnic Institute, Worcester, Massachusetts 01609*

TERRENCE L. GRAHAM (127), *Monsanto Agricultural Products Company, St. Louis, Missouri 63166*

JOHN IMSANDE (179), *Department of Genetics, Iowa State University, Ames, Iowa 50011*

JESSE M. JAYNES (105), *Department of Plant Pathology, Montana State University, Bozeman, Montana 59715*

ANDREW W. B. JOHNSTON (191), *John Innes Institute, Colney Lane, Norwich NR4 7UH, England*

ADAM KONDOROSI (191), *Institute of Genetics, Biological Research Center, Hungarian Academy of Sciences, P. O. Box 521, H-6701 Szeged, Hungary*

L. D. KUYKENDALL (299), *USDA, ARS Cell Culture and Nitrogen Fixation Laboratory, Beltsville, Maryland 20705*

THOMAS E. LOYNACHAN (311), *Department of Agronomy, Iowa State University, Ames, Iowa 50011*

L. W. MOORE (15), *Department of Botany and Plant Pathology, Oregon State University, Corvallis, Oregon 97331*

WILLIAM NEWCOMB (247), *Biology Department, Queen's University, Kingston, Ontario, Canada K7L 3N6*

C. E. PANKHURST (149), *Plant Physiology and Applied Biochemistry Divisions, D.S.I.R., Palmerston North, New Zealand*

RICK L. PESANO (83), *Department of Life Sciences, Worcester Polytechnic Institute, Worcester, Massachusetts 01609*

HAROLD L. PETERSON (311), *Department of Agronomy—Soils, Mississippi State University, Mississippi State, Mississippi 39762*

P. RUSSELL (225), *Department of Genetics, Iowa State University, Ames, Iowa 50011*

GARY A. STROBEL (105), *Department of Plant Pathology, Montana State University, Bozeman, Montana 59715*

W. D. SUTTON (149), *Plant Physiology and Applied Biochemistry Divisions, D.S.I.R., Palmerston North, New Zealand*

ROBERT TURGEON (59), *Biology Department, University of Dayton, Dayton, Ohio 45469*

Preface

The demands of an ever-increasing world population coupled with tightening energy supplies and increasing costs are forcing a global reappraisal of agricultural practice and philosophy. The use of chemically fixed nitrogen has been responsible for the major increases in crop productivity over the last fifty years; however, recently even industrialized countries have had to limit fertilizer use (and possibly ultimately eliminate their use in third world countries). In the last ten years applied fertilizer has almost eclipsed the biological fixation of nitrogen as an important agricultural concept. Only in the last decade has the potential importance of this phenomenon been appreciated. Coupled with this, the environmental hazards of herbicides and pesticides have revitalized interest in the biological control of crop diseases. The Rhizobiaceae embody both of these concepts since they include the crown gall disease bacteria and also the beneficial bacteria, *Rhizobium* spp., which symbiotically fix atmospheric nitrogen in conjunction with legumes. No general review of this area has appeared since the recent surge of interest has occurred, making this a timely contribution to the literature.

The first chapter in this volume discusses the taxonomy and identification of the Rhizobiaceae and suggests some new concepts that may have far reaching implications. The following five chapters contain a review on *Agrobacterium* and stress the concentration of research in the molecular biology of this important group of bacteria.

Subsequent chapters are on the genetics, molecular biology, agricultural, and morphological aspects of the rhizobia. Thus the reader can compare and contrast the recent findings in diverse areas of rhizobia research as well as compare the *Agrobacterium* with the *Rhizobia*. This book should be of specific interest to a wide range of readers, students, specialists, and individuals entering this field of research. It brings together modern thinking on this complicated and diverse bacterial family. We feel certain that it will be of special interest to investigators in agronomy, genetics, molecular biology, botany, and microbiology, but horticulturists and biochemists should also find certain aspects particularly relevant.

Naturally, the precise choice of topics and chapters strongly reflects our own areas of interest, but every attempt has been made to give a broad picture of this most interesting bacterial family. There is some overlap between certain chapters, but this was felt to be preferable to leaving gaps in explanations or discussions. Although some editing has been necessary, we have tried to retain the original sense and "flavor" of the individual authors.

We thank all our contributors for the painstaking preparation and attention to detail displayed in their individual chapters. For those who managed to meet deadlines our special thanks, and to those who took up the challenge at short notice we are greatly indebted. We have both learned much during the preparation of this volume, and we hope that the same will be true of its readers.

KENNETH L. GILES
ALAN G. ATHERLY

INTERNATIONAL REVIEW OF CYTOLOGY

SUPPLEMENT 1

Z. Hruban and M. Recheigl, Jr., Microbodies and Related Particles: Morphology, Biochemistry, and Physiology, 1969

SUPPLEMENT 2

Peter Luykx, Cellular Mechanisms of Chromosome Distribution, 1970

SUPPLEMENT 3

Andrew S. Bajer and J. Molè-Bajer, Spindle Dynamics and Chromosome Movements, 1972

SUPPLEMENT 4

G. H. Bourne, J. F. Danielli, and K. W. Jeon, eds., Aspects of Nuclear Structure and Function, 1974

SUPPLEMENT 5

G. H. Bourne, J. F. Danielli, and K. W. Jeon, eds., Aspects of Cell Control Mechanisms, 1977

SUPPLEMENT 6

G. H. Bourne, J. F. Danielli, and K. W. Jeon, eds., Studies in Ultrastructure, 1977

SUPPLEMENT 7

G. H. Bourne, J. F. Danielli, and K. W. Jeon, eds., Neuronal Cells and Hormones, 1978

SUPPLEMENT 8

G. H. Bourne, J. F. Danielli, and K. W. Jeon, eds., Aspects of Genetic Action and Evolution, 1978

SUPPLEMENT 9

J. F. Danielli and M. A. DiBerardino, eds., Nuclear Transplantation, 1979

SUPPLEMENT 10

Warren W. Nichols and Donald G. Murphy, eds., Differentiated Cells in Aging Research, 1979

SUPPLEMENT 11A

Indra K. Vasil, ed., Perspectives in Plant Cell and Tissue Culture, 1980

SUPPLEMENT 11B

Indra K. Vasil, ed., Perspectives in Plant Cell and Tissue Culture, 1980

SUPPLEMENT 12

A. L. Muggleton-Harris, ed., Membrane Research: Classic Origins and Current Concepts, 1981

INTERNATIONAL

REVIEW OF CYTOLOGY

SUPPLEMENT 13

Biology of the Rhizobiaceae

INTERNATIONAL REVIEW OF CYTOLOGY, SUPPLEMENT 13

The Taxonomy of the Rhizobiaceae

GERALD H. ELKAN

Department of Microbiology, North Carolina State University, Raleigh, North Carolina

I.	Introduction	1
II.	The Genus *Agrobacterium*	2
III.	The Genus *Rhizobium*	6
IV.	Relationship between *Agrobacterium* and *Rhizobium*	11
	References	12

I. Introduction

Beneke (1912) described a bacterial species as merely a collection of more or less similar clones that the taxonomist ties into a bundle which he chooses to call a species. How small or large this bundle is has depended upon the scientific insight of the individual taxonomist. The same principle has been applied to larger bundles that are called genera and families. The present state of bacterial taxonomy still largely depends upon this approach, even though we are now beginning to learn something of the phylogeny of bacteria (Fox *et al.*, 1980; DeLey, 1974).

Bergey's Manual of Determinative Bacteriology, eighth edition (Buchanan and Gibbons, 1974), describes the family Rhizobiaceae Conn 1938 as consisting of normally rod-shaped cells without endospores. The cells are aerobic, gram-negative, and motile having either one polar or subpolar flagellum or two to six peritrichous flagella. Many carbohydrates are utilized, with a usual production of considerable extracellular slime. There are two genera: *Agrobacterium,* all species of which with the exception of *A. radiobacter* incite cortical hypertrophies on plants, and *Rhizobium,* the species of which form nodules on the roots of Leguminosae.

The objective of this article is to summarize the current taxonomic status of the two genera within the family Rhizobiaceae and to examine their taxonomic interrelationships. Because the earlier works have been extensively reviewed elsewhere, and in view of the development of modern taxonomic techniques over the past two decades, most of the information discussed herein has appeared within the past 20 years. Additionally, because of the diversity within these genera, major emphasis will be placed upon studies involving a larger sampling of bacterial isolates.

1

II. The Genus *Agrobacterium*

The eighth edition of *Bergey's Manual* lists four species within the genus *Agrobacterium*. The key to these species appears in Table I, and the differentiating characteristics are listed in Table II. Since members of the genus *Agrobacterium* are soil inhabitants, they are a diverse group often hard to identify and classify. As an approach toward improving the taxonomy of the genus, DeLey *et al.* (1966) examined 45 diverse strains including representatives of all nomenspecies for the following features: DNA base composition, type of flagellation, 3-ketoglycoside formation, and phytopathogenicity for tomato and *Datura*. It was determined that, based upon the percent guanine plus cytosine (GC) of the bacterial DNAs, all strains of *Agrobacterium radiobacter, A. tumefaciens,* and *A. rhizogenes* had a very narrow percent GC range of 59.5–62.8. In conjunction with other morphological and physiological properties, this suggests a close genetic relationship.

All strains examined in these three species were peritrichously flagellated and frequently bore fimbriae-like structures which were longer, thinner, and more rigid than the flagella. All strains of *A. tumefaciens* were indeed pathogenic while, with only one exception, none of the *A. rhizogenes* or *A. radiobacter* strains were tumorigenic. The formation of 3-ketoglycosides appeared to be specific for *A. tumefaciens* and *A. radiobacter*.

According to *Bergey's Manual,* the difference between *Agrobacterium rubi* and *A. tumefaciens* is based on only a few minor features and one major one. That is, while *A. tumefaciens* causes crown gall in a wide number of hosts, *A. rubi* is tumorigenic only in *Rubrus* spp. (raspberries). Since *A. tumefaciens* has also been shown to cause crown gall in *Rubrus,* because the percent GC in these

TABLE I

KEY TO THE SPECIES OF THE GENUS *Agrobacterium*
ACCORDING TO BERGEY'S MANUAL, EIGHTH EDITION

A. Amino acids, nitrates, and ammonium salts used as sole source of nitrogen. 3-Ketolactose produced.
  1. Produce galls.

                                                 1. *A. tumefaciens*
  2. Do not produce galls.

                                                 2. *A. radiobacter*
B. Do not utilize amino acids, nitrates, and ammonium salts as sole source of nitrogen (require additional growth factors). 3-Ketolactose not produced.
  1. Produce hairy root of nursery stock.

                                               3. *A. rhizogenes*
  2. Produce galls on raspberries.

                                               4. *A. rubi*

TABLE II

DIFFERENTIAL CHARACTERISTICS OF THE SPECIES OF GENUS *Agrobacterium*
ACCORDING TO BERGEY'S MANUAL, EIGHTH EDITION

Characteristic	*A. tumefaciens*[a] and *A. radiobacter*	*A. rhizogenes* and *A. rubi*
Asparagine used as sole source of carbon and nitrogen	+	−
Nitrites produced from nitrates	+	−
Ammonium salts, nitrates, or amino acids utilized as sole nitrogen source	+	−
Growth factors and amino acids required	−	+
Calcium, or sodium, glycerophosphate mannitol nitrate agar		
Growth	+	−/S[b]
Halo or browning	+	−
Starr's and Lippincott's basal salts media, growth	+	−
Congo red and aniline blue mannitol agars		
Growth	+	M
Dye absorbed	P	F
Sodium selenite–yeast water–glucose agar		
Growth	+	S[c]
Selenite reduced	+	−
Litmus milk		
Reaction	N/Alk	A
Serum zone formation	+	F
3-Ketolactose production	+	−

[a] Since this table was prepared, it has been shown that there are two biotypes of these species, but the description given here applied only to biotype I (Keane *et al.*, 1970).

[b] S, Scant growth; M, moderate growth; F, faint; A, acid; Alk, alkaline; N, neutral; P, pronounced.

[c] Some strains of *A. rubi* cannot grow at all on this medium.

two species is indistinguishable, and since DNA transformations can readily be made between these two species, DeLey *et al.* (1966) proposed that they be combined. In a later paper, on the further basis of DNA base composition and hybridization, DeLey (1968) recommended that three species previously included in Bergey's classification be eliminated. Similar conclusions were made by a number of other workers (Heberlein *et al.*, 1967; Moffett and Colwell, 1968; Lippincot and Lippincot, 1969; Skyring *et al.*, 1971), and this recommendation was accepted; thus in the eighth edition *Agrobacterium gypsophilae*, *A. pseudotsugae,* and *A. stellulatum* are listed as species *incertae sedis*.

Keane *et al.* (1970) examined 50 *Agrobacterium* isolates, representing all species, using a series of biochemical tests, protein patterns following gel electrophoresis, serology, and pathogenicity. Initially, 90 different biochemical tests were screened, and 28 most likely to prove differential were selected for use. Biochemically, all isolates fell into two distinct biotypes. This was largely con-

firmed by serological reactions and by protein patterns following disk electrophoresis. Biotype 1 contains tumor-inducing, root-proliferating, and nonpathogenic forms. Biotype 2 contains both tumor-inducing and root-proliferating
forms. Results of the biochemical tests are summarized in Table III. These
workers concluded that division of the genus *Agrobacterium* into species according to pathogenicity was not satisfactory. Their study showed that biochemically,
serologically, and electrophoretically some isolates within the same species were
quite distinct, whereas it was not possible to distinguish among species in a
consistent manner. Kersters *et al.* (1973) conducted a numerical taxonomic

TABLE III
USEFULNESS OF BIOCHEMICAL TESTS
FOR THE DIFFERENTIATION OF *Agrobacterium*[a]

Biochemical tests used	Biotype 1	Biotype 2
Tests too variable for taxonomic use		
Acid from lactose	Variable	Variable
Acid from sorbose	Variable	Variable
Acid from dulcitol	Variable	Variable
Acid from glycerol	Variable	Variable
Test for oxidation–reduction	Variable	Variable
Reduction of nitrate to nitrite	Variable	Variable
Absorption of aniline blue	Variable	Variable
Results common to both biotypes		
Glucose oxidation	+	+
Catalase production	+	+
Gelatin hydrolysis	−	−
Lipolysis	−	−
Starch hydrolysis	−	−
Glucose hydrolysis	Acid	Acid
Rhamnose hydrolysis	Acid	Acid
Sucrose hydrolysis	Acid	Acid
Mannitol hydrolysis	Acid	Acid
Solicin hydrolysis	Acid	Acid
Tests differential for identification of biotypes		
Erythritol utilization	−	+
Growth factor requirements	+ or −	+
Citrate utilization	−	+
Malonate utilization	−	+
Ferric ammonium citrate utilization	+	−
Glycerophosphate utilization	+	−
Selenite	+	−
Oxidase	+	−
3-Ketolactose	+	−
Litmus milk	Alkaline	Acid

[a] Keane *et al.* (1970).

analysis of *Agrobacterium* isolates based on 92 biochemical, physiological, morphological, and mutational characters. They reported two main clusters of related organisms, which corresponded well with the two biotypes reported by Keane *et al.* (1970). In a less extensive study, White (1972) reported essentially similar results. Panagopoulos and Psallidas (1973) isolated *A. tumefaciens* from grapevine tumors in Russia and Greece and found seven heterogeneous isolates having a limited host range. These organisms are unique in that they do not fit either biotype 1 or 2. Therefore it is suggested that they be assigned to a new biotype 3. Whether this recommendation reaches acceptance depends upon further studies.

As early as 1969, Kerr (1969) showed that *A. radiobacter* could be converted to *A. tumefaciens* through genetic transfer of virulence. Pichinoty *et al.* (1977), using 200 determinative characters, also could not differentiate between *A. radiobacter* and *A. tumefaciens*. In an extensive study on host specificity in the genus *Agrobacterium*, Anderson and Moore (1979) examined 176 agrobacteria isolated from 26 plant species in 11 diverse families. Based upon pathogenicity, *A. rubi* strains were indistinguishable from *A. tumefaciens*, and five of eight *A. rhizogenes* strains formed tumors in some plants. They also concluded that speciation based on pathogenicity and host specificity was of little taxonomic use. In their numerical taxonomic study, Kersters *et al.* (1973) found representatives of all species in all groups and both major clusters. They concluded that separation of the genus *Agrobacterium* into species based on their state of phytopathogenicity was biologically unjustified and did not reflect the natural relationships among these bacteria.

Zaenen *et al.* (1974) found that tumor-inducing strains of *Agrobacterium* contained at least one large plasmid designated the tumor-inducing (Ti) plasmid. Curing of this plasmid leaves the bacterium avirulent. Ti plasmids fall into three groups based on DNA homology and restriction endonuclease analysis (Drummond and Chilton, 1978; Currier and Nester, 1976; Engler *et al.*, 1977; Van Larebeke *et al.*, 1974; Watson *et al.*, 1975). Recently, it has been suggested that the Ti plasmid has a major role in determining the host range (Thomashow *et al.*, 1980). This plasmid is quite mobile and can be easily transferred between isolates spontaneously or through laboratory manipulation. Kerr's (1969) observation of the transfer of virulence has now been well documented as being due to the Ti plasmid. Work with this plasmid then has shown rather conclusively the taxonomic problems resulting from the use of pathogenicity as the main determinant for species designation within this genus.

The conclusions based upon the body of literature point to a need for reclassifying this genus into two (perhaps three) species based mainly upon biochemical, physiological, serological, electrophoretic, DNA homology, and nutritional factors. Pathogenicity, being an unstable property, should not remain a major taxonomic determinant.

III. The Genus *Rhizobium*

As described in *Bergey's Manual*, eighth edition, the genus *Rhizobium* consists of two groups of species plus a miscellaneous grouping. The species and their major characteristics are summarized in Table IV. The current classification recognizes six species based upon host infectivity coupled with certain biochemical tests. This classification is historically based on the work of Baldwin and Fred (1927, 1929), Eckardt *et al.* (1931), and Fred *et al.* (1932). These workers proposed 16 cross-inoculation groups. Six of them were elevated to species level as currently listed in *Bergey's Manual*. The assumption that each species of *Rhizobium* nodulates only plants within a specified cross-inoculation group, and that within such a group all rhizobia from a host can inoculate all other numbers of the group, has lost credibility. In fact, D. C. Jordan and O. N. Allen, writing in *Bergey's Manual* (1974), state that the taxonomic position of *Rhizobium* is controversial and that the current classification can be regarded as only tentative. They conclude that a classification of *Rhizobium* based on the cross-inoculation (or plant affinity) concept, because of widespread anomalous cross-infections, is not very satisfactory. Graham (1976) summarizes the major limitations of the cross-inoculation concept of classification as follows:

TABLE IV

SPECIES OF *Rhizobium* AND THEIR CHARACTERISTICS AFTER *Bergey's Manual*, EIGHTH EDITION

Species	Serum zone in litmus milk	Acid reaction in litmus milk	Growth rate	Host nodulated	Common name of group
Group I: Two to six peritrichous flagella; GC content of the DNA 59.1–63.1 mole% (T_m)					
Rhizobium trifolii	+	−	Fast[a]	*Trifolium*	Clover
Rhizobium leguminosarum	+	−	Fast	*Pisum, Lens, Lathyrus, Vicia*	Pea
Rhizobium phaseoli	+	−	Fast	*Phaseolus vulgaris*	French bean
Rhizobium meliloti	+	−	Fast	*Melilotus, Medicago, Trigonella*	
Group II: Polar or subpolar flagellum; GC content of the DNA 61.6–65.5 mole% (T_m)					
Rhizobium japonicum	−	−	Slow[b]	*Glycine max*	Soybean
Rhizobium lupini	−	−	Slow	*Lupinus, Ornithopus*	Lupin
Rhizobium spp.	Variable	Variable	Variable	*Vigna, Desmodium, Arachis, Centrosema, Stylosanthes,* etc.	Cowpeal miscellany

[a] Colonies circular, convex, semitranslucent, raised, and mucilagenous; usually 2–4 mm in diameter on yeast extract–mannitol–mineral salts agar (YEM) within 3–5 days (turbidity in YEM broth in 2–3 days).

[b] Colonies circular, punctiform, opaque, rarely translucent, white, convex, and granular; do not exceed 1 mm in diameter within 5–7 days on YEM agar (turbidity in YEM broth in 3–5 days or longer).

1. *Cross-infection*. Cross-infection is the nodulation of plants from one affinity group by rhizobia from another. Each of the species has now been shown to be cross-infective to some degree.

2. *Insufficient nodulation data*. It is stated in *Bergey's Manual* (1974) that, of the 14,000 or so known species of legumes, only 8–9% have been examined for nodules and only 0.3–0.4% have been studied with respect to their symbiotic relationships with nodule bacteria.

3. *Scarcity of biochemical data*. Many of the biochemical studies involve only a few strains, and so it is difficult to generalize as to taxonomic meanings of biochemical differences. One taxonomic difference that does appear real is the designation of two groups of *Rhizobium* based upon growth rate.

The genus *Rhizobium* has traditionally been divided into two groups, as suggested by Lohnis and Hansen (1921). According to Allen and Allen (1950) the term "fast growers" commonly designates the rhizobia associated with alfalfa, clover, bean, and pea because in culture these grow much faster (less than one-half the doubling time) than the "slow growers," exemplified by soybean, cowpea, and lupine rhizobia. As a result of an extensive study on acid production by *Rhizobium* involving 717 strains, Norris (1965) hypothesized that the ancient form of the symbiont is represented by the slow growers which produce alkali and are commonly associated with legumes of tropical origin. Disputing this hypothesis, Graham (1964a) contended that the differences between slow- and fast-growing *Rhizobium* was too great to be based solely on evolutionary differentiation of root nodule bacteria from an organism similar to the present-day slow-growing type. Carbohydrate nutrition differences between fast and slow growers have been shown by many workers. In their review, Allen and Allen (1950) summarized that the slow-growing rhizobia were more specific in their carbohydrate requirements in every respect. The great difference in carbohydrate utilization between fast- and slow-growing root nodule bacteria was confirmed by Graham (1964), who concluded that carbohydrate utilization tests were clearly valid criteria for the subdivision of *Rhizobium*. For example, *Rhizobium trifolii* and *R. leguminosarum* can utilize 20 different carbohydrates and tricarboxylic acid cycle intermediates for growth, whereas *R. japonicum* can utilize only 8 of the 20.

The relative fastidiousness of the slow growers has been substantiated by more recent studies (Elkan and Kwik, 1968; Graham, 1964a,b). Specific enzymes may differ in the two groups; for example, Martinez-de-Drets and Arias (1972) propose an enzymic basis for differentiation between these groups based on the presence of an NADP-6-phosphogluconate dehydrogenase found in fast growers but absent in slow growers. While the major pathways seem to be similar, there is evidence that the preferred pathways may be different. However, most of the metabolic studies were done with only one or very few *Rhizobium* strains (Elkan and Kuykendall, 1981).

Based on the great similarities between the species of fast-growing rhizobia and equally great differences between them and the slow growers (such as *R. japonicum*), Graham (1964a) proposed the consolidation of *R. trifolii, R. leguminosarum,* and *R. phaseoli* into a single species and creation of a new genus, *Phytomyxa,* to contain strains of slow-growing *Rhizobium* (type species: *Phytomyxa japonicum*).

Vincent (1977) surveyed the literature and concluded that fast or slow growth on yeast mannitol agar, where the fast growers lower the pH of the medium and the slow growers produce an alkaline end point, is a major acceptable distinction between species of rhizobia. This differentiation has also been utilized in *Bergey's Manual,* eighth edition (Table IV), but Vincent includes in the fast growers some members of the cowpea miscellany. The slow growers include two reasonably well-defined subgroups, *Rhizobium lupini* and *R. japonicum,* plus a large number of relatively poorly defined members of the cowpea miscellany.

As the cross-inoculation approach became more unsatisfactory, alternative taxonomic methods were applied to these organisms. These involved nutrition, Adansonian analysis, serology, DNA base ratio studies, DNA hybridization studies, and lately protein fingerprinting or analysis.

Graham (1964a) studied 100 features in an Adansonian analysis and concluded:

1. That *R. leguminosarum, R. trifolii,* and *R. phaseoli* should be pooled to form a single species.

2. That *R. meliloti* should be maintained as a separate species.

3. That *R. japonicum* and *R. lupini* should be pooled with organisms of the cowpea miscellany to form a single species and perhaps, with additional studies, a genus.

Moffett and Colwell (1968), in a different Adansonian analysis found substantial agreement with the above findings of Graham. They also substantiated the fact that the major separation of *Rhizobium* species is into fast and slow growers and that this separation is apparently at the generic level. Using a different method of numerical taxonomy, 't Mannetje (1967) has reexamined the findings of Graham (1964a). He proposes that *Rhizobium* be considered as consisting of two sections, one with *R. japonicum* as the type species and the other with *R. leguminosarum* as the type species. He further recommends that the genus *Rhizobium* not be split at the generic level until much more information is available.

DeLey and Rassel (1965) examined the DNA base composition and flagellation as taxonomic tools. Utilizing 35 strains of *Rhizobium* they found a correlation between the DNA base composition and the type of flagellation. The peritrichously flagellated fast-growing rhizobia had a low percent GC composition ranging from 58.6 to 63.1%. These organisms were found in all cross-inoculation groups examined. The subpolarly flagellated, slow-growing strains ranged from 62.8 to 65.5% GC. From this study and a literature comparison, they tentatively proposed that there were two subgroups of fast growers (*R. leguminosarum* and *R. meliloti*), while the slow growers consisted of one genetic species (*R. japonicum*). Thus they proposed three species within the genus *Rhizobium*.

DeLey (1968), in an extensive review, critically analyzed the data from the various numerical taxonomy, DNA base ratios, and DNA hybridization studies. Based on these data he proposed a revision of the genus *Rhizobium* as summarized in Table V.

Graham (1969) reviewed the analytical serology of the Rhizobiaceae. Based on the studies of many workers, three broad serological groups have been defined as follows:

1. *Rhizobium trifolii, R. leguminosarum,* and *R. phaseoli*
2. *Rhizobium meliloti*
3. *Rhizobium japonicum* and *R. lupini.*

These serological groupings are in agreement with the numerical taxonomic studies of Graham (1964a). DNA homologies of the above groups of organisms (*R. trifolii, R. leguminosarum,* and *R. phaseoli*) were determined by Jarvis *et al.* (1980). Based on the data obtained they proposed that *R. trifolii* and *R. leguminosarum* be combined in one species and within this species various biovars be designated according to plant specificity. It is recommended that *R. phaseoli* be retained at present as a separate species pending more study.

Using three different methods of nucleic acid hybridization, Gibbons and Gregory (1972) concluded that *R. leguminosarum* and *R. trifolii* could not be distinguished. *Rhizobium lupini* and *R. japonicum* were also shown to be closely related, and a close relationship also appeared between *R. meliloti* and *R. phaseoli* (but with less certainty). In a recent study, Hollis *et al.* (1981) were not able to differentiate between the *Bergey's Manual* reference strains of *R. japonicum* and *R. lupini* by DNA–DNA hybridization.

Protein analysis by two-dimensional gel electrophoresis was used by Roberts *et al.* (1980) to classify *Rhizobium* strains. They reported that all the slow-growing strains (group II) were closely related and quite distinct from the group I isolates.

Rhizobium meliloti strains, as previously shown using other methods, formed a distinct group, while the rest of the group I organisms seemed to be more diverse. All the previously discussed studies were devoted to taxonomic relationships among species of rhizobia. In many cases, because of the number of species (plus the cowpea miscellany), there was commonly a relatively small representation for each group or species.

Few studies have been concerned with the taxonomic status within a species.

TABLE V

Species of *Rhizobium* and Their Characteristics[a]

Species	Relation to the current species	Flagellation	Percent GC	Serum zone litmus milk	Acid reaction litmus milk	Growth rate	Nodule-forming characteristics, special features
Rhizobium leguminosarum	*R. phaseoli* +	Peritrichous	59.0–63.5	+	–	Fast	Forms nodules on one or more of *Trifolium, Phaseolus vulgaris, Vicia, Pisum, Lathyrus, Lens*
Rhizobium meliloti	Unchanged	Peritrichous	62.0–63.5	+	+	Fast	Forms nodules with *Melilotus, Medicago, Trigonella*
Rhizobium rhizogenes	*A. rhizogenes*	Peritrichous	61.0–63.0	+	–	Fast	Causes hairy root disease of apples and other plants
Rhizobium radiobacter	*A. tumefaciens* + *A. radiobacter* + *A. rubi*	Peritrichous	59.5–63.0	+	–	Fast	Frequently produces galls on angiosperms; produces 3-keto-glycosides
Rhizobium japonicum	*R. japonicum* + *R. lupini* + cowpea miscellany	Subpolar	59.5–65.5	–	–	Slow	Nodulates many different legumes including one or more of *Vigna, Glycine, Lupinus, Ornithopus, Centrosema*, etc.

[a] After DeLey (1968).

In one such study, Hollis *et al.* (1981) examined the genetic interrelationships, using DNA–DNA hybridization, of 29 diverse isolates of *R. japonicum*. The results indicate that *R. japonicum* strains fall into two widely divergent DNA homology groups plus one subgroup. These two groups are well out of the limits defining a species and thus illustrate the diversity within this so-called species.

Graham (1976) expresses his great concern about the decline of *Rhizobium* taxonomic papers published since 1968. At present the situation has not changed. In view of the ever-increasing importance of *Rhizobium* it appears vital that the taxonomy of this genus be clarified. Detailed studies within each of the current species and especially the cowpea miscellany need to be carried out. There appears to be a dearth of biochemical and broad-spectrum serological tests useful for classification and identification of rhizobia. More nodulation studies, involving additional plant species, need to be conducted. Only if considerably more of such data are obtained can this taxonomy be properly codified.

IV. Relationship between *Agrobacterium* and *Rhizobium*

Since 1888 investigators have remarked upon the similarities between these two groups of organisms. The early works were summarized by Fred *et al.* (1932) who also reaffirmed the close relationship between the crown gall organism and the root nodule bacteria. This relationship has been the basis for their inclusion in the family Rhizobiaceae through several editions of *Bergey's Manual* including the eighth edition. On the other hand, some authors have felt that the two genera are unrelated and have separated them into different families (i.e., Krassilnikov, 1959; Prevot, 1961).

Recently, there have been a number of investigations using DNA base ratio studies and hybridization and Adansonian analyses, all designed to investigate the relationships between *Agrobacterium* and *Rhizobium*. In one of the first of these, Graham (1964a) studied 100 features of 121 strains of *Rhizobium, Agrobacterium, Beijerinckia, Bacillus,* and *Chromobacterium*. Among other conclusions, he found a close relationship between fast-growing rhizobia and agrobacteria and concluded that *Agrobacterium radiobacter* and *A. tumefaciens* should be united and included as *R. radiobacter* in the genus *Rhizobium*. This was confirmed in a later serological study (Graham, 1971) which, however, did not include *Agrobacterium rhizogenes* or *A. rubi*. DeLey *et al.* (1966) reported that *Agrobacterium, R. leguminosarum,* and *R. meliloti* all had a percent GC value within the same range, again possibly indicating a close relationship. These conclusions were corroborated by Skyring and Quadling (1969) as a result of a principal component analysis study. 't Mannetje (1967), using a different numerical analysis, suggested that *A. radiobacter* and *A. tumefaciens* be continued but not included in the genus *Rhizobium* pending further studies.

DeLey (1968) reviewed the major extant studies based on DNA base ratios (Wagenbreth, 1961; DeLey and Rassel, 1965; DeLey *et al.* 1966), hybridization data (Heberlein *et al.*, 1967), and computer analysis (Graham, 1964a; 't Mannetje, 1967; Moffet and Colwell, 1968). He proposed the formation of a single genus *Rhizobium* which would contain both *Agrobacterium* and *Rhizobium* species. DeLey's revision is summarized in Table V.

This system appears to be a logical improvement over the current scheme but, according to Graham (1976), does not address two important problems: (1) It does not consider the rhizobia associated with the many unexamined legumes; and (2) there still is nothing inherent in this scheme that would allow identification of a strain in the absence of plant host data. Even with these limitations, however, DeLey's proposal offers an interim improvement in the taxonomy of the Rhizobiaceae.

A somewhat different grouping was recommended by White (1972) based on a numerical analysis of 70 metabolic and nutritional tests and involving 23 *Rhizobium* and 32 *Agrobacterium* isolates. His data confirmed the idea that *Agrobacterium* and *Rhizobium* be combined and that four species of fast-growing rhizobia be recognized, the slow-growing rhizobia being separate. The resultant reorganization would include: group I, *R. radiobacter;* group II, *R. meliloti;* group III, related to *A. rhizogenes* and biotype-2 strains of Keane *et al.* (1970); group IV, *R. leguminosarum.* Crown gall-forming agrobacteria were associated with groups I and III. These groupings were made based on biochemical tests, without regard to nodulating ability or pathogenicity. White recognized the problem mentioned previously, that is, the lack of easy or precise methods for identifying these organisms in the absence of a plant host or plant response.

In summary, the data indicate a close relationship between fast-growing rhizobia and some agrobacteria. Slow-growing rhizobia are not closely related to agrobacteria (and probably not closely related to fast-growing rhizobia, as well). A reorganization of this family to combine the two genera seems warranted. Again, however, there arises the problem of not having enough information for more than an interim reorganization which would be an improvement on the current scheme but still would serve only as a holding situation.

REFERENCES

Allen, E. K., and Allen, O. N. (1950). *Bacteriol. Rev.* **14,** 273–330.
Anderson, A. R., and Moore, L. W. (1979). *Phytopathology* **69,** 320–323.
Baldwin, I. L., and Fred, E. B. (1927). *Soil Sci.* **24,** 217–230.
Baldwin, I. L., and Fred, E. B. (1929). *J. Bacteriol.* **17,** 141–150.
Beneke, W. (1912). "Bau Und Leben Der Bakterien." Berlin Univ., Leipzig.
Buchanan, R. E., and Gibbons, N. E. (1974). "Bergey's Manual of Determinative Bacteriology," 8th ed. Williams & Wilkins, Baltimore.
Currier, T. C., and Nester, E. W. (1976). *J. Bacteriol.* **126,** 157–165.

DeLey, J. (1968). *Annu. Rev. Phytopathol.* **6**, 63–90.

DeLey, J. (1974). *Taxonomy* **23**, 291–311.

DeLey, J., and Rassel, A. (1965). *J. Gen. Microbiol.* **41**, 85–91.

DeLey, J., Bernaerts, M., Rassel, A., and Guilmot, J. (1966). *J. Gen. Microbiol.* **43**, 7–17.

Drummond, M. H., and Chilton, M.-D. (1978). *J. Bacteriol.* **136**, 1178–1183.

Eckhardt, M. M., Baldwin, I. L., and Fred, E. B. (1931). *J. Bacteriol.* **21**, 273–285.

Elkan, G. H., and Kwik, I. (1968). *J. Appl. Bacteriol.* **31**, 399–404.

Elkan, G. H., and Kuykendall, L. D. (1981). *In* "Ecology of Nitrogen Fixation" (W. J. Broughton, ed.), Oxford Univ. Press, London and New York, in press.

Engler, G., Van Montagu, M., Zaenen, I., and Schell, J. (1977). *Biochem. Soc. Trans.* **5**, 930–931.

Fox, G. E., Stackebrandt, E., Hespell, R. B., Gibson, J., Maniloff, J., Dyer, T. A., Wolfe, R. S., Balch, W. E., Tanner, R. S., Magrum, L. J., Zablen, L. B., Blakemore, R., Gupta, R., Bonen, L., Lewis, B. J., Stahl, D. A., Luehrsen, K. R., Chen, K. N., and Woese, C. R. (1980). *Science* **209**, 457–463.

Fred, E. B., Baldwin, I. L., and McCoy, E. (1932). "Root Nodule Bacteria and Leguminous Plants." Univ. of Wisconsin Press, Madison.

Gibbons, A. M., and Gregory, K. F. (1972). *J. Bacteriol.* **111**, 129–141.

Graham, P. H. (1964a). *J. Gen. Microbiol.* **35**, 511–517.

Graham, P. H. (1964b). *Antonie van Leeuwenhoek* **30**, 68–72.

Graham, P. H. (1969). *In* "Analytical Serology of Microorganisms" (J. B. Kwapinski, ed.), Vol. 2, pp. 353–378. Wiley, New York.

Graham, P. H. (1971). *Arch. Mikrobiol.* **78**, 70–75.

Graham, P. H. (1976). *In* "Symbiotic Nitrogen Fixation in Plants" (P. S. Nutman, ed.), Vol. 7, pp. 99–112. Cambridge Univ. Press, London and New York.

Heberlein, G. T., DeLey, J., and Tijgat, R. (1967). *J. Bacteriol.* **94**, 116–124.

Hollis, A. B., Kloos, W. E., and Elkan, G. H. (1981). *J. Gen. Microbiol.* **123**, 215–222.

Jarvis, B. D. W., Dick, A. G., and Greenwood, R. M. (1980). *Int. J. Syst. Bacteriol.* **30**, 42–52.

Keane, P. J., Kerr, A., and New, P. B. (1970). *Aust. J. Biol. Sci.* **23**, 585–595.

Kerr, A. (1969). *Nature (London)* **178**, 703.

Kerr, A., and Panagopoulos, C. G. (1977). *Phytopathol. Z.* **90**, 172–179.

Kersters, K., DeLey, J., Sneath, P. H. A., and Sackin, M. (1973). *J. Gen. Microbiol.* **78**, 227–239.

Krassilnikov, N. A. (1959). "Diagnostik der Bakterien und Actinomyceter." Fischer, Jena.

Lippincott, J. A., and Lippincott, B. B. (1969). *J. Gen. Microbiol.* **59**, 57–75.

Löhnis, F., and Hansen, R. (1921). *J. Agric. Res.* **20**, 543–546.

't Mannetje, L. (1967). *Antonie Van Leeuwenhoek* **33**, 477–491.

Martinez de Drets, G., and Arias, A. (1972). *J. Bacteriol.* **109**, 467–470.

Moffett, M. L., and Colwell, R. R. (1968). *J. Gen. Microbiol.* **51**, 245–266.

Norris, D. O. (1965). *Plant Soil* **22**, 143–166.

Panagopoulos, C. G., and Psallidas, P. G. (1973). *J. Appl. Bacteriol.* **36**, 233–240.

Pichinoty, F., Mandel, M., and Garcia, J.-L. (1977). *Ann. Microbiol.* **128A**, 303–310.

Prevot, A. R. (1961). "Traite de Systematique Bacterienne." Dunod, Paris.

Roberts, G. P., Leps, W. T., Silver, L. E., and Brill, W. J. (1980). *Appl. Environ. Microbiol.* **39**, 414–422.

Skyring, G. W., and Quadling, C. (1969). *Can. J. Microbiol.* **15**, 142–158.

Skyring, G. W., Quadling, C., and Rowatt, J. W. (1971). *Can. J. Microbiol.* **17**, 1299–1311.

Thomashow, M. F., Panagopoulos, C. G., Gordon, M. P., and Nester, E. W. (1980). *Nature (London)* **283**, 794–796.

Van Larebeke, N., Engler, G., Holsters, M., van den Elsacker, S., Zaenen, I., Schilperoort, R. A., and Schell, J. (1974). *Nature (London)* **252**, 169–170.

Vincent, J. M. (1977). *In* "A Treatise on Dinitrogen Fixation" (R. W. F. Hardy and W. S. Silver, eds.), pp. 277–366. Wiley, New York.

Wagenbreth, D. (1961). *Flora (Jena)* **151,** 219–230.

Watson, B., Currier, T. C., Gordon, M. P., Chilton, M. D., and Nester, E. W. (1975). *J. Bacteriol.* **123,** 255–264.

White, L. O. (1972). *J. Gen. Microbiol.* **72,** 565–574.

Zaenen, I., Van Larebeke, N., Teuchy, H., Van Montague, M., and Schell, J. (1974). *J. Mol. Biol.* **86,** 109–127.

Biology of *Agrobacterium tumefaciens:* **Plant Interactions**

L. W. MOORE AND D. A. COOKSEY

*Department of Botany and Plant Pathology, Oregon State University,
Corvallis, Oregon*

I.	Introduction	15
II.	Disease Cycle	16
	A. Inoculum	16
	B. Dissemination and Inoculation	24
	C. Penetration and Conditioning of Host Tissue	25
	D. Infection	26
	E. Multiplication of *Agrobacterium tumefaciens*	33
III.	Control of Crown Gall Disease	35
	A. Biological Methods	35
	B. Chemotherapy and Soil Fumigation	38
	C. Sanitation and Cultural Management	40
IV.	Conclusions	41
	References	42

I. Introduction

Agrobacterium tumefaciens infects plants from 93 families (DeCleene and Delay, 1976), causing crown gall disease. This may be the broadest host range of any plant disease pathogen. The host range includes some unusual plants such as cactus, juniper, baby's breath, and conifer, but most of the damage occurs on stone and pome fruits. Losses occur primarily at the nursery where galled plants are culled and discarded, cullage often amounting to 80% or more. Such epidemics represent severe financial losses to the nursery grower. Orchard trees and landscape plants also are damaged when inferior root systems developed around the galls and in some instances from wood decay fungi that enter the tree through the gall. Infected plants can be stunted and subject to blow-over during wind storms.

Interest in crown gall disease has been at two levels. Historically the causal agent and factors affecting disease development were studied for disease control. Recently, the molecular biology of the infection process and the genetics of agrobacteria have been emphasized, somewhat to the exclusion of biological studies.

Despite early efforts to unravel the biology of the bacterium–host interaction, there is still much that is unknown. Contradictions are found readily in the

15

literature for any given parameter such as the role of pH, moisture, temperature, or antagonistic microorganisms on survival of *A. tumefaciens* in soil. All this underscores the current need for a better understanding of the biology of the pathogen in the soil and its interaction with the host. This article will focus on the specific events in the disease cycle of crown gall, contrasting current findings with those of the past and emphasizing the areas needing research.

II. Disease Cycle

Agrobacterium tumefaciens can induce tumorlike galls on the roots, crowns, leaves, and stems of plants. The galls may be a millimeter in diameter or up to 100 lb in weight (Smith, 1920). Small galls can be confused with other overgrowths such as excessive callus formation around a wound, nematode and insect galls, *Plasmodiphora brassicae* root malformations, *Rhizobium* nodules on legumes, and nitrogen-fixing nodules formed by actinomycetes on plants such as Russian olive and alder. The tumors generally appear 2–4 weeks after a plant is infected but may be delayed 18 months on roses (Munnecke *et al.*, 1963). This incubation period is lengthened below 17°C, and abnormally long incubation periods constitute latent infections. Bacteria released from the tumors either infect or colonize nearby plants. A number of factors aid in dissemination of the inoculum, and these will be discussed later in detail. Once a new infection is established, the cycle is completed and the disease is perpetuated. The simplified cycle of events that perpetuates the disease will now be examined in detail beginning with the inoculum.

A. INOCULUM

1. *Detection and Classification*

Improved methods for detecting low populations of agrobacteria and studying their biology in soil and on the plant are still needed despite numerous earlier studies (Patel, 1928a,b; Hildebrand, 1941a; Weinhold, 1970; Schroth *et al.*, 1971).

Agrobacteria are now classified into two major groups (Kersters *et al.*, 1973; White, 1972), generally designated biotypes 1 and 2 (Keane *et al.*, 1970), even though the diagnostic tests separating the two biotypes are sufficient to give them species status (Anderson, 1977; Lippincott *et al.*, 1981) (Table I). Both groups contain pathogens and nonpathogens. A new strain isolated only from grape plants and intermediate between biotypes 1 and 2 has been named biotype 3 (Kerr and Panagopolous, 1977). Selective media (Table II) separate agrobacteria into biotypes and discriminate against the myriad of other soil microbes that otherwise overwhelm *Agrobacterium* strains on nonselective media. Even so, *Ag-*

TABLE I

MAJOR DIAGNOSTIC TESTS

DISTINGUISHING AMONG THE BIOTYPES OF AGROBACTERIA[a]

Test	Biotype 1	Biotype 2	Biotype 3	Biotype 1-2
3-Ketoglycoside	+[b]	−	−	−
Ferric ammonium citrate	+	−		
Sodium chloride tolerance	3–4%	1%	>2%	
Citrate utilization	−	+		+
Oxidase reaction	+	−		−
Utilization of L-tyrosine	−	+		
Litmus milk	Alkaline	Acid	Alkaline	Alkaline
Acid from:				
Erythritol	−	+	−	+
Ethanol	+	−	−	
Melezitose	+	−	−	
Alkali from:				
Malonate	−	+	+	−
L-(+)-Tartrate	V	+	+	
Propionate	V	−	−	
Mucic acid	−	+	−	
Selective media				
Schroth *et al.* (1965)	+	−	−	
New and Kerr (1971)	−	+	−	
Clark[c]	+	V−	V−	

[a] Adapted from: Anderson (1977), Keane *et al.* (1970), Panagopoulos *et al.* (1978), and Spiers (1979).

[b] +, −, and V indicate positive, negative, and variable reactions, respectively.

[c] Spiers (1979) states that some biotype-2 strains grow on Clark's medium but only poorly. M. N. Schroth (personal communication) has isolated biotype 3 on Clark's medium.

robacterium strains on most selective media can be swamped by other more competitive bacteria at soil dilutions less than 10^{-2} (Lippincott *et al.*, 1981).

2. *Geographical Distribution of Biotypes*

Biotype distribution varies from a predominant biotype in some geographic regions to mixtures in others (Table III). Only one biotype of *Agrobacterium* was isolated from California soils (Schroth *et al.*, 1971), and a single biotype predominated in southern Australia (Kerr, 1974). Obviously, use of a medium selective for only one biotype can cause erroneous conclusions about the predominance of a particular biotype.

The evolutionary and ecological significance of biotype distribution is unknown. In the Netherlands, biotype-1 strains predominate, are insensitive to

TABLE II

SELECTIVE MEDIA FOR THE ISOLATION OF *Agrobacterium* STRAINS

Medium	Major biotype	Habitat	Results	Additional comments
Clark (1969)	1	Soil and galls	A few biotype 2 grew, but poorly	Biotype 3 recovered (M. N. Schroth, personal communication)
Heskett–Kado (Kado and Heskett) (1970)	1 and 2	Soil and galls	Most biotype 2 grew (Spiers, 1979)	
Patel (1928b)	1 and 2	Soil and galls	Discriminated poorly against the normal soil microflora	
New–Kerr (New and Kerr, 1971)	2	Soil and galls	Most biotype-1 strains could not utilize erythritol	Nitrate reportedly prevented growth of *A. rhizogenes*; *Rhizobium trifolii* was unable to grow; however, Anderson (1977) found that 26/27 *Rhizobium* strains grew, as did *A. rhizogenes*
Schroth *et al.* (1965)	1	Soil and galls	Excluded 99% of other species growing on nonselective media; efficiency of recovery, 38%; 18 of 28 soils yielded *A. tumefaciens*; 27/28 soils yielded *A. radiobacter*	Reducing antibiotics increased efficiency of recovery; some biotype 2 grew when ammoniacal nitrogen was used; use of selective medium as a liquid enrichment medium allowed detection of *A. radiobacter* in all 28 soils
Burr (unpublished combination of Schroth *et al.* and New–Kerr)	1, 2, 3	Galls	10% of the recovered biotype 3 were virulent	Could not recover *virulent* agrobacteria from grape galls on Schroth *et al.* or New–Kerr media

agrocin 84, and are not controlled biologically with strain K84 (H. Miller, personal communication). Similarly, the biotype-3 strains from grape are not sensitive to agrocin 84 or controlled biologically. In contrast, agrocin 84-sensitive biotype-2 pathogens predominate in Australia, and biological control has been excellent (A. Kerr, personal communication). Further characterization of these strains according to opine catabolism and tumor-inducing (Ti) plasmid homology would aid our understanding of the significance of biotypes and their distribution.

TABLE III

PREDOMINANT BIOTYPES OF *Agrobacterium* ACCORDING TO GEOGRAPHICAL LOCATION

Location	Biotype	Pathogens	Nonpathogens	Selective media employed[a]
		Number of strains		
Australia (Kerr, 1969)	1	0	112	D, E
	2	85	15	D, E
Canada (Dhanvantari, 1978)	1	3	—	D
	2	27	—	D
Greece (Panagopolus and	1	13	6	Nutrient agar
Psallidas, 1973)	2	23	2	Nutrient agar
	Intermediate	4	3	Nutrient agar
New Zealand (Spiers, 1979)	1	123	64	A, B, C
	2	1	39	A, B, C
	1–2	0	36	A, B, C
The Netherlands (H. Miller,	1	90	Rare	A, C, E
(personal communication)	2	0	Rare	A, C, E
Hungary (Sule, 1978)	1	6	—	
	2	38	—	
	3	25	—	
United States				
Oregon (Anderson, 1977)	1	30	48	Kado's DIM (unpublished), C, E
	2	60	27	Kado's DIM (unpublished), C, E
	Intermediate	7	9	Kado's DIM (unpublished), C, E
(Moore and Allen, 1977)	1	62	119	Kado's DIM (unpublished), C, E
	2	400	709	Kado's DIM (unpublished), C, E
Maryland (Alconero, 1980)	1	0	9	A, B, C, D, E
	2	2	13	A, B, C, D, E
Pennsylvania (L. Forer,	1	6	—	A, C, E
(personal communication)	2	60	—	A, C, E
	Intermediate	3	—	A, C, E
South Carolina	1	2	38	A, B, C, D, E
(Alconero, 1980)	2	43	102	A, B, C, D, E
Tennessee (Alconero, 1980)	1	37	25	A, B, C, D, E
	2	123	15	A, B, C, D, E

[a] A, Clark (1969); B, Kado and Heskett (1970); C, New and Kerr (1971); D, Patel (1928a,b); and E, Schroth *et al.* (1965).

3. Source and Survival

a. *Soil.* Evidence exists for short- and long-term survival of *A. tumefaciens* in soil (Table IV), which has lead to conflicting descriptions of *A. tumefaciens* as a soil invader (Dickey, 1961) and a soil inhabitant (Schroth *et al.*, 1971). This conflict is due in part to comparisons of data from fallow soil with data from soils bearing host and nonhost plants. Certain factors increase the persistence of *A.*

TABLE IV

SOIL FACTORS FAVORING CROWN GALL DISEASE OR SURVIVAL OF *Agrobacterium tumefaciens*

Investigator	Test location	Helpful	Hindering
Dickey (1961)	Laboratory greenhouse	Sterile soil, direct relationship between population and soil moisture in upper 10 cm of unsterile soil	Unsterile soil (especially at 34°C), coarse-textured, addition of sulfur to give pH of 4.2 (below 10 cm)
Patel(1929)	Laboratory	Sterilized soils (736 days maximum survival)	Unsterile soil, survival best in sand and least in clay soil (480 days maximum)
Smith (1912, 1925)	Field	Sandy soil	—
Chamberlain(1962)	Field	Light-textured soil, previous crop of asparagus plus heavy fertilization	Heavy soils
Lelliott (1971)	Field	Peat soils (pH 5.9)	—
Moore (personal communication and unpublished)	Field	Low, wet spots in field, heavier-textured soils	—
Deep *et al.* (1968)	Field	Fumigation	—
Ross *et al.* (1970)	Field	—	Fumigation
Sherbakoff (1925)	Field	Abundant humus (especially barnyard manure well supplied with lime)	—
Patrick (1955)	Field	Straw refuse	Antagonistic microflora

tumefaciens in soil (Table IV), but survival does not exceed 800 days in laboratory studies using fallow, sterilized soils and is less in unsterile soils (Dickey, 1961).

In contrast to quantitative measurements from controlled laboratory studies, evidence for the long-term persistence of naturally occurring *A. tumefaciens* in field soil is based largely on the occurrence and severity of the disease (Banfield and Mandanberg, 1935; Townsend, 1915; Hildebrand, 1941a,b). The problem with such inferences is that plant susceptibility can be altered by physical, organic, and microbial environmental factors. Thus an increase in disease may not be correlated with higher populations of *A. tumefaciens* in the soil, as predicted by Weinhold (1970), but with changes in host susceptibility or populations of antagonistic microorganisms (Patrick, 1955; Cooksey and Moore, 1980).

Certain edaphic factors affected survival of *A. tumefaciens* in fallow soil under controlled laboratory and greenhouse conditions (Dickey, 1961). However, the soil was infested with strains of *A. tumefaciens* that may not have been adapted naturally to that soil. Different strains of *A. tumefaciens* apparently can adapt to opposing environments such as sandy soil versus heavy-textured soil and acid versus alkaline soil, especially in the plant rhizosphere (Patrick, 1955). Lelliot (1971) concluded that crown gall in apple rootstock beds in England could not be related to the type of bed, age of bed, soil pH, or type of loam or silt loam.

Survival of *A. tumefaciens* in field soil is also inferred, because the incidence of disease following soil fumigation was reduced but not eliminated even at double the rate of fumigant (Ross *et al.*, 1970). In contrast, more crown gall occurred on mazzard cherry seedlings grown from seed planted in fumigated than on those from nonfumigated nursery soil (Deep *et al.*, 1968). The seedlings were free of crown gall when harvested after the first growing season but developed galls during the second growing season following transplanting. Since the seedlings were transplanted to pots of autoclaved soil, *A. tumefaciens* was probably a component of the rhizosphere. The plants grew more vigorously in the fumigated soil which may have increased their susceptibility to infection, as reported for grape plants grown on heavily fertilized soil (Chamberlain, 1962).

The population of *A. tumefaciens* in natural or cultivated soil is low to nondetectable except in the vicinity of infected plants (Patel, 1928a, 1929; New and Kerr, 1972; Schroth *et al.*, 1971), and the ratio of *A. tumefaciens* to *Agrobacterium radiobacter* is less than 1. This ratio was lowest (1:13) in soils where stone fruits had been planted previously and highest (1:500) in grass pasture soil that had never been cultivated or supported host plants other than dicotyledonous weeds (Schroth *et al.*, 1971). Weinhold (1970) predicted less than a 1% incidence of crown gall with ratios of 1:500 and a 10–20% incidence at 1:13. The enigma is that *A. tumefaciens* exists in the soil in low numbers, yet epidemics occur involving 80% or more infected plants. Does this mean that the host plant enriches selectively for *A. tumefaciens?* If so, what qualifies the pathogen to be enriched selectively over the more numerous nonpathogenic agrobacteria?

The seemingly contradictory reports relative to survival of *Agrobacterium* in soils of different types and conditions is probably a reflection of the ability of a specific strain of *A. tumefaciens* to adapt to a unique habitat. For example, galling was more severe on apple seedlings at one planting site than at another, even though all the seedlings were produced at one nursery site (Moore, unpublished), suggesting strain adaptation.

b. *Contaminated Water.* Water is nearly indistinguishable as a source of *A. tumefaciens* and a means of dissemination. Serious losses occurred when stratified peach seeds were planted on land with no history of orchard or nursery crops, apparently from soil-borne *A. tumefaciens* (Smith and Cochran, 1944). Later it was concluded that irrigation water was contaminated by the pathogen in winter rain runoff that had passed through several orchards containing galled trees. Smith and Cochran soaked galls from peach roots overnight in water which was then poured over canned soil that had been fumigated with chloropicrin and planted with peach seeds. Fifty-four percent of the seedlings were infected at harvest versus 7% in the noninfested control soil.

Although Smith and Cochran concluded that *A. tumefaciens* did not persist in soil for long periods, Schroth *et al.*, (1971) isolated *A. tumefaciens* from pasture soils that had never been tilled or cultivated. However, the pasture soils may have received *A. tumefaciens* from runoff water during winter rains, since agrobacteria can move considerable distances in currents of soil water (Patel, 1928a; Hedgecock, 1910).

c. *Infected or Colonized Plants.* Galled plants are the most common source of *A. tumefaciens*. The bacterium is usually present in high numbers, partially protected by the tumor tissue, and liberated into surrounding soil or water. For example, volunteer shoots from buried roots of a previous orchard planting were galled, thus providing a source of inoculum for newly planted trees (Ark, 1954). Symptomless cuttings from galled grape plants resulted in a large percentage of the rooted vines becoming galled (Garcia and Rigne, 1913). Dormant peach trees heeled in with infected trees for 1 month developed galls during the growing season (Ness, 1917). Ness also observed that wild vegetation such as blackberry patches frequently were galled and cautioned against planting directly on freshly cleared land.

Since the rhizosphere is a favorite habitat of agrobacteria (Starkey, 1931), plants may be colonized with *A. tumefaciens* but not infected until wounded. *Agrobacterium tumefaciens* from galls of infected mother-block trees in rootstock beds often colonize the newly propagated rootstock, which develop galls upon outplanting (Lelliot, 1971). When symptomless red raspberry plants from fields of low crown gall incidence (0.1%) and high incidence (84%) were transplanted, 2.6% of the plants from the low-incidence soil were infected after one growing season, whereas 94% of the plants from soils with a high incidence of infection were galled (Banfield and Mandanberg, 1935). Cuttings from chrysanthemum plants contaminated with the pathogen can develop crown gall during the rooting process (New and Milne, 1976). Even cherry seed collected from the soil under a galled tree can be contaminated and infect the subsequent seedlings (Toumey, 1900).

Regardless of the source of the planting stock, the planting site must be considered. Seedlings from the same nursery source planted at two locations in Oregon and two in New York State showed a 70% incidence of galling at one of the Oregon production site versus 3% at the other Oregon site and 12 and 6% at the New York sites (Moore, 1976). Similarly, when symptomless cherry seedlings from one production site were lined out besides those from another site, one lot was severely galled at harvest but not the other (Young, 1954). Alternatively, symptomless apple seedlings from a single source were inoculated with *A. tumefaciens* and planted at different planting sites. The incidence of galled seedlings varied according to the planting site, suggesting an interaction of the pathogen with other microorganisms (L. Moore, unpublished). Each of two *A. radiobacter* strains coinoculated with a mixture of pathogenic strains prevented infection of mazzard cherry seedlings *in situ,* whereas three other *A. radiobacter* strains coinoculated with these pathogens enhanced the disease (Moore, 1977).

Agrobacteria were recovered from the roots of several unidentified weeds growing in nursery soils at a ratio of 1 *A. tumefaciens* to 28 *A. radiobacter.*

The incidence of crown gall can increase with cultivation of susceptible crops (Patel, 1928a; New and Kerr, 1972; Schroth *et al.,* 1971; Townsend, 1915). The agrobacteria may overwinter in or on old roots or plant debris from the previous crop. In contrast, a nursery in Washington has planted apple seedlings continuously for years with nominal occurrence of crown gall (S. Lochrie, personal communication). The soil contains 10^6 colony-forming units (CFU)/gm of a *Penicillium* sp. that produces large zones of inhibition against *Agrobacterium* strains *in vitro* (D. A. Cooksey, and L. W. Moore, unpublished).

4. *Loss of Pathogenicity*

Agrobacterium radiobacter predominates over *A. tumefaciens* in soil and in the rhizosphere of uninfected plants, yet 80% or more of the plants in a field may be infected by *A. tumefaciens.* Kerr (1969) suggested that a loss of virulence following infection or swamping of *A. tumefaciens* by *A. radiobacter* might explain the difficulty in isolating virulent *Agrobacterium* from natural soils. His later experiments demonstrated that neither loss of virulence nor swamping of *A. tumefaciens* by *A. radiobacter* occurred. Nevertheless, Kerr showed that tumor development was markedly restricted or completely inhibited when 100 times more *A. radiobacter* than *A. tumefaciens* was applied to a wound. Ratios of 100–500 *A. radiobacter* to 1 *A. tumefaciens* are common in some soils (Schroth *et al.,* 1971), and Weinhold (1970) predicts less than 1% infection from such ratios. Hypothetically, a susceptible host could enrich selectively for *A. tumefaciens,* but if this is true, why doesn't the disease occur every year rather than sporadically? Does *A. tumefaciens* mutate to an avirulent form? Is *A. radiobacter* better equipped to survive in the soil and rhizosphere? Is there virulence conversion of nonpathogens to pathogens?

In the laboratory, loss of virulence can occur during routine subculturing of *A. tumefaciens* (Smith, 1912), which is strain-dependent (Abo-El-Dahab *et al.,* 1978) and irreversible (Kado, 1976a). However, Van Lanen *et al.* (1952) concluded that the virulence of *A. tumefaciens* was remarkably stable when compared with that of most plant and animal pathogens, since 20–30 serial transfers in glycine media were required to cause a loss of infectivity.

Following dichlone fungicide treatment of cherry seedlings, the incidence of crown gall after inoculation with an avirulent strain was as high as with the virulent strain (Deep and Young, 1965). Deep and Young suggested that the avirulent strain was converted to a virulent strain. However, without dichlone, the avirulent strain induced a low percentage of infection. Instead of virulence conversion, dichlone may have eliminated antagonists to naturally occurring virulent *Agrobacterium* in the field soil, which were then complemented (Lippincott and Lippincott, 1978b) by the avirulent strain, thereby increasing the incidence of crown gall.

B. DISSEMINATION AND INOCULATION

Dissemination of *A. tumefaciens* to a susceptible plant can be passive or active, with the former playing the major role especially from plant-propagating material.

1. *Passive*

a. *Propagating Material.*   Propagating materials in the form of cuttings, seedlings, and root stocks produced vegetatively in layering beds can all be colonized by the pathogen or carry latent infections. Detecting latent infections or colonized tissues is a problem because of the absence of symptoms. However, there is good evidence that symptomless planting stock from disease areas will have a greater percentage of galled plants. Symptomless trees and rootstocks derived from apple rootstock beds in England became infected after outplanting (Lelliot, 1971). Symptomless first year raspberry plants were selected from fields where 0.1–84% of the plants were galled, and during the following growing season galls developed on 3 and 94% of the plants from the lightly and heavily infected plantings, respectively (Banfield and Mandanberg, 1935). Young (1954) planted mazzard cherry seedlings from one nursery source beside seedlings from another nursery source. At harvest there was a high incidence of galled trees from one but not from the other. Siegler and Piper (1931) planted symptomless apple seedlings that were neither washed nor disinfected with mercuric chloride and observed 48% infection compared to 8% infection on grafts from seedlings that were thoroughly washed before grafting, and 1% when the seedlings were washed and disinfected. Even seed collected under a mother-block tree may be contaminated by the pathogen and produce seedlings that become infected (Toumey, 1900). Grape cuttings removed from plants with galls at the base developed more galls after rooting and planting out than cuttings from symptomless plants (Hedgecock, 1910). *Agrobacterium tumefaciens* was isolated from sap oozing from a surface-sterilized, decapitated grape vine. Apparently, the bacteria were translocated systemically in the plant; thus cuttings from diseased plants could carry the pathogen internally and become infected when the vine was wounded.

b. *Water and Rain.*   Surface water moving through orchards with galled trees became contaminated with *A. tumefaciens* and flowed into water used to irrigate soil where peach seeds were planted. Seedlings harvested from these fields were nearly 100% diseased (Smith and Cochran, 1944). Subsequently, peach roots bearing crown gall were soaked overnight in water, and then this water was poured over chloropicrin-sterilized soil in cans planted with peach seeds; 54% of the seedlings were galled at harvest, compared to 8% of the seedlings from the water control. Hedgecock (1910) observed that, when water used to irrigate terraced grapes broke through one of the upper terrace dikes and cascaded down through the lower terraces, crown gall developed wherever the floodwater had passed. Patel (1928a) also showed that *A. tumefaciens* could be moved over considerable distances in currents of soil water.

c. *Other Methods.* Machinery and tools used in the cultivation and propagation of nursery stock can be contaminated as they pass through galled tissue, thus carrying the pathogen to new tissues. *Lippia canescens* is a dicotyledonous plant used in Arizona for lawn cover; infections were spread throughout the entire planting with continued mowing (M. Stanghellini, personal communication). Nematodes provide wounds for infection by *A. tumefaciens* and may carry the organism on their body or within. Root-chewing insects also provide wounds for infection (Riker and Hildebrand, 1934) and may carry the pathogen to the wound site.

2. Active

Agrobacterium tumefaciens bacteria became concentrated after 0.5 hour at the end of a capillary tube containing expressed sap from tomato plants (Riker, 1923). Conversely, when a capillary tube containing *A. tumefaciens* in sterile water was placed in a hanging drop of expressed sap, the bacteria migrated into the droplet, indicating that a chemotactic stimulus could attract *A. tumefaciens* to the wound of a plant. The role of chemotaxis in the dissemination of *A. tumefaciens* is unknown, but the strength and specificity of the chemical gradient could affect such a chemotactic response (MacNab and Koshland, 1972) and may explain the erratic chemotactic response of *A. tumefaciens* to excised roots (Schroth *et al.*, 1971). The amount of free water available in soil pores would influence the ability of *A. tumefaciens* to respond to the chemotactic stimulus. Narrow capillaries similar to those found in soil reduced the growth rate and cell size of bacteria, and organic matter was inaccessible when concentrated in capillaries less than 1 μm in diameter (Zvyagintsev, 1970).

C. Penetration and Conditioning of Host Tissue

Wounds are required for penetration of the host by the pathogen and for altering the host cells to a susceptible state (Braun, 1952). Wounding may occur during seed germination (Schroth *et al.*, 1971), by insects and nematodes (Griffin *et al.*, 1968; Nigh, 1966; Orion and Zutra, 1971), and from cultural practices such as root pruning of nursery stock and cultivation.

Agrobacterium tumefaciens introduced into tissues by vacuum infiltration in the absence of a wound does not produce tumors. Wounds apparently condition plant cells to become susceptible to tumor induction in addition to providing a point of entry. Wounded *Kalanchoe* stems required 6–12 hours to become susceptible to tumor induction, and susceptibility increased to a maximum 2–3 days after wounding (Braun, 1952). The nature of the conditioning response is still not clear, but it is accompanied by changes in cell metabolism, such as increased DNA synthesis (Klein *et al.*, 1953; Klein, 1958) and changes in cell membrane permeability (Braun, 1952).

An exception to the wound requirement for tumor formation is penetration of *A. tumefaciens* through lenticels (Kerr, 1972). Kerr has suggested that the bacteria penetrate the loosely compacted cells at the surface of lenticels and contact the susceptible meristematic cells beneath. Although meristematic cells of apical buds usually are not susceptible to infection (Braun, 1954), the meristematic cells in lenticels may have something in common with wound-conditioned cells.

D. INFECTION

Tumor induction in wounded plants involves two major steps: attachment of the bacterium to host cell walls and transfer of bacterial plasmid DNA into the plant cell (Chilton *et al.*, 1977). There are several environmental and host factors that may influence tumor induction and development, but in most cases the specific steps of the infection process being affected are unknown.

1. The Infection Process

a. *Attachment.* Tumor induction in bean leaves was inhibited when avirulent or heat- and ultraviolet-killed *A. tumefaciens* strains were coinoculated with virulent bacteria, suggesting that the avirulent or killed bacteria excluded the pathogen from essential attachment sites (Lippincott and Lippincott, 1969a). Lipopolysaccharides isolated from strains active in pathogen exclusion also blocked tumor induction (Whatley *et al.*, 1976). Ti plasmids from strains K14 and B6S3 coded for the lipopolysaccharide active in pathogen exclusion, but lipopolysaccharide from some strains lacking Ti plasmids also excluded pathogens. Site-active lipopolysaccharide may thus be determined by either plasmid or chromosomal genes (Whatley *et al.*, 1978).

More direct evidence for site attachment by *A. tumefaciens* was obtained by Matthysse *et al.* (1978), using suspensions of cultured cells from tobacco and carrot. When a virulent strain was added to suspensions of plant cells, the number of free bacteria in the suspension gradually decreased over a 2-hour period, presumably because the bacteria were binding to plant cells. Light and scanning electron microscopy confirmed the attachment to carrot cells. Only bacteria with a Ti plasmid attached to plant cells, although lipopolysaccharide from avirulent strain NT1 was able to exclude attachment of the pathogen as it did in the bean leaf assay (Whatley *et al.,* 1976). These data suggest that some other substance coded for by the Ti plasmid is required for attachment to the tissue culture cells in addition to site-active lipopolysaccharide (Matthysse *et al.,* 1978).

Further work is needed to determine requirements for attachment to other hosts and to plants under field conditions. Strains of *A. tumefaciens* mutated specifically for loss of the site-active lipopolysaccharide, for example, might be used for this purpose. Liao and Heberlein (1979) found that polymixin-resistant *A. tumefaciens* B6 mutants were less virulent on carrot and potato disks than on bean leaves. Crude cell envelope preparations from the polymixin mutants were less active in pathogen exclusion than preparations from wild-type B6, suggesting that site-specific lipopolysaccharide was lost from the cell envelope upon acquisition of polymixin resistance. It would be useful to identify the specific cell wall modifications in these mutants and determine whether the mutations occur in plasmid or chromosomal genes.

Equally important in the attachment process is the nature of the host cell attachment site. The number of tumors formed in the bean leaf assay was greatly reduced when *A. tumefaciens* B6 was incubated with isolated cell walls from bean leaves, but not when incubated with isolated cell membranes (Lippincott *et al.,* 1977b). It was further shown that pectin and polygalacturonic acid mimicked the effect of plant cell walls in this test. However, cell walls from tumor tissues, young, healthy bean seedlings, or several species of monocots did not bind *A. tumefaciens,* although all three cell wall types were active in binding when treated with pectinesterase (Lippincott and Lippincott, 1978a). Since pectinesterase catalyzes hydrolysis of the methyl ester bonds of pectic compounds, the data suggest that the degree of methylation of cell wall pectins is important for the attachment of *A. tumefaciens* to a particular host.

b. *Transformation.* The oncogenic nature of crown gall tumors was first proven by White and Braun (1942), who found that tumor tissue could be grown in culture independently of the causative organism. Subsequently Braun (1947) hypothesized that a tumorigenic substance, known as the tumor-inducing principle, passed from the bacterium to the plant cell, causing the transformation. The tumor-inducing principle was still unidentified in 1975 (Kado, 1976b), but the most appealing hypothesis was that it was a nucleic acid capable of self-replication once introduced into the plant cell. This hypothesis was confirmed with the discovery that virulent *A. tumefaciens* strains harbored large plasmids (Zaenen *et al.,* 1974), and that a portion of the plasmid DNA was stably incorporated into crown gall tumor cells (Chilton *et al.,* 1977) and integrated into the plant DNA (Thomashow *et al.,* 1980a).

The mechanism of transfer of plasmid DNA to plant cells is not yet known, but certain requirements of the bacterium are related to this stage in the infection process. First, tumor induction is temperature-sensitive; temperatures above

30°C prevent tumor formation (Riker, 1926), although the plant, bacterium, and fully induced tumors can grow above 30°C. Transfer of plasmid DNA from the bacterium to the host may be inhibited during the temperature-sensitive stage in tumorigenesis, because plasmid transfer between *Agrobacterium* strains is also inhibited above 30°C (Tempe *et al.*, 1977). However, these plasmid transfer functions are not identical, since methionine represses plasmid transfer between bacteria but does not affect tumor induction (Hooykaas *et al.*, 1979).

The second requirement is active metabolism by the bacterium at the wound site. Mutants of *A. tumefaciens* B6 auxotrophic for adenine, methionine, or asparagine form fewer tumors per viable cell than the wild-type B6 (Lippincott *et al.*, 1965). Infectivity was enhanced when the auxotrophs reverted to the prototrophic form, or by inoculating the leaf with the mutant plus the required amino acids (Lippincott and Lippincott, 1966). In addition, rifampicin (an inhibitor of bacterial RNA polymerase) inhibited tumor induction on *Kalanchoe* leaves when applied to wounds during the first 24 hours after inoculation, whereas fully induced tumors were not affected (Beiderbeck, 1970a). When rifampicin-resistant strains of *A. tumefaciens* were used, rifampicin did not hinder tumor induction, indicating that RNA synthesis by the bacterium was essential during tumor induction (Beiderbeck, 1970b). This metabolic activity may be required for replication and transfer of the Ti plasmid or needed to maintain viability of bacteria during the conditioning of host cells to a susceptible state after wounding (Lippincott and Lippincott, 1966).

c. *Role of Hormones in Tumor Development.*   Higher levels of auxins and cytokinins are found in crown gall tissues than in normal tissues (Klein, 1965), suggesting that these hormones increase cell enlargement and cell division during tumorigenesis. *Agrobacterium tumefaciens* produces indoleacetic acid and a number of cytokinins in culture (Kaper and Veldstra, 1958; Upper *et al.*, 1970; Kaiss-Chapman and Morris, 1977), and synthesis of indoleactic acid is reportedly coded by the Ti plasmid in strain C58 (Liu and Kado, 1979). Because crown gall tissues synthesize auxins and cytokinins in the absence of the bacterium (Braun, 1958), the bacterial genes for auxin and cytokinin synthesis probably are contained in the T-DNA region of the Ti plasmid, which is incorporated into the plant cell. When the T-DNA is lost from tumor cells under certain conditions (see Section II,D,1,e), the tissue is no longer hormonally independent (Yang *et al.*, 1980b).

d. *Invasion of Host.*   Tumors are typically found on the crown or on the taproot of the plant, but they also develop on aerial tissues of plants such as raspberry and grape. Tumors that develop some distance from inoculated wound sites are called secondary tumors. The occurrence of secondary tumors suggests that the bacterium may move some distance within the plant tissues from the point of infection.

Smith *et al.*, (1911, 1912) reported that *A. tumefaciens* was an intracellular parasite, and that secondary tumors were formed from "tumor strands" connected to primary tumors. Others reported that the bacteria were in the intercellular spaces, spreading through xylem vessels or through the intercellular spaces to initiate secondary tumors (Riker, 1923; Hill, 1928). Stonier (1956) confirmed by autoradiography the intercellular location of *A. tumefaciens* in the pith and cortex and also inside xylem vessels of sunflower and *Kalanchoe,* suggesting that the bacteria might spread through the plant by the vascular system. The spread of *A. tumefaciens* through xylem vessels of grapevine (Lehoczky, 1968) and chrysanthemum (Miller, 1975) has been reported, but efforts to repeat this work with chrysanthemum (R. Hartman, personal communication) and grape (T. J. Burr, personal communication) have been unsuccessful.

Although the bacteria may be translocated through xylem vessels in some hosts, the infection process leading to secondary tumors is not well studied. A requirement for wounding in secondary tumor formation is not well established. Secondary tumors may develop in grapevine where fissures develop from frost damage (Lehoczky, 1968), but in other hosts wounds at the site of secondary tumor formation are not as obvious. Secondary tumors on castor bean leaves were inhibited when an avirulent *A. tumefaciens* strain was inoculated with a virulent strain (El Khalifa and Yousif, 1974), suggesting that bacterial binding to a specific site was essential in secondary and primary tumor formation. The involvement of Ti plasmids in the formation of secondary tumors has not been studied.

e. *Tumor Reversal.* Tumors induced by nopaline-degrading strains of *A. tumefaciens* that develop abnormal shoots and leaves are referred to as teratomas. Teratoma shoots grafted to a normal plant can develop into morphologically normal shoots which eventually flower and produce viable seed (Braun, 1959; Braun and Wood, 1976). However, tissues from plants regenerated from teratomas can revert to the tumorous state when planted on a basic culture medium (Turgeon *et al.*, 1976), and such tissues have been shown to contain T-DNA (Yang *et al.*, 1980a). Regeneration of plants from teratoma tissues is thus a reversible suppression of the tumorous state, but not a true reversal to the normal state.

A true reversal can occur when cells of the regenerated teratoma undergo meiosis. Haploid tissue derived from anthers and tissue from the F_1 generation of seed did not revert to tumorous growth on a basic culture medium (Turgeon *et al.*, 1976) and were free of T-DNA (Yang *et al.*, 1980a,b). The mechanism by which the T-DNA is lost during meiosis is unknown.

Reversion of unorganized tumors incited by several octopine-degrading and null-type strains has also been reported (Sacristan and Melchers, 1977; Einset and Cheng, 1979). The regenerated plant tissues from these tumors did not revert to the tumorous state when planted on a basic culture medium, indicating a true reversal to the normal state. Plants regenerated from unorganized tumors have not yet been examined for the presence of T-DNA.

2. Factors Influencing Infection

a. *Susceptibility of Hosts.* In a controlled environment, wounded hosts may be consistently and uniformily infected, but field plants are subject to many factors that influence their susceptibility. Therefore environmental and host variables that lead to fluctuations in infection and severity of the disease have been studied.

Temperature limitations are the best studied of the factors influencing host susceptibility. Tomato seedlings maintained at 30°C or above did not develop tumors, although the plant and bacterium could both grow above this temperature (Riker, 1926). Similarly, periwinkle and *Kalanchoe* plants developed tumors at 25°C but not at 32°C (Braun, 1947, 1952). Once initiated at 25°C, however, tumor development was not affected by growing the plants at 32°C. The higher temperatures apparently inhibit tumor initiation rather than development and maintenance of the tumorous state.

The temperature-sensitive step in tumorigenesis may involve inhibition of transfer of the Ti plasmid from the bacterium to the plant (Tempe *et al.,* 1977); however, some hosts have a critical temperature higher than 30°C. Initiation of crown gall on mazzard cherry seedlings occurred at temperatures up to 37°C, which was near the limiting temperature for growth of that host (Deep and Hussin, 1965). Tumor initiation also occurred on radish seedlings kept at day and night temperatures of 38 and 34°C, respectively (Moore and Tingey, 1976). These data show that transfer of the Ti plasmid to plants apparently can occur at temperatures much higher than 30°C, suggesting that the temperature-sensitive step may be related to processes other than Ti plasmid transfer or may be strain-dependent. *Agrobacterium tumefaciens* strains differ considerably in the temperature at which virulence is lost (Lin and Kado, 1977), and the plasmid transfer functions in different strains may vary in their temperature sensitivity.

Host susceptibility is also dependent on the age of the plant and the site of infection. Younger tissues generally are more susceptible than older tissues, although very young "embryonic" tissues lack adherence sites for the bacterium (Lippincott and Lippincott, 1978a). Tomato seedlings were more susceptible to infection in the upper internodes than in basal internodes, and 10-week-old seedlings were more susceptible than 15-week-old seedlings (Link *et al.,* 1953). In addition, tumors grew larger on the younger tomato seedling tissues. The severity of plant stunting caused by tumors in radish seedlings was also greatest when younger tissues were infected (Moore and Tingey, 1976).

Both the growth of tumors and the stunting of galled plants are influenced by nutrient availability. Tumors on tomato seedlings supplied with low nitrate nitrogen levels developed almost as fast as seedlings supplied with high nitrogen levels (Link *et al.*, 1953). Relative to the size of the plant, tumors grew faster at lower than at higher nitrogen levels. Infected low-nitrogen tomato plants also showed greater stunting and premature leaf abscission. Hussin and Deep (1965) reported that the relative tumor size on both tomato and mazzard cherry seedlings decreased as the level of nitrate nitrogen increased. Tomato tumors were also reduced in size when the potassium level was increased. An increase in phosphorus levels alone did not affect tumor size, but an increase in both nitrogen and phosphorus led to a greater reduction in tumor size than an increase in nitrogen alone. Tumors were initiated but did not develop on tomato and cherry seedlings grown in Hoagland's solution lacking boron; the growth of tumors appeared to be more dependent on boron than in normal tissues. The level of boron in tomato tissues was shown to decrease with increasing nitrogen levels, suggesting that the nitrogen effects may be related to the boron deficiency. Low nitrate nitrogen levels may also be involved in the observed increase in crown gall in cherry seedlings following soil fumigation (Deep *et al.*, 1968), because nitrification is inhibited following soil fumigation (Huber and Watson, 1974).

b. *Host Specificity.* Although the genus *Agrobacterium* has a wide host range (DeCleene and DeLey, 1976), individual strains usually exhibit some degree of host specificity (Anderson and Moore, 1979). Several early reports established that there were strains that were pathogenic on some hosts but not on others (Smith *et al.*, 1911; Siegler, 1928; Wormald, 1945; Braun, 1950; McKeen, 1954). The recently described biotype-3 strains (Kerr and Panagopoulos, 1977) have been isolated only from grapevine and can infect only a limited number of test hosts (Panagopoulos *et al.*, 1978). Some strains from grapevine (Loper and Kado, 1979), *Lippia,* and mazzard cherry (Anderson and Moore, 1979) are extremely host-specific and form tumors only on the host from which they were isolated.

The molecular basis of host specificity has been investigated in several *A. tumefaciens* strains. An early demonstration by Klein and Klein (1953) that the virulence and host specificity traits were transmitted together from one strain to another in transformation experiments suggests that these two traits are genetically linked. Because we now know that genes for tumor induction are located on the Ti plasmid, it appears likely that genes for host specificity were also plasmid-borne. Transfer of the Ti plasmid from a broad-host-range pathogen to narrow-host-range strains from grapevine extended the host range of the grapevine strains to that of the donor strain (Sonoki *et al.*, 1978; Loper and Kado, 1979). The Ti plasmids from narrow-host-range strains have also been transferred to avirulent strains, resulting in virulent transconjugants with the host range of the donor (Thomashow *et al.*, 1980b; Moore unpublished). These experiments clearly establish that host specificity is associated with the Ti plasmid in these strains, but how this trait is determined by the plasmid genes is unknown. It has been suggested that host specificity might be involved in the binding of *Agrobacterium* to the host (Thomashow *et al.*, 1980b). This is an attractive hypothesis, considering recent evidence that the degree of methylation of host cell wall pectins may partly determine whether the bacterium will bind to the host (Lippincott and Lippincott, 1978a). The expression of host specificity might involve a differential response to the degree of methylation of host cell wall pectins.

c. *Complementation between Strains.*   When a mixture of *A. tumefaciens* and certain other *Agrobacterium* strains is inoculated into pinto bean leaves, tumor growth is greater than with *A. tumefaciens* alone. This complementation between strains involves either an increase in tumor size without an increase in tumor initiation (J. A. Lippincott and B. B. Lippincott, 1969) or an increase in tumor initiation accompanied by an increase in tumor size (Lippincott and Lippincott, 1970).

The first type of complementation has been correlated with the synthesis of a tumor growth factor (TGF-1) in leaves inoculated with the tumor-enhancing strains (Lippincott *et al.*, 1968). TGF-1 has been identified as a carnosine-like derivative of histidine (Lippincott *et al.*, 1972). Its synthesis is induced by many virulent and avirulent *Agrobacterium* strains, but induction requires site attachment by the bacterium (B. B. Lippincott and J. A. Lippincott, 1969). It is interesting that some avirulent strains show both the site attachment and tumor growth factor-inducing abilities that might normally be associated with virulent strains. Although these avirulent strains apparently lack functional Ti plasmids, the possibility of plasmid involvement and even genetic exchange with the host cannot be excluded.

The second type of complementation increases tumor growth by increasing tumor initiation. Several avirulent *Agrobacterium* strains increased the number of tumors formed on bean leaves when coinoculated with a virulent strain (Lippincott and Lippincott, 1970). Tumor growth increased directly with the number of tumors initiated (El Kahlifa and Lippincott, 1968). Tumor initiation complementation is apparently dependent on plasmid DNA and has a requirement for site attachment (Lippincott *et al.*, 1978a,b; Lippincott *et al.*, 1977a). Complementation may involve plasmid-coded products responsible for transfer of the Ti plasmid (Lippincott *et al.*, 1978a,b).

Complementation between strains has also been observed in field-grown stone fruit seedlings. Moore (1977) found that several avirulent *Agrobacterium* strains significantly increased both the number and size of tumors on mazzard cherry seedlings when inoculated *in situ* with virulent strains. Inoculation of avirulent strain K84 with *A. tumefaciens* also increased the number of tumors on peach seedlings in one test (Miller *et al.*, 1979). Tumor initiation complementation probably occurred with strain K84 on peach, since strain K84 enhanced tumor initiation of bean leaves (Lippincott *et al.*, 1978).

E. Multiplication of *Agrobacterium tumefaciens*

1. *Growth in Soil, Tumors, and Rhizosphere*

Precise measurements have not been made of the growth rates of agrobacteria in soil, the rhizosphere, or galls. Agrobacteria generally can be recovered from these three habitats and in some cases show an increase in population over an earlier assay. Agrobacteria also colonize the roots of nonhost plants such as corn (Starkey, 1931) and weeds (L. W. Moore, unpublished). However, there is no evidence that they multiply in fallow, natural soil. In fumigated soil held in the laboratory there was an initial increase in the population after inoculation, followed by a decline to nondetectable levels (Patel, 1928a; Dickey, 1961; Munnecke and Ferguson, 1960). In contrast to laboratory studies, *A. tumefaciens* and *A. radiobacter* are not eliminated from fumigated field soils (Deep *et al.*, 1968; Schroth *et al.*, 1971) even at high fumigation rates (Ross *et al.*, 1970).

The rhizosphere, where the population per gram of roots is 100- to 1000-fold higher than per gram of soil, offers protection to agrobacteria. Although generation times are unavailable for agrobacteria in the rhizosphere, the population of *Agrobacterium* inoculated into host roots increases over time (Fig. 1) (Moore, 1973; New and Kerr, 1972), especially on small lateral roots (Moore, 1977). The population attains a maximum of about 10^6 cells/gm of root regardless whether a natural or an artifical inoculum is introduced.

Microbial competition and antagonism generally suppress agrobacteria, especially *A. tumefaciens*, and the number of *A. radiobacter* recovered from soil or

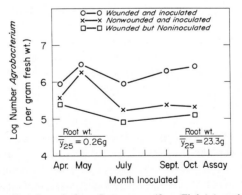

Fig. 1. Recovery of agrobacteria from *Prunus cerasifera* (Ehrh.) (myrobalan plum) seedlings. Different groups of seedlings were inoculated with *A. tumefaciens* at periodic intervals during the growing season, beginning shortly after germination and ending 2 weeks before the roots were assayed. Mean root weights are given for the beginning (April) and end (assay) of the test. Regardless of the inoculation date, all the seedlings were assayed the second week of October by macerating individually three roots from each treatment in a Waring blender and plating serial dilutions of the root suspension onto the selective media of Kado and Heskett (1970) and New and Kerr (1971).

noninfected plants exceeds that of *A. tumefaciens*. *Agrobacterium radiobacter* has a faster growth rate, or it survives better, or *A. tumefaciens* mutates rapidly to avirulence. Despite statements of virulence stability for *A. tumefaciens* (Van Lanen *et al.*, 1952), there is a high rate of spontaneous mutation to avirulence in at least four *Agrobacterium* strains (Cooksey and Moore, unpublished). Alternatively, genes essential to the infection process may be transposed from the Ti plasmid to the chromosome, rendering the pathogen avirulent. Under certain environmental conditions, a reversal of this process may occur, with the transposon moving back to the Ti plasmid, thus restoring virulence.

Although the tumor harbors large populations of agrobacteria, the generation time of *A. tumefaciens* in the gall is unknown and probably varies with the type of host, strain of pathogen, and environmental conditions. Single cells of *A. tumefaciens* inoculated into wounded tomato stems multiplied into thousands of cells within 5 days, but the rate of multiplication was not determined (Hildebrand, 1942). New and Kerr (1972) showed that *A. tumefaciens* inoculated into wounded tomato stems increased in number the first 3 days and then remained constant over the next 7 or 8 days. Thus *A. tumefaciens* in the rhizosphere or in a wound attain a constant population density because of limited nutrients. However, nutrient availability could increase over time as the tumor enlarges.

There is a need for research in the area of population dynamics of *Agrobacterium* in the soil, the rhizosphere, and the tumor, with emphasis on generation times. Tools are now available for performing these studies without extensively disrupting the natural environment. These tools include the use of antibiotic markers in agrobacteria and fluorescent antibody techniques modified according to Schmidt's (1974) procedure. *Agrobacterium* strains could be added alone and in combination with other naturally occurring microbes to the rhizosphere of plants produced aseptically from tissue culture to study the interaction of these microbes. These interactions will be influenced by root exudates and environmental challenges to the plant from herbicides, other plants species, nutrient levels from fertilizers, water potential, and fumigation.

2. Role of Opines

Opines are unusual α-N-substituted amino acid derivatives produced by the tumor cells but have no known function in these cells even though they constitute 0.5–7% by dry weight of the tumor (Gordon, 1981). Genes on the infecting Ti plasmid enter the host cell and induce a tumor that synthesizes a specific opine (Tempe *et al.*, 1979). Opines are growth substrates for oncogenic bacterial strains *in vitro* and induce conjugative activity of Ti plasmids. It is proposed that opines play the same role in nature as *in vitro*.

Tempe *et al.* (1979) suggest that opine production by crown gall cells serves the unique function of promoting Ti plasmid dissemination through bacterial multiplication and conjugation. Since *A. tumefaciens* utilizes opines that it has directed the host tissue to produce following infection, the opine plasmid is hypothesized to create a unique ecosystem for itself, rich in metabolites useful to itself but not to the plant or to other microorganisms. Although in part true, this may be an oversimplification because other microbes are harbored within the tumor. We regularly isolate *A. radiobacter* along with *A. tumefaciens* from naturally infected plants, and certain pseudomonads and fungi are isolated which utilize octopine as a sole carbon and nitrogen source (Moore, unpublished).

Despite the many similarities between agrobacteria and fast-growing rhizobia, octopine and nopaline utilization is uncommon in *Rhizobium* (Lippincott *et al.*, 1973). However, a spontaneous mutant of *Rhizobium trifolii* grew on minimal medium containing octopine as the sole carbon and nitrogen source (Sknotnicki and Rolfe, 1978).

III. Control of Crown Gall Disease

The economic importance of crown gall disease has encouraged many attempts to develop effective control methods, both preventive and therapeutic, but preventive measures are the most effective. We will discuss the various control methods in relation to the specific stages of the disease cycle to which they are directed.

A. BIOLOGICAL METHODS

1. *Biological Control with Microbial Antagonists*

New and Kerr (1972) showed that crown gall in Australia was strongly correlated with a high ratio of virulent to avirulent *Agrobacterium* strains in soil near the infected plants. In contrast, a high proportion of avirulent strains was usually found near healthy plants. Biological control was later achieved by artificially increasing the proportion of avirulent strain K84 on roots of peach seedlings or on peach seed (Kerr, 1972). Strain K84 is now used commercially for the prevention of crown gall on a variety of crops in Australia and the United States and has been tested extensively in many other countries (Moore and Warren, 1979). It has been highly successful in many areas but unsuccessful in others.

The mechanism of biological control by strain K84 has been attributed to the production of a bacteriocin (agrocin 84) that inhibits certain pathogenic strains of *A. tumefaciens* (Kerr and Htay, 1974). Sensitivity to agrocin 84 *in vitro* appears to involve active transport of the molecule into sensitive strains by means of a periplasmic binding protein (Murphy and Roberts, 1979) coded for by Ti plasmid genes (Engler *et al.*, 1975).

Another mechanism for agrocin-84 activity was proposed by Smith and Hindley (1978) who found that incubation of the pathogenic strain C58 in the presence of agrocin 84 reduced its ability to bind to host cells. It was proposed that agrocin 84 inhibited tumor induction by causing a disruption of the cell envelope structure required for binding to the host.

However, biological control in the field is not always correlated with the production of agrocin 84. Three reports (Schroth and Moller, 1976; Moore, 1977; Cooksey and Moore, 1980) have shown that some agrocin 84-resistant *A. tumefaciens* strains can be controlled by strain K84 on field-grown *Prunus* seedlings. In addition, agrocin-resistant pathogens were controlled by strain K84 on pear seedlings but not on apple (Moore, 1979), suggesting that the host species could influence the effectiveness of strain K84.

In greenhouse tomato tests, strain K84 can prevent infection by agrocin-resistant strains when the antagonist is applied 24 hours before the pathogen (Cooksey and Moore, 1980). Infection of agrocin-sensitive strains also was reduced when an agrocin mutant of strain K84 was inoculated into tomato stems 24 hours before the pathogen (Cooksey and Moore, 1980). Strain K84 was thus able to reduce infection independently of agrocin production on *Prunus* and tomato seedlings, but the action was nonspecific and required a 24-hour delay on tomato seedlings. Because strain K84 was shown to have wound site-binding activity (Whatley *et al.*, 1976), it is possible that physical blockage by attachment to infection sites is another mechanism by which strain K84 prevents infection. It might bind readily to *Prunus* wound sites but not as readily to apple or tomato seedlings. As mentioned earlier, variable attachment to different host species has been proposed as a mechanism of host specificity in *A. tumefaciens* (Thomashow *et al.*, 1980b).

Because strain K84 is ineffective in some areas and on certain crops, hundreds of other bacteriocin-producing *Agrobacterium* strains have been tested for biological control (Kerr and Panagopolous, 1977; Moore, 1977; Garrett, 1979), but they have been ineffective. Moore (1977) isolated three *A. radiobacter* strains from Oregon that prevented crown gall on mazzard cherry seedlings growing *in situ* when inoculated with *A. tumefaciens*, but they did not show antibiotic activity *in vitro* against the pathogens. These strains may have prevented infection by blocking infection sites. Because the importance of antibiotic production in biological control of crown gall in the field even with strain K84 is not clear, the search for new biological control agents should not be limited to strains that produce antibiotics.

The use of microbes other than antagonistic agrobacteria has only recently been explored. The presence of natural fungal competitors in nursery soils was suggested by the finding by Deep and Young (1965) that preplanting fungicide treatments increased the incidence of crown gall of cherry seedlings. When a suspension of unidentified fungi was later used as a preplanting treatment, the incidence of galled seedlings was reduced about 25%. Cooksey and Moore (1980) isolated species of *Penicillium, Aspergillus,* and *Trichoderma* that produced potent antibiotics against species of *Agrobacterium*. When mixed with *A. tumefaciens,* some of these antagonists prevented infection of radish and tomato seedlings in greenhouse experiments. In field tests, several strains also reduced the incidence of galling on mazzard cherry seedlings. Species of *Pseudomonas* and *Bacillus* were also effective antagonists. In some cases, there was a low correlation between *in vitro* activity and biological control. One *Pseudomonas* antagonist prevented tumor formation on tomato when inoculated with strains of *A. tumefaciens* that were insensitive *in vitro* to its antibiotic activity. Other antagonists failed to prevent galling by pathogenic strains that were sensitive *in vitro*. These data suggest that mechanisms other than antibiotic production can be important in biological control with microbial antagonists.

2. *Wound Callusing*

Since wounds were necessary for optimum infection of three *Prunus* species (Moore, 1976), tests for reducing the susceptibility of these wounds were conducted. Wounds on stored dormant mazzard cherry seedlings remained susceptible for 107 days, contrasted to 6 days for wounds made in June on mazzard seedlings growing in the field (Moore, 1976). Below-ground wounds on red raspberry remained susceptible to infection for 7 weeks (Hildebrand, 1941), and *A. tumefaciens* remained alive for 12 weeks in noninfected wounds of *Cupressus arizonica* (Greene and Melhus, 1919; Brown and Evans, 1933); thus wounds may remain open to infection for varying degrees of time. The question is whether we can alter cultural management to induce more rapid healing and therefore lessen susceptibility.

Braun and Mandle (1948) completely halted tumorigenesis in *Kalanchoe daigremontana* when these plants were placed at 32°C shortly after inoculation. Bacteria were capable of transforming normal cells to tumor cells at 25°C only during a 4-day period immediately following wounding of the host. These authors reported that wound healing progressed in *Kalanchoe* at about the same rate as 25°C as at 32°C. Wound healing was accompanied by active cell division between the second and third days after wounding. However, Lipetz (1965) maintained *K. daigremontana* plants at 32°C after wounding and observed that wound cambium appeared 12 hours before comparable plants maintained at 25°C, indicating a faster rate of wound healing at 32°C.

We hypothesized that acceleration of wound healing by higher temperatures could protect root-pruned cherry seedlings from infection by *A. tumefaciens*. Mazzard cherry seedlings were root-pruned and placed at 18° and 24°C in a rooting medium, and the tops were kept at 4°–7°C. Samples of seedlings were removed periodically over a 3-week period, inoculated with *A. tumefaciens,* potted in natural soil, and grown in the greenhouse. After 1 week at 24°C the plants were reduced from 60% incidence of infection at zero time to 5%, whereas 3 weeks was needed at 18°C. A commercial grower heat-treated root-pruned mazzard cherry for 3 weeks before planting, and 66% of the nonheated seedlings were galled from natural inoculum compared to 6% of those heated; thus heat treatment is one means of preventing crown gall.

3. *Host Resistance*

Resistant germplasm has been used extensively in plant pathology as a deterrent to diseases. According to Smith (1923), some plant species appeared immune to *A. tumefaciens* (olive, onion, and garlic), others were slightly susceptible (avocado and most monocots), and others were highly susceptible (raspberry, blackberry, rose, peach, almond, and grape). Some apple varieties show marked differences in susceptibility to *A. tumefaciens* (Brown, 1929), and Smith (1925) listed a number of resistant *Prunus* species that might be of value as dwarfing rootstocks. Fourteen out of 83 grape varieties tested were completely free of crown gall; a few were badly diseased, but most were only slightly injured (Garcia and Rigne, 1913). Almond and peach were the most susceptible stone fruits in California, and propagation of walnuts on *Juglands regis* rootstock was abandoned in favor of the less susceptible *J. califorinica* or *J. hindsii* rootstock (Wilson and Ogawa, 1979). There appears to be a scion effect on susceptibility, because some cultivars are much more susceptibile upon the same rootstock than others (Greene and Melhus, 1919).

Little is known about the mechanism of resistance to crown gall. The plant sap of a begonia immune to crown gall had extreme acidity (approximately 0.1 *N*), and this was given as a possible reason for its resistance (Smith and Quirk, 1926). Recently, cowpea plants aquired resistance to *A. tumefaciens* after they were inoculated with cowpea mosaic virus and failed to develop tumors (Saedi *et al.*, 1979). Attempts to increase resistance by thermal neutron irradiation of seeds of peach, almond, and cherry was unsuccessful (DeVay, as cited in Wilson and Ogawa, 1979).

A. CHEMOTHERAPY AND SOIL FUMIGATION

Numerous chemicals for protecting wounds from infection by *A. tumefaciens* and eradicating existing galls from plants have been tested. Rootstocks from a cherry layer bed containing galled mother plants were dipped in a 0.25% sodium hypochlorite solution prior to planting, but they were not protected and often were damaged by the chemical treatment (Garrett, 1971). In some instances, there has been a marked increase in crown gall on mazzard cherry trees and other woody roseaceous hosts after a preplanting treatment with the fungicide dichlone. Similar results occurred following the use of captan and certain dosages of terramycin (Deep and Young, 1965). Mazzard cherry seedlings inoculated with *A. tumefaciens* before dipping in Agristrep (streptomycin sulfate), copper sulfate, or a captan–sodium hypochlorite solution exhibited a higher percentage of galled seedlings than inoculated controls (Moore, 1977).

Investigators do not always agree on results relative to chemical treatments with the same compound. For example, the incidence of infected grafts was reduced from 48 to 8% when apple seedlings were surface-sterilized with mercuric chloride before grafting (Siegler and Piper, 1931). When the seedlings were washed thoroughly and surface-sterilized, the infection was less than 1%. In contrast, Sherbacoff (1925) found that a 20-minute preplanting dip of apple trees in mercuric chloride did not prevent infection. Part of the disagreement could have arisen because Sherbacoff soaked his experimental trees overnight in water prior to chemical treatment. Since *A. tumefaciens* is easily disseminated in water, any galled trees present during the overnight period of soaking could have contaminated the others. Siegler and Bowman (1940) observed that stratified peach seed treated with mercuric chloride before planting had much less infection than untreated seed. However, almond seed treated before planting with 1 ppm mercuric chloride plus a wetting agent, with 1% sodium hypochlorite plus a wetting agent, was infected approximately equally to untreated controls (Schroth *et al.*, 1971).

Efforts to eradicate tumors on orchard trees have met with varying degrees of success. Galls on severely infected trees were removed surgically, and the tree roots dipped in different concentrations of mercuric chloride, hydrogen peroxide, Formalin, potassium permanganate, copper sulfate, methyl violet, and salycylic acid (Ness, 1917). By the end of the growing season, galls had usually returned where they had initially been cut off. Most of the trees treated with mercuric chloride were killed. Application of a sodium dinitrocresol (Elgetol)–methonal mixture during July through December 1939 gave 100% eradication of galls 3–10 in. in diameter on almond trees, and no new tumors were observed in September of 1940 (Ark, 1941). Schroth and Hildebrand (1968) tested a series of aromatic hydrocarbons that selectively eradicated crown gall from a wide range of plant species. This selective toxicity was attributed to enhanced penetration of the tumor cell walls and tissue (Moore and Schroth, 1976). The absence of an epidermal layer on the tumors also allowed rapid loss of water from the tissue following treatments. In orchards, roots of galled trees were exposed by spraying with water under high pressure, and tumors were treated with the aromatic hydrocarbons (Ross *et al.*, 1970).

The incidence of crown gall disease on almonds and roses was reduced about 70% when soils were fumigated before planting with a mixture of chloropicrin and methyl bromide (Ross *et al.*, 1970). Complete control was not obtained even when the amount of fumigant was doubled. In contrast, the disease was enhanced on mazzard cherry seedlings following fumigation of the soil in which the seeds were planted (Deep *et al.*, 1968). These conflicting reports regarding the effect of soil fumigation on crown gall disease could arise from inadequate soil moisture which is essential for good fumigation (Munnecke and Ferguson, 1960). In these experiments, bacteria were killed relatively easily by fumigating the soil. Dickey (1962) reported that the population of *Agrobacterium* in soil was reduced more than 99% at depths of 3, 9, and 15 in. using a methyl bromide–chloropicrin fumigant, but agrobacteria present in buried galls were not affected. In soil, the greatest reduction was from 10^8 to 10^4 CFU/gm. However, a 99% reduction of 10^8 CFU/gm still leaves 10^6 cells, which is ample to infect plants.

C. SANITATION AND CULTURAL MANAGEMENT

Ideally, only pathogen-free, symptomless plant propagules should be planted, because disease-free planting stock may be colonized by *A. tumefaciens* or have latent infections. However, current methods of detection are laborious and seldom used except with chrysanthemum cuttings that are culture-indexed for several different pathogens.

Sanitation practices are aimed at preventing contamination of healthy plants and equipment by soil particles, plant debris, and water which can harbor *A. tumefaciens*. Materials used to disinfect pruning tools, benches, and equipment include steam, 70% denatured alcohol, 5% Formalin, and various commercial brands of disinfectant. Contaminated irrigation water should be chlorinated, and plant debris destroyed to provide clean working areas and tools.

Dormant planting stock commonly is heeled in through the winter in sawdust or soil. However, when diseased and healthy apple and peach trees were heeled in together for 1 month prior to planting, nearly 100% of the plants were infected at the end of the growing season (Ness, 1917). Similarly, raspberry plants that were heeled in with infected plants showed a much higher percentage of galling than plants that were heeled in with healthy plants (Banfield and Mandanberg, 1935). It was also shown that plants selected from areas in the field with a low incidence of crown gall produced much cleaner plants than symptomless plants from a field with many diseased trees. Thus culture procedures should be modified to prevent contamination of healthy trees in this manner.

Nonsusceptible plants should be grown for several years in soil where infected trees or plants have been grown before replanting with a susceptible crop (Ark, 1954; Townsend, 1915; Ness, 1917). Patel (1928a) stated that *A. tumefaciens* generally was not recovered from soils where susceptible plants had not been grown. These findings are in agreement with those of New and Kerr (1972) who found *A. tumefaciens* in soil only near a galled tree. However, Schroth *et al.* (1971) recovered low ratios of *A. tumefaciens* to *A. radiobacter* from pastureland never planted with susceptible crops.

Crown gall has been reduced by growing plants in soils ammended with sulfur (600 lb/acre) to make the soil more acid (Ark, 1941; Sherbakoff, 1925). Siegler

(1938) added lime in every other row of acid soil before planting peach seeds. After one growing season, 32 and 3% of the peach seedlings from limed (pH 6.8) and nonlimed (pH 5.0) soils, respectively, were galled. However, in the coastal regions of Oregon and Washington, crown gall can reach epidemic proportions in soils that are naturally acid because of normally high precipitation. Conversely, high levels of disease can occur in the relatively alkaline soils east of the Cascades in Washington and in the Sacramento Valley (Siegler, 1938). We conclude that infection is possible wherever a plant can grow satisfactorily, probably because of strain adaptation by agrobacteria to the specific environment.

IV. Conclusions

Agrobacteria can be classified into two major groups called biotypes 1 and 2. Strains with characteristics intermediate between biotypes 1 and 2 have been found and in some cases named biotype 3 and 1-2. More intermediates are likely to be discovered. The distribution and significance of these biotypes with reference to the biology of agrobacteria is understood poorly. Combinations of selective media must be used to study the distribution of biotypes, since they often occur together.

The source of *A. tumefaciens* inoculum in field soil in the absence of host plants is still not clear. Plants can become infected when planted in soil with low to nondetectable levels of the pathogen, yet large numbers (10^5–10^6) of the pathogen are required to infect artificially a high percentage of plants. Usually, the ratio of pathogenic to nonpathogenic strains in soil is 1:500 or greater in the absence of a galled plant, or no pathogenic strains of *Agrobacterium* are recovered. The plant does not appear to enrich selectively for the pathogen, since both pathogens and nonpathogens colonize roots equally well. We know little about the influence of root exudates on *A. tumefaciens* or its interaction with other microbes in the rhizosphere. Such interactions may favor the multiplication of *A. tumefaciens* to a level at which infection can occur. Another hypothesis is that the Ti plasmid functions as an episome; it could become integrated into the bacterial chromosome until some stimulus from plant roots or the associated microflora triggered its excision and restored its autonomous plasmid functions. However, there is no evidence to support this hypothesis at present. Avirulent *Agrobacterium* strains also carry large plasmids (Merlo and Nester, 1977; Satikkholeslam *et al.*, 1979) which may be converted to Ti plasmids in nature. However, hybridization of plasmid DNA from these avirulent strains with that from *A. tumefaciens* strains C58 and A6 revealed only 10–50% homology with strain C58 and about 10% with strain A6.

We now have an understanding of the basic molecular events that occur during crown gall tumorigenesis: Bacterial plasmid DNA is transferred to the plant cell and integrated into the plant chromosome where it directs the development of the tumorous state. However, our understanding of events that precede and follow the transfer of plasmid DNA is limited.

Attachment of the bacterium to the host cell wall is accepted as an essential stage in tumorigenesis, but these experiments have been performed on a limited number of hosts and only in the laboratory or greenhouse. It would be worthwhile to extend this work to economically important plant species. The possibility should be examined also that differential wall attachment is involved in host specificity.

References

Abo-El-Dahab, M. K., El-Goorani, M. A., and El-Wakil, M. A. (1978). *Phytopath. Z.* **91**, 14–22.

Alconero, R. (1980). *Plant Dis.* **64**, 835–838.

Anderson, A. R. (1977). Ph.D. thesis, Oregon State University.

Anderson, A. R., and Moore, L. W. (1979). *Phytopathology* **69**, 320–323.

Ark, P. A. (1941). *Phytopathology* **31**, 956–957.

Ark, P. A. (1954). *Plant Dis. Rep.* **38**, 207–208.

Banfield, W. M., and Mandanberg, E. C. (1935). *Phytopathology* **25**, 5–6. (Abstr.)

Beiderbeck, R. (1970a). *Z. Naturforsch. B* **25**, 735–738.

Beiderbeck, R. (1970b). *Z. Naturforsch. B* **25**, 1458–1460.

Braun, A. C. (1943). *Am. J. Bot.* **30**, 674–677.

Braun, A. C. (1947). *Growth Symp.* **11**, 325–337.

Braun, A. C. (1950). *Phytopathology* **40**, 1058–1060.

Braun, A. C. (1952). *Growth* **16**, 65–74.

Braun, A. C. (1954). *Annu. Rev. Plant Physiol.* **5**, 133–162.

Braun, A. C. (1958). *Proc. Natl. Acad. Sci. U.S.A.* **44**, 344–349.

Braun, A. C. (1959). *Proc. Natl. Acad. Sci. U.S.A.* **45**, 932–938.

Braun, A. C., and Mandle, R. J. (1948). *Growth* **12**, 255–269.

Braun, A. C., and Wood, H. N. (1976). *Proc. Natl. Acad. Sci. U.S.A.* **73**, 496–500.

Brown, J. G., and Evans, M. M. (1933). *Phytopathology* **23**, 97–101.

Brown, N. A. (1929). *J. Agric. Res.* **39**, 747–766.

Chamberlain, G. C. (1962). *Can. Plant Dis. Surv.* **42**, 208–211.

Chilton, M. D., Drummond, M. H., Merlo, D. J., Sciaky, D., Montoya, A. L., Gordon, M. P., and Nester, E. W. (1977). *Cell* **11**, 263–271.

Clark, A. G. (1969). *J. Appl. Bacteriol.* **32**, 348–354.

Cooksey, D. A., and Moore, L. W. (1980). *Phytopathology* **70**, 506–509.

DeCleene, M., and DeLey, J. (1976). *Bot. Rev.* **42**, 389–466.

Deep, I. W., and Hussin, H. (1965). *Plant Dis. Report.* **49**, 734–735.

Deep, I. W., and Young, R. A. (1965). *Phytopathology* **55**, 212–216.

Deep, I. W., McNeilan, R. A., and MacSwan, I. C. (1968). *Plant Dis. Rep.* **52**, 102–105.

Dhanvantari, B. N. (1978). *Can. J. Bot.* **56**, 2309–2311.

Dickey, R. S. (1961). *Phytopathology* **51**, 607–614.

Dickey, R. S. (1962). *Plant Dis. Rep.* **46**, 73–76.

Einset, J. W., and Cheng, A. (1979). *In Vitro* **15**, 703–708.

El Khalifa, M. D., and Lippincott, J. A. (1968). *J. Exp. Bot.* **19**, 749–759.

El Khalifa, M. D., and Yousif, A. M. (1974). *Physiol. Plant Pathol.* **4**, 435–442.

Engler, G., Holsters, M., Van Montagu, M., and Schell, J. (1975). *Mol. Gen. Genet.* **138**, 345–349.

Garcia, F., and Rigney, J. W. (1913). *New Mexico Agric. Exp. Sta. Bull.* **85.**
Garrett, C. M. E. (1971). *Rep. E. Malling Res. Sta. for 1970,* p. 104.
Garrett, C. M. E. (1979). *Ann. Appl. Biol.* **91,** 221–226.
Gordon, M. (1981). *In* "The Biochemistry of Plants" (A. Marcus, ed.), Vol. 6, pp. 531–570. Academic Press, New York.
Greene, L., and Melhus, I. E. (1919). *Iowa State College Agric. Exp. Sta. Res. Bull.* **50,** 147–176.
Griffin, G. D., Anderson, J. L., and Jorgenson, E. C. (1968). *Plant Dis. Rep.* **52,** 492–493.
Hedgecock, G. G. (1910). Field studies of the crown gall of the grape. *USDA Bull.* No. 183. 33 pp.
Hildebrand, E. M. (1941a). *Plant Dis. Rep.* **25,** 200–202.
Hildebrand, E. M. (1941b). *N. SP. J. Agric. Res.* **61,** 685–696.
Hildebrand, E. M. (1942). *J. Agric. Res.* **65,** 45–59.
Hill, J. B. (1928). *Phytopathology* **18,** 553–564.
Hooykaas, P. J. J., Roobol, C., and Schilperoort, R. A. (1979). *J. Gen. Microbiol.* **110,** 99–109.
Huber, D. M., and Watson, R. D. (1974). *Annu. Rev. Phytopathol.* **12,** 139–165.
Hussin, H., and Deep, I. W. (1965). *Phytopathology* **55,** 575–578.
Kado, C. I. (1976a). *Beltsville Symp. Agric. Res. Virol. Agric.* **1,** 247–266.
Kado, C. I. (1976b). *Annu. Rev. Plant Pathol.* **14,** 265–308.
Kado, C. I., and Heskett, M. G. (1970). *Phytopathology* **60,** 969–976.
Kaiss-Chapman, R. W., and Morris, R. O. (1977). *Biochem. Biophys. Res. Commun.* **76,** 453–459.
Kaper, J. M., and Veldstra, H. 1958. *Biochim. Biophys. Acta* **30,** 401–420.
Keane, P. J., Kerr, A., and New, P. B. (1970). *Aust. J. Biol. Sci.* **23,** 585–595.
Kerr, A. (1969). *Aust. J. Biol. Sci.* **22,** 111–116.
Kerr, A. (1972). *J. Appl. Bacteriol.* **35,** 493–497.
Kerr, A. (1974). *Soil Sci.* **118,** 168–172.
Kerr, A., and Htay, K. (1974). *Physiol. Plant Pathol.* **4,** 37–44.
Kerr, A., and Panagopoulos, C. G. (1977). *Phytopathol. Z.* **90,** 172–179.
Kersters, K., DeLey, J., Sneath, P. H. A., and Sackin, M. (1973). *J. Gen. Microbiol.* **78,** 227–239.
Klein, D. T., and Klein, R. M. (1953). *J. Bacteriol.* **66,** 220–228.
Klein, R. M. (1958). *Proc. Natl. Acad. Sci. U.S.A.* **44,** 350–354.
Klein, R. M. (1965). *Handb. Pflanzenphysiol.* **15,** 209–235.
Klein, R. M., Rasch, E. M., and Swift, H. H. (1953). *Cancer Res.* **13,** 499–502.
Lehoczky, J. (1968). *Phytopathol. Z.* **63,** 239–246.
Lelliott, R. A. (1971). *Plant Pathol.* **20,** 59–63.
Liao, D. H., and Heberlein, G. T. (1979). *Can. J. Microbiol.* **25,** 185–191.
Lin, B.-C., and Kado, C. I. (1977). *Can. J. Microbiol.* **23,** 1554–1561.
Link, G. K., Wilcox, H. W., Eggers, V., and Klein, R. M. (1953). *Am. J. Bot.* **40,** 436–444.
Lipetz, J. (1965). *Science* **149,** 865–867.
Lippincott, B. B., and Lippincott, J. A. (1966). *J. Bacteriol.* **92,** 937–945.
Lippincott, B. B., and Lippincott, J. A. (1969). *J. Bacteriol.* **97,** 620–628.
Lippincott, B. B., Margot, J. B., and Lippincott, J. A. (1977a). *J. Bacteriol.* **132,** 824–831.
Lippincott, B. B., Whatley, M. H., and Lippincott, J. A. (1977b). *Plant Physiol.* **59,** 388–390.
Lippincott, J. A., and Lippincott, B. B. (1969). *J. Bacteriol.* **99,** 496–502.
Lippincott, J. A., and Lippincott, B. B. (1970). *Infect. Immun.* **2,** 623–630.
Lippincott, J. A., and Lippincott, B. B. (1978a). *Science* **199,** 1075–1078.
Lippincott, J. A., and Lippincott, B. B. (1978b). *Phytopathology* **68,** 365–370.
Lippincott, J. A., Webb, J. H., and Lippincott, B. B. (1965). *J. Bacteriol.* **90,** 1155–1156.
Lippincott, J. A., Lippincott, B. B., and El Khalifa, M. D. (1968). *Physiol. Plant.* **21,** 731–741.
Lippincott, J. A., Lippincott, B. B., and Chi-Cheng Chang. (1972). *Plant Physiol.* **49,** 131–137.
Lippincott, J. A., Beiderbeck, R., and Lippincott, B. B. (1973). *J. Bacteriol.* **116,** 378–383.
Lippincott, B. B., Margot, J. B., and Lippincott, J. A. (1977a). *J. Bacteriol.* **132,** 824–831.
Lippincott, B. B., Whatley, M. H., and Lippincott, J. A. (1977b). *Plant Physiol.* **59,** 388–390.

Lippincott, J. A., Chang, C., Creaser-Pence, V. R., Birnberg, P. R., Rao, S. S., Margot, J. B., Whatley, M. H., and Lippincott, B. B. (1978). *Proc. Int. Conf. Plant Pathol. Bacteriol., 4th, Angers, France* pp. 189–197.

Lippincott, J. A., Lippincott, B. B., and Starr, M. P. (1981). *In* "The Prokaryotes" (M. P. Starr, N. Stolp, A. Valows, H. G. Schlegel, and H. G. Truper, eds.). Springer-Verlag, Berlin and New York, in press.

Liu, S.-T., and Kado, C. I. (1979). *Biochem. Biophys. Res. Commun.* **90,** 171–178.

Loper, J. E., and Kado, C. I. (1979). *J. Bacteriol.* **139,** 591–596.

MacNab, R. M., and Koshland, D. E., Jr. (1972). *Proc. Natl. Acad. Sci. U.S.A.* **69,** 2509–2512.

McKeen, W. E. (1954). *Phytopathology* **44,** 651–655.

Matthysse, A. G., Wyman, P. M., and Holmes, K. V. (1978). *Infect. Immun.* **22,** 516–522.

Merlo, D. J., and Nester, E. W. (1977). *J. Bacteriol.* **129,** 76–80.

Miller, H. N. (1975). *Phytopathology* **65,** 805–811.

Miller, R. W., Brittain, J. A., and Watson, T. (1979). *Fruit South. July* pp. 30–32.

Moore, L. W. (1973). *Int. Congr. Plant Pathol., 2nd, Minneapolis, Minn.*

Moore, L. W. (1976). *Am. Nurseryman* 1954.

Moore, L. W. (1977). *Phytopathology* **67,** 139–144.

Moore, L. W. (1979). *In* "Soil-borne Plant Pathogens" (B. Schippers and W. Gams, eds.), pp. 533–568. Academic Press, New York.

Moore, L. W., and Allen, J. (1977). *Proc. Am. Phytopathol. Soc.* **4,** PA-40 (Abstr.).

Moore, L. W., and Schroth, M. N. (1976). *Phytopathology* **66,** 1460–1465.

Moore, L. W., and Tingey, D. T. (1976). *Phytopathology* **66,** 1328–1333.

Moore, L. W., and Warren, G. (1979). *Annu. Rev. Phytopathol.* **17,** 163–179.

Munnecke, D. E., and Ferguson, J. (1960). *Plant Dis. Rep.* **44,** 552–555.

Munnecke, D. E., Chandler, P. A., and Starr, M. P. (1963). *Phytopathology* **53,** 788–799.

Murphy, P. J., and Roberts, W. P. (1979). *J. Gen. Microbiol.* **114,** 207–213.

Ness, H. (1917). *Texas Agric. Exp. Sta. Bull.* No. 211.

New, P. B., and Kerr, A. (1971). *J. Appl. Bacteriol.* **34,** 233–236.

New, P. B., and Kerr, A. (1972). *J. Appl. Bacteriol.* **35,** 279–287.

New, P. B., and Milne, K. S. (1976). *N. Z. J. Exp. Agric.* **4,** 109–115.

Nigh, E. L. (1966). *Phytopathology* **56,** 150. (Abstr.).

Orion, D., and Zutra, D. (1971). *Isr. J. Agric. Res.* **21,** 27–29.

Panagopoulos, C. G., and Psallidas, P. G. (1973). *J. Appl. Bacteriol.* **36,** 233–240.

Panagopoulos, C. G., Psallidas, P. G., and Alivizatos, A. S. (1978). *Proc. Int. Conf. Plant Pathog. Bact., 4th, Angers, France* pp. 221–228.

Patel, N. K. (1928a). *Phytopathology* **18,** 331–343.

Patel, M. K. (1928b). *Phytopathology* **16,** 577.

Patel, M. K. (1929). *Phytopathology* **19,** 295–300.

Patrick, Z. A. (1955). *Can. J. Bot.* **32,** 705–735.

Riker, A. J. (1923). *J. Agric. Res.* **25,** 119–132.

Riker, A. J. (1926). *J. Agric. Res.* **32,** 83–96.

Riker, A. J., and Hildebrand, E. M. (1934). *J. Agric. Res.* **48,** 887–912.

Ross, N., Schroth, M. N., Sanborn, R., O'Reilly, H. J., and Thompson, J. P. (1970). *Calif. Agric. Exp. Sta. Bull.* No. 845.

Sacristan, M. D., and Melchers, G. (1977). *Mol. Gen. Genet.* **152,** 111–117.

Saedi, D., Bruening, G., Kado, C. I., and Dutra, J. C. (1979). *Infect. Immun.* **23,** 298–304.

Satikholeslam, S., Lin, B.-C., and Kado, C. I. (1979). *Phytopathology* **69,** 54–58.

Schmidt, E. L. (1974). *Soil Sci.* **118,** 141–149.

Schroth, M. N., and Hildebrand, D. C. (1968). *Phytopathology* **58,** 848–854.

Schroth, M. N., and Moller, W. J. (1976). *Plant Dis. Rep.* **60,** 275–278.

Schroth, M. N., Weinhold, A. R., McCain, A. H., Hildebrand, D. C., and Ross, N. (1971). *Hilgardia* **40,** 537-552.

Schroth, M. N., Thompson, J. P., and Hildebrand, D. C. (1965). *Phytopathology* **55,** 645-647.

Sherbakoff, C. D. (1925). *Phytopathology* **15,** 105-109.

Siegler, E. A. (1928). *J. Agric. Res.* **37,** 301-313.

Siegler, E. A. (1938). *Phytopathology* **28,** 858-859.

Siegler, E. A., and Bowman, J. J. (1940). *Phytopathology* **30,** 417-426.

Siegler, E. A., and Piper, R. V. (1931). *J. Agric. Res.* **43,** 985-1002.

Sknotnicki, M. L., and Rolfe, B. G. (1978). *J. Bactriol.* **133,** 518-526.

Smith, C. O. (1912). *Univ. Calif. Agric. Exp. Sta. Tech. Bull.* No. 235, pp. 531-557.

Smith, C. O. (1925). *J. Agric. Res.* **31,** 957-971.

Smith, C. O., and Cochran, L. C. (1944). *Plant Dis. Rep.* **28,** 160-162.

Smith, E. F. (1920). "Bacterial Diseases of Plants." Saunders, Philadelphia, Pennsylvania.

Smith, E. F. (1923). *J. Radiology* **4,** 295-317.

Smith, E. F., and Quirk, A. J. (1926). *Phytopathology* **16,** 491-508.

Smith, E. F., Brown, N., and Townsend, C. O. (1911). *US Dept. Agric. Bur. Plant Indust. Bull.* No. 213.

Smith, E. F., Brown, N. A., and McCulloch, L. (1912). *US Dept. Agric. Bur. Plant Indust. Bull.* No. 225.

Smith, V. A., and Hindley, J. (1978). *Nature (London)* **276,** 498-500.

Sonoki, S., Ireland, C. R., Loper, J. E., Baraka, M., and Kado, C. I. (1978). *Proc. Int. Conf. Plant Pathol. Bacteriol., 4th, Angers, France* pp. 133-142.

Spiers, A. G. (1979). *N. Z. J. Agric. Res.* **22,** 631-636.

Starkey, R. L. (1931). *Soil Sci.* **32,** 367-393.

Stonier, T. (1956). *Am. J. Bot.* **43,** 647-655.

Süle, S. (1978). *J. Appl. Bacteriol.* **44,** 207-213.

Tempe, J., Petit, A., Holsters, M., van Montagu, M., and Schell, J. (1977). *Proc. Natl. Acad. Sci. U.S.A.* **74,** 2848-2849.

Tempe, J., Guyon, P., Tepfer, D., and Petit, A. (1979). *In* "Plasmids of Medical, Environmental and Commercial Importance" (K. N. Timmis and A. Phler, eds.), pp. 353-363. Elsevier, Amsterdam.

Thomashow, M. F., Nutter, R., Montoya, A. L., Gordon, M. P., and Nester, E. W. (1980a). *Cell* **19,** 729-739.

Thomashow, M. F., Panagopoulos, C. G., Gordon, M. P., and Nester, E. W. (1980b). *Nature (London)* **283,** 794-796.

Toumey, J. W. (1900). *Univ. Ariz. Agric. Exp. Sta. Bull.* No. 33.

Townsend, C. O. (1915). *USDA Bull.* No. 203.

Turgeon, R., Wood, H. N., and Braun, A. C. (1976). *Proc. Natl. Acad. Sci. U.S.A.* **73,** 3562-3564.

Upper, D., Helgeson, J. P., Kemp, J. D., and Schmidt, C. J. (1970). *Plant Physiol.* **45,** 543-547.

Van Lanen, J. M., Baldwin, I. L., and Riker, A. J. (1952). *J. Bacteriol.* **63,** 715-721.

Weinhold, A. R. (1970). *In* "Root Disease and Soilborne Pathogens" (T. A. Toussoun, R. V. Bega, and P. E. Nelson, eds.), pp. 22-24. Univ. of Calif. Press, Berkeley, California.

Whatley, M. H., Bodwin, J. S., Lippincott, B. B., and Lippincott, J. A. (1976). *Infect. Immun.* **13,** 1080-1083.

Whatley, M. H., Margot, J. B., Schell, J., Lippincott, B. B., and Lippincott, J. A. (1978). *J. Gen. Microbiol.* **107,** 395-398.

White, L. O. (1972). *J. Gen. Microbiol.* **72,** 565-574.

White, P. R., and Braun, A. C. (1942). *Cancer Res.* **2,** 597-617.

Wilson, E. E., and Ogawa, J. M. (1979). "Fungal, Bacterial, and Certain Nonparasitic Diseases of Fruit and Nut Crops in California." Agric. Sci. Publ., Univ. of Calif., Berkeley, California.

Wormald, H. (1945). *Trans.Br. Mycol. Soc.* **28**, 134–146.

Yang, F., Merlo, D. J., Gordon, M. P., and Nester, E. W. (1980a). *Mol. Gen. Genet.* **179**, 223–226.

Yang, F., Montoya, A. L., Merlo, D. J., Drummond, M. H., Chilton, M.-D., Nester, E. W., and Gordon, M. P. (1980b). *Mol. Gen. Genet.* **177**, 707–714.

Young, R. A. (1954). *Plant Dis. Rep.* **38**, 417–420.

Zaenen, I., Van Larebeke, N., Teuchy, H., Van Montague, M., and Schell, J. (1974). *J. Mol. Biol.* **86**, 109–127.

Zvyagintsev, D. G. (1970). *Microbiologiya* **39**, 161–165.

INTERNATIONAL REVIEW OF CYTOLOGY, SUPPLEMENT 13

Agrobacterium tumefaciens in Agriculture and Research

FAWZI EL-FIKI AND KENNETH L. GILES

Department of Life Sciences, Worcester Polytechnic Institute, Worcester, Massachusetts

I.	Introduction	47
II.	Crown Gall Disease around the World	48
III.	Host Range	50
IV.	Factors Affecting Development of the Disease	50
V.	Virulence and Avirulence of *Agrobacterium tumefaciens*	52
VI.	Agricultural Control of *Agrobacterium tumefaciens*	54
VII.	Agrocin and Its Mode of Action on *Agrobacterium tumefaciens*	54
VIII.	Chemical Treatments	55
IX.	Evaluation of Present Methods of Control	56
	References	56

I. Introduction

Crown gall disease, caused by the bacterium *Agrobacterium tumefaciens,* causes major crop losses in agriculture throughout the world. These losses are most marked in horticultural stone fruits and other ornamental crops, though many dicotyledonous plants can be affected. The pathogen has a wide host range and is an economic problem in all countries exporting fruit, since until recently there had been no truly effective method for controlling the disease. Even cutting the trees and replanting with new cultivars has not stopped the advance of losses caused by this disease. Crown gall disease is soil-borne, and once infected, trees suffer a decrease in yield which becomes worse year after year. One method of containing the disease has been to change to the cultivation of monocotyledonous plants, which are not hosts for *Agrobacterium,* and some resistant dicotyledonous varieties do exist.

From another point of view, that of plant pathological research, *A. tumefaciens* is most attractive, because it represents the first case of well-defined genetic parasitism in plants. The plant tumors formed as a result of crown gall disease make a useful parallel to animal tumors and have been used as such in cancer research. The method of integration of the tumor-inducing (Ti) plasmid has also attracted the attention of would-be genetic engineers of plants, since it might provide some lessons about the types of vectors necessary for the transfer of DNA into foreign cells.

II. Crown Gall Disease around the World

It was in 1897 that Cavara first isolated a bacterial organism from galls formed on grape stems. He found that these organisms could reproduce the disease by subsequent inoculation into uninfected stems. In 1907, Smith and Townsend described the organism in detail and named it *Bacterium tumefaciens*. Once described and characterized, it was soon recognized as a major problem in horticultural nurseries throughout Europe and, with the increasing transportation of stock plants and cultivars around the world, it soon became evident as a worldwide problem whose economic implications should not be underestimated.

In 1953 Wood stated that financial losses due to the disease in the United States of America were somewhere on the order of 3 billion dollars. Riker and Hilderbrandt (1953) noted that crown gall disease in the United States affected particularly peach, apple, pear, rose, and sugar beet. They also recognized the fact that the galls appeared more frequently at ground level. These galls often became sites of subsequent infection by other bacteria and fungi, the secondary infection occurring particularly during wet weather. Dochinger (1969) reported aerial galls on the branches and stems of three hybrid poplars in Iowa. Tumors were also produced in 16 species of *Acer, Aesculus, Alnus, Juglans, Quercus, Robinia,* and *Tilia.* It was also noted, as commonly found, that infection rates were higher in nurseries and intensively cultivated areas than under field conditions. This is particularly true of fruit tree stocks. This observation has been confirmed by Kerr (1972) in Australia, who described the disease as most significant for almonds, plums, and prunes and again noted that the problem began with nursery stocks and varied from year to year with climate.

Garret (1978) reported a commercial blackberry plantation in England which had become infected with crown gall within a few years of its establishment at a nursery and noted that there was really no safe time for the cutting out of infected canes, old canes remaining susceptible for several days at their cut surfaces. Although the cuts were susceptible throughout the year, infection occurred most readily during the season of rapid growth.

The susceptibility of nursery stocks to infection by crown gall is confirmed by the report of Dhanyantari (1978) from Canada, where it was found that 30 pathogenic isolates of *Agrobacterium* were made from 68 plum and cherry nurseries in Niagara and southern Ontario. Durgpal (1971) described a bacterial disease of stone fruits, apricots, etc., with the characteristic symptoms of crown gall as being a major disease in certain parts of India, and El-Helaly noted that the disease existed in Egypt and caused problems with many fruit trees in that country (El-Helaly *et al.,* 1963).

Estimates of losses caused by crown gall on stone fruit crops are given for the world in Table I and for Australia in Table II (A. Kerr, personal communication). Although the data were derived from 1963 figures for California, there is no

TABLE I

ESTIMATED LOSSES CAUSED BY CROWN GALL IN VARIOUS STONE FRUIT CROPS
GROWN THROUGHOUT THE WORLD[a]

Crop	Production (tons × 103)	Price/ton ($A)	Value ($A × 106)	Loss (%)	Estimated loss ($A × 106)
Almond	669	730	488	4.0	20
Apricot	1160	320	371	2.0	7
Cherry	1580	1337	2112	1.0	21
Peach and nectarine	6055	293	1774	4.0	71
Plum and prune	4574	410	1875	1.0	19
Estimated total loss					138

[a] Percentage losses for the different crops were obtained from 1963 California data. There is no reason to believe that percentage losses have altered since then, nor that the disease is any more or less severe in California than in other stone fruit growing areas of the world. World production of the various crops is for 1974 and price per ton is for 1974-1975.

reason to believe the percentage losses have altered since then or that the disease is any more or less severe in California than in other stone fruit-growing areas of the world. It can be concluded from the tables that, if the losses from crown gall could be prevented in these stone fruit crops, savings of up to 138 million dollars could be achieved in the industry. Obviously this is of no small concern to plant pathologists involved with the disease, and the rest of this chapter will be devoted to the symptomology, biology, and control methods available to counteract the disease.

TABLE II

ESTIMATED LOSSES CAUSED BY CROWN GALL IN VARIOUS STONE FRUIT CROPS GROWN IN AUSTRALIA

Crop	Production (tons × 103)	Price/ton ($A)	Value ($A × 106)	Loss (%)	Estimated loss ($A × 106)
Almond	1.4	730	1.0	4.0	0.04
Apricot	35.0	320	11.2	2.0	0.22
Cherry	10.0	1337	13.4	1.0	0.13
Peach and nectarine	79.0	293	23.1	4.0	0.92
Plum and prune	24.0	410	9.8	1.0	0.10
Estimated total loss					1.41

[a] Percentage losses for the different crops were obtained from 1963 California data. There is no reason to believe that percentage losses have altered since then, nor that the disease is any more or less severe in California than in other stone fruit growing areas of the world. Australian production of the various cros is for 1974 and price per ton is for 1974-1975.

III. Host Range

DeCleen and DeLey (1976) surveyed the susceptibility of 1193 species belonging to 588 genera and 138 families selected from throughout the plant kingdom.
They determined that 60% of gymnosperms and dicotyledonous angiosperms
were sensitive to crown gall disease. All the plants concerned were infected with
a selection of *Agrobacterium* strains B6, ChrIIb, Hop, and B23, and they found
that none of these strains infected fungi, bryophytes, or pteridophytes. Although
their data point to the fact that not all strains of *A. tumefaciens* have the same
host range, insufficient statistical data are available and not enough plant taxa
have been infected with sufficient strain selections of *A. tumefaciens* to come to
any strong conclusions based on this work.

Horner (1945) confirmed that 19 species failed to become infected after
contact with the Hop strain of *A. tumefaciens*. Braun (1953) studied host factors
which determined tumor morphology with Turkish tobacco and sunflower infected with both T37 and B6 strains of *A. tumefaciens*. He concluded that three
major factors affected tumor morphology and tumor induction: (1) the strain of
A. tumefaciens used, (2) the position of infection on the plant, and (3) inherent
genetic factors in both the plant and the bacterium. Obviously these considerations put previous results in a different light, suggesting that the developmental
state of the plant and infection site may be factors influencing positive or negative tumorigenesis.

Speiss (1974) reported that *A. tumefaciens* might be ineffective on mosses
because of the different responses to cytokinins of *Funaria* and other bryophytes.
He also suggested that the ecology of bryophytes might in fact separate them
from the usual saprophytic environment of free-living *Agrobacterium* cells.

Anderson and Moore (1979) isolated 176 *Agrobacterium* strains, principally
from locations within the United States, from 26 host species representing 11
plant families. These strains were inoculated into 11 predetermined crown gall
hosts to examine host specificity. It was found that 66% of pathogenic strains
induced a response in between 6 and 8 of the 11 host species. Twenty-seven of
the strains were nonpathogenic on the new hosts, and 3% of them infected only
the host species from which they had been originally isolated. This suggests that
host specificity and host origin may in some cases be connected, and this raises
the possibility of the host plant modifying the infective ability of the *Agrobacterium* plasmid.

IV. Factors Affecting Development of the Disease

Agrobacterium tumefaciens can live as a saprophyte in the soil for several
years. Infection of the host plant occurs through a lesion or wound in the plant's

surface into which the bacterium, or a number of bacteria, are introduced. The bacterium appears to become closely associated with the cell wall of the damaged cell and, in a manner as yet unidentified, passage of the bacterial plasmid takes place through the cell wall of the damaged cell into the living neighboring cells of the surrounding lesion. The plasmid, again in a manner unknown, is transported to the nucleus of the living cell, where approximately 15% of its linear length becomes covalently integrated into the plant chromosomal DNA. This results in hormone independence of the infected cell, subsequent rapid division, and the production of characteristic tumors. From an agricultural point of view, any factor which damages the intact plant or causes a lesion to occur on the outer surfaces of the plant is likely to increase the rate of infection by *Agrobacterium*. Many horticultural techniques rely on controlled wounding, cutting, and grafting of stock and scion to produce rooted plants of a desired cultivar. Such a grafting technique is frequently used for the production of stone fruit plants, such as pear and apricot, and the method is also used in vine and apple tree production. However carefully the grafting is done, with tools being flamed and sterilized, there is still a risk of contamination by the bacterium, both in the air and, subsequent to planting, from the soil. There is little doubt that, in nursery situations, grafting is the main cause of infection by crown gall organisms. However, apart from this obvious direct factor, there are a number of indirect and less obvious phenomena which can also bring about lesions on the plant surface, thereby allowing infection.

Soil and air pollution from industrial wastes can cause changes in soil microbiology and the biological balance among organisms in cultivated soils. McCallan and Weedon (1940) pointed out some of the effects of sulfur dioxide, chlorine, and ammonia on soil; Serat *et al.* (1966) reported that ozone inhibited bacteria, and certainly ozone causes stress on plants, which may change their status from resistant to susceptible. Ridley and Sims (1966) pointed out that ozone toxicity produced lesions on plants and that these lesions could provide sites for infection with any pathogen, leading to an increase in infection of the plant population. Costonis and Sinclair (1967, 1972) and Yarwood and Middleton (1954) found that ozone could protect plants when applied in conjunction with gasoline vapor. Such an observation has been confirmed by Menser and Chaplin (1969) and Howell (1970). These authors found that ozone could affect host plants by increasing the production of phenolic substances, often in correlation with resistance to plant pathogens. Nevertheless, the balance of the literature seems to point to the fact that air pollutants appear to stimulate, or at least modify, the metabolic activity of plants, favoring the development of invasive, parasitic organisms growing actively in the tissue.

Certain common agricultural practices often interact with air pollutants, changing their overall effect on host plants. Heck *et al.* (1965) reported that, in low-fertility soils, plants became more sensitive to air pollutants than plants grown under higher-fertility conditions. On this particular point, there is considerable conflict in the literature. Dunning *et al.* (1972) stated that a deficiency in potassium increased ozone injury in both beans and soybeans, and Brennan *et al.* (1950) found a similar effect after increasing the application of phosphates, in which case tomatoes were shown to become more sensitive to pollutants. On the other hand, Adedipe *et al.* (1972) reported a decrease in ozone sensitivity caused by increasing the sulfur in applications of fertilizer to beans. Excess nitrogen decreased the sensitivity while excess sulfur increased the sensitivity to injury by sulfur dioxide in both tobacco and tomato (Leone and Brennan, 1972).

Pesticides, whether herbicides, insecticides, fungicides, bacteriocides, or fumigants, have a marked pollutant effect on both the plant and bacterial ecosystem. Some of these chemicals have been shown to have a mutagenic effect which may be seen in either the plant host or the pathogen. Such genetic changes may be reflected in dramatic differences in host relationships and interactions with disease symptomology (Kappas, 1978). Insecticides especially can affect soil cultivation by changing the soil microbiology. Lichenstein and Schulz (1961) demonstrated that they could produce an effect either by volatilizing into the air or by remaining in residual amounts in the soil (Cliath and Spencer, 1971).

Other, more obvious factors, which are liable to cause injury or wounding to the outer surfaces of the plant—factors such as insects, nematodes, rodents, and birds—also have a profound effect on the rate of infection and cause lesions much in the same way as grafting or budding, as described earlier. The time of planting and the specific cultivars used can also affect the severity of the disease.

Climatic factors, too, play a part in development of the symptoms, since the third step of the infection, intiation of the plant tumor due to transformation of the plant cells, is temperature-sensitive. Temperatures from 20° to 27°C favor the transformation process; temperatures over 30°C decrease the transformation rate. It has been reported by Tempe *et al.* (1977) that, at 37°C, total inhibition of transformation in tomato, grapevine, rose, and kalanchoe occurs. Such a temperature-sensitive step in the transformation explains some of the variation which occurs from year to year in the severity of the disease. It is of interest to note that Gurlitz and Mathysse (1979) found that attachment of the bacterium to the plant cell wall occurred at all temperatures between 20° and 35°C, with an optimum between 27° and 30°C. Attachment of the bacterium begins within 15–60 minutes after infection in carrots.

V. Virulence and Avirulence of *Agrobacterium tumefaciens*

Since 1907, the nature of the virulence of *A. tumefaciens* has been under investigation. It became obvious initially that strains of the bacterium could be either virulent or avirulent. The problem showed itself most clearly when mixed infections containing virulent and avirulent bacteria were used to initiate a tumor. After development of the tumor, the bacteria elaborated within the tumor tissues all appeared to be virulent. It took a long time for this puzzle to be solved, and further development of techniques in molecular biology was required before a satisfactory answer could be found.

In 1947, Braun indicated that there was a tumorigenic inducing factor in *A. tumefaciens* responsible for crown gall initiation. Subsequent to this report, attempts to isolate and characterize the tumor-inducing factor were made. Manocha (1970) stated that a bacteroid type of *A. tumefaciens* existed in the nucleolus of crown gall cells as shown by electron microscopy in sunflower and *Datura stramonium*. Hohl (1961) had previously found the elaboration of intact bacteria subsequent to infection. Much of the research on the tumor-inducing factor has focused on the molecular biology of infection, and Beljansky *et al.* (1974) succeeded in purifying the bacterial DNA. Beardsley (1972) claimed that he had found bacteriophage DNA in the tumors of sunflower and tobacco. To add to the confusion, Braun and Wood (1966) concluded that the tumor-inducing principle was RNA, or that RNase entered the bacteria or host cell and selectively activated some component essential for tumor inception; and Swain and Rier (1972) reported that RNA could produce tumors directly, using the strain B6. Discovery of the Ti plasmid in *A. tumefaciens* by Van Larebeke *et al.* (1974) and Schell *et al.* (1976) marked the turning point in the search for the virulence factor. Their results were confirmed by Chilton *et al.* (1974) and Watson *et al.* (1975). These investigators showed that, when the Ti plasmid was lost, virulence was lost from the bacterial strain, and that virulence was acquired by avirulent strains when the plasmid was introduced by conjugation (Bomhoff *et al.*, 1976; Gordon *et al.*, 1976; Schell *et al.*, 1976).

The secondary function of the Ti plasmid in tumorigenesis is the production of certain specific arginine derivatives: octopine, nopaline, agropine, etc. These opines are characteristic for the individual strain of bacterium and form two major classes. Octopine-producing plasmids develop unorganized tumors resembling normal plant callus, and nopaline plasmids develop a teratoma type of tumor in which small shoots develop on the surface of the tumor. These arginine derivatives are not found in normal tissue (Gordon *et al.*, 1976). The indication is that these opines serve as a nutritional source for the bacteria within the tumor and stimulate the occurrence of conjugation between neighboring bacteria, thus explaining the apparent anomaly of the mixed infection phenomenon mentioned earlier. This idea of plant metabolites being sequestered by invading bacteria has given rise to the concept of genetic parasitism, a new concept in plant pathology, suggesting naturally occurring genetic engineering and also exciting a ferment of molecular biology.

Such a concept of disease-promoted transfer of virulence eliminates the possibility of using heavy inoculations of avirulent strains in the permanent control of

Agrobacterium in soil. Although heavy inoculations of avirulent strains give some respite from the disease, within a few years the plasmid successfully transfers itself to the whole population.

VI. Agricultural Control of *Agrobacterium tumefaciens*

Chater and Holmes (1980) summarized the useful methods of disease control as follows:

1. Use healthy stock and uninfested soil. Rotate infested fields at least once every 4 years by planting a replacement monocotyledonous plant such as corn, wheat, or barley.
2. Avoid wounding the stem or roots during planting or cultivation.
3. Remove and destroy gall-bearing young as soon as they appear.
4. Sterilize in alcohol, or with a flame, the tools, knives, and tables used in grafting. Mercuric chloride can also be used as a surface sterilizant.
5. Wipe with a good disinfectant all boxes, benches, and work storage areas after use with a germicide or Formalin.
6. When the gall is established, paint it with a solution of Elgetol–methanol, or with a penicillin or steptomycin solution.
7. Before making a cutting, immerse the stem in a 0.5% sodium hypochlorite solution.

VII. Agrocin and Its Mode of Action on *Agrobacterium tumefaciens*

Kerr and New (1972) and Kerr and Htay (1974) found that some oncogenic strains of *A. tumefaciens* were sensitive to a bacteriocin produced by *Agrobacterium radiobacter* strain 84. The bacteriocin has been called agrocin 84. Colonies resistant to agrocin are no longer oncogenic, and this correlation between resistance to agrocin and oncogenicity has been attributed to curing of the Ti plasmid. These authors indicate that both characters are controlled by the genome of the Ti plasmid present in the virulent *A. tumefaciens* strains.

Thompson *et al.* (1977) purified and characterized agrocin 84. They indicated that it was resistant to all proteases and peptidases. Biological activity was not destroyed by ninhydrin, dansyl chloride, or succinic anhydride. Agrocin 84 was sensitive to periodate and acetic anhydride treatment, but this result was not supported by the findings of McGardell (1976), who found K84 was a peptide of molecular weight 2500, but confirmed the report of Robert *et al.* (1977) that it was a 6-*N*-phosphoamitole nucleotide with a molecular weight of about 100–1300. Das *et al.* (1979) studied the relationship between agrocin-resistant strains and their lipid profiles. They found that agrocin-resistant strains caused an increase in the amount of unsaturated fatty acids and a decrease in the saturated fatty acid component.

Sule (1978) stated that agrocin 84 may survive at least 6 months in beet culture in field experiments, and in peat culture the compound effectively controlled crown gall in roseberry. According to the results of Kerr (1980), dipping or soaking nursery stock in agrocin 84 acts as an effective antibiotic measure in treating crown gall. He also used terramycin, vancomycin, and auromycin; however, none of these effectively controlled the pathogen. Kerr noted that the effect of agrocin 84 was particularly marked on the oncogenic strains of *Agrobacterium*.

As well as stimulating the use of agrocin 84 as a control method, these findings have reemphasized the search for alternative control methods. Beiderbeck (1972) found that α-amatin inhibited the growth of *A. tumefaciens in vitro* and reduced tumor initiation. Rennert (1978) found that hydroxyurea strongly inhibited the formation of tumors of *A. tumefaciens* on sunflower stems and suggested that it acted mainly on the host cell at the time of induction of tumors.

There is considerable agricultural research extending the selection of resistant plant varieties. For example, American grape varieties appear to be more resistant to crown gall than European varieties. Apple trees seem to do better on Mahaleb rootstock, and nut trees better on black walnut rootstock, than on others in combatting the disease. On the other hand, most research has taken into account only heavily contaminated nursery soils, which can be more easily treated, sterilized, or fumigated than is the case under field conditions. Any generalized application of a biocide must take into account not only the soil condition, water content, and temperature, etc., but also the stage of plant development at which it is applied. Moje *et al.* (1957) and Katznelson and Richardson (1943) found that bacteria were more resistant to the effect of biocides than fungi and tended to increase in soil fumigated with chloropicrin.

Soil treatment with heat, using steam, is the most effective and efficient method of soil sterilization. It leaves no toxic residue and is easily applied in the greenhouse (Baker and Roistacher, 1957). In practice, to be efficient, the soil temperature must be raised to 100°C for 30 minutes; however, this is an expensive procedure in terms of fuel and other factors. Since the energy crisis and the rise in gas and oil prices, it has become more expensive than formerly. Newhall (1955) estimated that 40 tons of coal were required per acre of soil to steam to a depth of 1 ft. It is considerations such as these which have made the search for alternative methods of control necessary.

VIII. Chemical Treatments

A number of organic and inorganic compounds, such as phosphorus and halogenated nitrobenzenes, have biocidal and biological properties that can be used to control *A. tumefaciens*. These chemicals include both volatile and non-volatile components which break down in the soil to volatile biocidal products such as sodium methyldithiocarbamate. These volatile compounds require a period of several days between treatment and subsequent planting. Volatile water-soluble compounds, however, such as formaldehyde and allyl alcohol, are effective in aqueous solutions against surface pathogens, but the residual effect with all these compounds with a broad spectrum of toxicity is a major problem. El-Helaly *et al.* (1963) found that Chlorthion was the only phosphorus compound capable of inhibiting the growth of *A. tumefaciens*. Parathion was ineffective. Boyle and Price (1963) reported that vanomycin was effective in inhibiting crown gall on tomato and rose cutting stock.

Systemic fungicides can also be used to minimize crown gall effects. Helton and Williams (1968) used cycloheximide thiosemicarbazone, cyclohexamide acetate, and 8-quinolinol phosphate, each at concentrations of 50, 500, or 5000 ppm, to minimize crown gall damage on 3-year-old cherry trees, without any apparent loss of production that might have been associated with clearing and fumigation of the soil. These compounds can be used as both curative and preventative agents, and cyclohexamide thiosemicarbazone achieves these effects without any chemical injury to the trees or the environment.

IX. Evaluation of Present Methods of Control

Crown gall is a significant agronomic factor in causing yield losses in stone fruit crops. Estimates of 138 million dollars lost in this industry alone cannot be dismissed lightly. It is likely that total agronomic loss from this disease far exceeds these estimates. With the present energy crisis, it is unfortunate that the costs of pesticides have reached such levels as to make their use obstructively expensive; also, their residual actions in the soil have become ecologically undesirable. The only existing methods which seem to cause little ecological disruption are the inoculation of infected areas with avirulent strains of the bacterium—only temporary relief can be gained by this method—and more effective and exacting control of applications of agrocin 84, which seems to have no undesirable side effects and is both inexpensive and effective.

REFERENCES

Adedipe, N. O., Hofstra, G., and Ormrod, D. P. (1972). *Can. J. Bot.* **50,** 1789–1793.
Anderson, A. R., and Moore, L. W. (1979). *Phytopathology* **69,** 320–323.
Baker, K. F., and Roistacher, C. N. (1957). *Calif. Univ. Agric. Exp. Sta. Man.* **23,** 123–196.
Beardsley, R. E. (1972). *Prog. Exp. Tumor. Res.* **15,** 482–509.
Beiderbeck, R. (1972). *Z. Natureforsch. Teilb.* **27,** 1393–1394.

Beljansky, M., Aaron-Da gunha, M. I., Beljanski, M. S., Manigult, P., and Bourgarel, P. (1974). *Proc. Natl. Acad. U.S.A.* **71**, 1585-1589.

Bomhoff, G., Klapwijk, P. M., Kester, H. C. M., Shilperoort, R. A., Hernalsteens, J. P., and Schell, J. (1976). *Mol. Gen. Genet.* **145**, 177-181.

Boyle, A. M., and Price, R. M. (1963). *Phytopathology* **53**, 1272-1275.

Braun, A. C. (1947). *Am. J. Bot.* **34**, 234-240.

Braun, A. C. (1953). *Bot. Gaz.* **114**, 363-371.

Braun, A. C., and Wood, H. N. (1976). *Proc. Natl. Acad. Sci. U.S.A.* **73**, 496-500.

Brennan, E. G., Leone, I. A., and Daines, R. H. (1950). *Plant Physiol.* **25**, 736-747.

Cavara, F. (1897). *Stn. Sper. Agric. Ital. Modena* **30**, 482-509.

Chater, C. S., and Holmes, F. W. (1980). "Insect and Disease Control for Trees and Shrubs." Massachusetts Dept. of Environmental Management, U. Mass., USDA, and C.E.S. (C-148).

Chilton, M. D., Currier, T. C., Farrand, S. K., Bendich, A. J., Gordon, M. P., and Nester, E. W. (1974). *Proc. Natl. Acad. Sci. U.S.A.* **71**, 3672-3676.

Cliath, M. M., and Spencer, F. W. (1971). *Soil Sci. Soc. Am. Proc.* **35**, 791-795.

Costonis, A. C., and Sinclair, A. W. (1967). *Phytopathology* **57**, 807. (Abstr.).

Costonis, A. C., and Sinclair, A. W. (1972). *Eur. J. Forest Pathol.* **2**, 65-73.

Das, P. K., Mitali, B., and Gora, C. C. (1979). *J. Gen. Appl. Microbiol.* **25**, 1-10.

DeCleen, M., and DeLay, J. (1976). *Bot. Rev.* **42**, 388-403.

Dhavantari, B. N. (1978). *Can. J. Bot.* **56**, 2309-2311.

Dochinger, L. S. (1969). *Phytopathology* **59**, 1024. (Abstr.)

Dunning, J. A., Tingey, D. T., and Heck, W. W. (1972). *N.C. Univ. Agric. Exp. Sta. J. Ser. Pap.* **3931**, 1-22.

Durgpal, J. C. (1971). *Indian Phytopathol.* **24**, 379-382.

El-Helaly, A. F., Abu-El-Dahab, M. K., and Zeitoun, F. N. (1963). *Phytopathology* **53**, 726-764.

Engler, G., Holsters, M., Van Montogu, M., Schell, J., Hernalsteens, J. P., and Shilperoort, R. A. (1975). *Mol. Gen. Genet.* **138**, 345-349.

Garret, C. M. E. (1978). *Plant Pathol.* (*London*) **27**, 182-186.

Gordon, M. P., Farrand, S. K., Sciaky, D., Montoya, A. L., Merlo, D. J., and Nester, E. W. (1976). *Crown Gall Symp. Univ. Minnesota.*

Gurlitz, R. H. C., and Matthysse, A. G. (1979). *Annu. Meet. Am. Soc. Plant Physiol.* p. 314.

Heck, W. W., Dunning, J. A., and Hindawi, I. J. (1965). *Air Pollut. Control Assoc.* **15**, 511-515.

Helton, A. W., and Williams, R. E. (1968). *Phytopathology* **58**, 782-787.

Hohl, H. R. (1961). *Phytopathol. Z.* **40**, 317-356.

Horner, G. M. (1945). *Plant Dis. Rep.* **29**, 98-110.

Howell, R. K. (1970). *Phytopathology* **60**, 1626-1629.

Kappas, A. (1978). *Mutat. Res.* **51**, 189-197.

Katznelson, H., and Richardson (1943). *Can. J. Res.* **C21**, 249-255.

Kerr, A. (1972). *J. Appl. Bacteriol.* **35**, 493-497.

Kerr, A. (1980). *Plant Dis.* **64**, 25-30.

Kerr, A., and Htay, K. (1974). *Physiol. Plant Pathol.* **4**, 37-44.

Leone, I. A., and Brennan, E. (1972). *J. Air Pollut. Control Assoc.* **22**, 544-550.

Lictenstein, E. P., and Schulz, K. R. (1961). *J. Econ. Entomol.* **54**, 517-522.

Loubser, J. J. (1978). *Plant Dis. Rep.* **62**, 730-731.

McCallan, S. E. A., and Weedon, F. R. (1940). *Contrib. Boyce Thomp. Inst.* **11**, 331-342.

McCardell, B., and Pootjes, C. F. (1976). *Antimicrob. Agents Chemother.* **10**, 498-502.

Manocha, M. S. (1970). *Can. J. Microbiol.* **17**, 819-820.

Menser, H. A., and Chaplin, S. F. (1969). *Tob. Sci.* **13**, 169-170.

Moje, W., Martin, J. P., and Baines, R. C. (1957). *J. Agric. Food Chem.* **5**, 32-36.

Moore, L. W. (1977). *Phytopathology* **67**, 139-144.

New, P. B., and Kerr, A. (1972). *J. Appl. Bacteriol.* **35,** 279-287.

Newhall, A. G. (1955). *Bot. Rev.* **21,** 198-250.

Rennert, A. (1978). *Acta Soc. Bot. Pol.* **47,** 51-64.

Ridley, J. D., and Sims, E. T., Jr. (1966). *S.C. Agric. Exp. Sta. Res. Ser.* **70,** 1-12.

Riker, A. J., and Hilderbrandt, A. C. (1953). "Plant Disease." USDA Year Book, pp. 10-14.

Roberts, W. P., Tate, M. E., and Kerr, A. (1977). *Nature (London)* **265,** 379-381.

Saedi, D., Bruening, G., Kaob, C. I., and Dutra, J. (1979). *Infect. Immun.* **23,** 298-304.

Schell, J., Van Montagu, M., DePicker, A., DeWaele, D., Engler, G., Genetello, C., Hernalsteens, J. P., Holsters, M. E., Silva, B., Van den Elsacker, S., Van Larebeke, N., and Zaenen, I. (1976). *Crown Gall Symp. Minnesota Univ.*

Serat, W. F., Budinger, F. E., Jr., and Muller, P. K. (1966). *Atoms Environ.* **1,** 21-32.

Smith, E. F., and Townsend, C. O. (1907). *Science* **25,** 671-673.

Speiss, L. D. (1974). *Plant Physiol.* **55,** 583-585.

Sule, S. (1978). *Phytopathol. Z.* **91,** 273-275.

Swain, L. W., and Rier, J. P., Jr. (1972). *Bot. Gaz.* **133,** 318-324.

Tempe, J. A., Petit, M., Holsters, M., Montagu, V., and Schell, J. (1977). *Proc. Natl. Acad. Sci. U.S.A.* **74,** 2848-2849.

Thompson, R. J., Hamilton, R. H., and Pootjes, C. F. (1977). *Annu. Meet. Am. Plant Physiol. Can. Soc. Plant Physiol., Univ. Wisconsin* p. 601.

Van Larebeke, N., Engler, G., Holsters, M., Van den Elsaker, S., Zaenen, I., Schilperoort, R. A., and Schell, J. (1974). *Nature (London)* **252,** 169-170.

Watson, B., Currier, T. C., Gordon, M. P., Chilton, M. D., and Nester, E. W. (1975). *J. Bacteriol.* **23,** 255-264.

Wood, J. I. (1953). "Plant Disease." USDA Year Book, pp. 1-5.

Yarwood, C. E., and Middleton, J. T. (1954). *Plant Physiol.* **29,** 393-395.

INTERNATIONAL REVIEW OF CYTOLOGY, SUPPLEMENT 13

Suppression of, and Recovery from, the Neoplastic State

ROBERT TURGEON

Biology Department, University of Dayton, Dayton, Ohio

I.	Introduction	59
II.	The Potential of the Tumor Cell	60
III.	Crown Gall Teratoma	62
	A. Tumorigenesis	62
	B. Definition of Teratomas	63
	C. Description of Teratomas	64
	D. Control of Tumor Morphology	65
IV.	Recovery from Crown Gall Tumor Disease	71
V.	Suppression and Recovery in Other Neoplastic Diseases	73
	A. Habituation	73
	B. Chemical Modification of the Tumor Phenotype *in Vitro*	74
	C. Temperature-Sensitive Tumor Growth	76
	D. Teratocarcinoma	77
	References	78
	Note Added in Proof	81

I. Introduction

Certain members of the Rhizobiaceae are well known for their infectious interactions with plant hosts. These interactions have been widely studied, especially the formation of nodules in the legume–*Rhizobium* spp. symbiosis and the initiation of non-self-limiting tumors in crown gall disease caused by *Agrobacterium tumefaciens*. In both cases the growth potential of the plant is engaged to produce an overgrowth which harbors the bacteria. The initiation of plant cell division benefits both the prokaryotic and eukaryotic symbionts in the legume–*Rhizobium* spp. interaction, but in crown gall disease the overall effect on the physiology and growth of the host is detrimental. In the latter case the plant loses control over cell growth and proliferation; tumors are formed which may eventually reach enormous proportions.

This article will be devoted primarily to a discussion of the growth potential of crown gall tumors and, more specifically, to an examination of the crown gall teratoma. It will become evident that transformation does not necessarily lead to a complete relaxation of growth control, and that the phenotypic properties

59

associated with this neoplastic disease may be suppressed. In addition, the reversion of crown gall cells to the untransformed condition will be described. Other aspects of crown gall disease will be analyzed as they pertain to the central theme of growth control. Since the principles of tumor growth suppression and reversion are as applicable to the tumors of animals as to those of plants, the discussion will be extended, for comparative purposes, to a limited number of animal tumor types in which modification of the neoplastic phenotype has been demonstrated. The reader's attention is directed to a number of reviews on crown gall and the tumor problem in general (Braun, 1969, 1974, 1977, 1978; Meins, 1977; Pierce et al., 1978; Turgeon, 1981).

II. The Potential of the Tumor Cell

Neoplastic transformation leads to an apparently stable alteration in cellular phenotype. "Once a tumor cell, always a tumor cell" is a common belief. Certainly the durability of the transformed state has impressed many with clinical experience who witness too often the steady, irrevocable progression of the overwhelming majority of human malignancies.

However, there is evidence that transformation is not always irreversible; certain tumor cells respond to controls which regulate growth, and as a result the neoplastic potential of the cells declines and may even be lost. The inhibition of neoplastic growth is often the cause, or the result, of cell differentiation. Armin C. Braun has, over a long career, championed the view that both plant and animal tumor cells have a greater potential for expressing the differentiated phenotype than commonly supposed and, furthermore, that the neoplastic state is reversible in many circumstances. Braun's major thesis has been that neoplasia is an aberrant expression of the cell's normal capacity for cell division; under proper conditions the tumor cell may respond to controls which regulate growth and differentiation.

During the normal growth of multicellular organisms cells divide and subsequently differentiate in a coordinated pattern; there is a proper balance between increase in cell number and differentiation. In a sense, continuously dividing stem cell populations are as distinctive as their differentiated progeny. They are specialized for the function of proliferation, and they respond to controls which restrict this potential. Neoplastic transformation results in continued division of cells even in the presence of controls which moderate the proliferation of normal cells. Thus the tumor cell is said to be *autonomous*.

Host control of neoplastic growth has been demonstrated in a number of situations. For example, adenocarcinoma cells in the prostate gland may require testosterone for continued growth. Huggins and Hodges (1941) were able to improve the condition of many patients by castration and the administration of estrogens. Another well-known example of host control is found in the mammary carcinoma of mice, which develops rapidly during pregnancy but sometimes disappears between pregnancies (Foulds, 1969). A more recently described example of host influence on tumor growth is the transformation of mouse cells by spleen focus-forming virus. Although these cells are able to proliferate in wild-type mice, they cannot do so in mice carrying two mutant alleles at the *Steel* locus (Mak *et al.*, 1979). The mutant mice have the normal number of hematopoietic stem cells, but the stem cells are unable to undergo extensive erythroid differentiation, and as a result the mice are severely anemic (McCulloch *et al.*, 1965). Both normal and transformed cells are subject to the regulatory influence of the *Steel* locus.

The dependency of transformed cells on the regulatory influence of the host may be reduced with time; this phenomenon is known as tumor progression (Pierce *et al.*, 1978). Tumors which are relatively benign may gradually lose certain characteristics of their tissue of origin and acquire the ability to metastasize; i.e., they become more anaplastic (Pierce *et al.*, 1978). Therefore, tumors which arise at the same site may develop different growth potentials; some grow slowly and are evidently restricted by the host, while others are able to divide and metastasize rampantly.

The restriction of tumor growth may be due to controls which are either intrinsic or extrinsic to the transformed cell, or both. In the cases cited above tumor cells responded to the physiological influence of the host (extrinsic control), but the *ability* to respond is an intrinsic capacity which may be of limited duration and extent. Intrinsic limitations on growth were demonstrated in an early series of experiments on crown gall disease. Braun (1943, 1947) interrupted the transformation of *Vinca* plants and observed that the growth rates of the tumors depended upon the duration of transformation. When these tumor cells were cultured, they continued to grow at rates proportional to their growth rate on the plant, proving that cell proliferation in slow-growing tumors *in vivo* was not a host-mediated response. The growth rate of slowly growing *Vinca* tumors increased to that of fully transformed cells when additional hormones and nutrients were supplied in the culture medium (Wood and Braun, 1961).

Plant tumors are also subject to extrinsic growth limitations. For example, inoculation of different sites on the plant yields tumors of varying sizes. Undoubtedly the best example of host control over plant tumor growth involves crown gall teratoma. This tumor has the intrinsic capacity to produce shoots, but the degree of organization of the shoots is largely host-mediated, i.e., extrinsic. The crown gall teratoma is therefore an excellent system for studying the developmental potential of tumor cells.

III. Crown Gall Teratoma

A. TUMORIGENESIS

Crown gall disease affects a wide range of dicotyledonous species including at least 142 genera in 61 families (Braun, 1977). The disease is initiated by *A. tumefaciens* which enters the plant through a wound site. Once transformation is complete, neoplastic cells proliferate in the absence of the inciting pathogen (White and Braun, 1942; Braun, 1943); the overgrowth is a true tumor rather than a simple hyperplastic reaction to infection. The neoplastic phenotype is stable through mitotic divisions, and bacteria-free tumor cells are transplantable into appropriate hosts. Crown gall cells produce growth regulators in quantities which exceed normal physiological requirements and, as a result, they are able to grow *in vitro* on media not supplemented by growth hormones (Braun, 1958). Unlike normal plant tissues, crown gall tumors synthesize amino acid metabolites known as opines. Opines of the octopine group are synthesized by the reductive condensation of basic amino acids with pyruvate. The known opines of this group, with the corresponding amino acids in parentheses, are octopine (arginine) Ménagé and Morel, 1964), lysopine (lysine) (Biemann *et al.*, 1960), histopine (histidine) (Kemp, 1977), and octopinic acid (ornithine) (Ménagé and Morel, 1964). Opines of the nopaline group are synthesized by the condensation of α-ketoglutarate with amino acids; they are nopaline (arginine) (Goldmann *et al.*, 1969) and nopalinic acid (ornithine) (Firmin and Fenwick, 1977; Kemp, 1978).

Natural isolates of *A. tumefaciens* fall into three groups based on the ability to utilize opines as carbon and nitrogen sources. Octopine and nopaline strains use the corresponding groups of opines, while null strains are unable to utilize either. Octopine and nopaline strains of *A. tumefaciens* induce tumors which synthesize octopine and nopaline, respectively.

Oncogenic strains of *A. tumefaciens* harbor a large plasmid $(90-150 \times 10^6$ daltons) (Van Larebeke *et al.*, 1974, 1975; Watson *et al.*, 1975; Bomhoff *et al.*, 1976) known as the tumor-inducing (Ti) plasmid. Transformation is associated with, and apparently accomplished by, transfer of a stable, replicating portion of Ti plasmid DNA to the plant cell (Chilton *et al.*, 1977). At least some of this foreign DNA is transcribed (Dummond *et al.*, 1977; Gurley *et al.*, 1979; Yang *et al.*, 1980a). The molecular biology of crown gall is considered elsewhere in this volume. In morphological terms, tumors induced by *A. tumefaciens* are of two general types: unorganized tumors and teratomas. While scattered vascular elements may differentiate within unorganized tumors, they do not produce recognizable tissues or organs. Teratomas, on the other hand, are tumors in which transformed cells have a discernible capacity for organogenesis.

B. Definition of Teratomas

In plant biology the term "teratoma" is used according to its original definition; "terato" is a combining form meaning "wonder" or "monster," and the suffix "oma" means a morbid growth or tumor. A plant teratoma is a tumor which produces malformed, abnormal structures. The medical definition which describes a teratoma as a multiplicity of tissues derived from more than one primitive germ layer (Bolande, 1977) has no application in plant biology.

Teratomas were recognized as a distinct form of crown gall tumor from the earliest days of research on the subject. At that time they were usually referred to as embryomas. E. F. Smith (1917, 1921) and Levin and Levine (1918, 1920) described the growth of teratomas on a number of plant species. Smith (1917, 1921) argued that teratoma shoots were composed of normal cells stimulated to grow and organize by the undifferentiated cells of the tumor. Smith's view was disputed by Levin and Levine (1918, 1920) who proposed that the shoots were composed of true tumor cells. This is of course a fundamental point. Many plants have a pronounced capacity for regeneration, and therefore it is not surprising that organs, either shoots or roots, are often formed near the vicinity of a wound or tumor. Tumors produce excessive amounts of growth regulators which influence nontransformed cells and may stimulate them to generate shoots or roots. The occurrence of normal organs near growing tumors is widespread, and the organs may actually be generated through, or from within, the tumor mass itself.

It is more important to realize that transformed cells are able to participate in organogenesis. To do so, tumor cells must cooperate with neighboring cells and consequently lose a certain degree of autonomy; the neoplastic phenotype of the cell is suppressed. We now know that Levin and Levine (1918, 1920) were correct; crown gall teratomas are composed of transformed cells (Braun, 1948; Braun and Wood, 1976; Turgeon et al., 1976; Huff and Turgeon, 1981).

A plant tumor is not considered a teratoma unless the transformed cells, of which it is composed, are able to organize (Turgeon, 1981). The presence of shoots or roots emanating from a tumor or from its vicinity does not necessarily indicate that the tumor is a teratoma. The best test for determining whether tissue is composed of tumor cells is to grow a cell or a piece of the tissue in vitro. Under these circumstances hormone dependency can be tested directly, and tissue from culture may then be grafted into an appropriate host, if so desired, to determine whether it is transplantable. Since the majority of teratomas are induced by nopaline strains, these tumors typically produce nopaline (Goldman et al., 1969) and nopalinic acid (Firmin and Fenwick, 1977; Kemp, 1978). However, the presence of opines in a presumed teratoma is not unequivocal proof that the structure contains tumor cells unless it can be shown that the opines have not been imported from their site of synthesis in the unorganized portion of the tumor mass. The presence of octopine or nopaline dehydrogenase is much more reliable proof that tumor cells are present. Whenever possible, it is advisable to evaluate the tissue in question using the following criteria: hormone autonomy *in vitro,* transplantability, resistance to reinfection by *A. tumefaciens* (Braun, 1948; Turgeon *et al.,* 1976), opine synthesis (where applicable), and of course the presence of T-DNA.

It must be recognized that the above definition of the teratoma suffices at the present time but is unlikely to do so in the future. Current research trends place heavy emphasis on manipulation of the Ti plasmid. When cells are transformed (in the broad sense of "changed") by altered T-DNA, they will undoubtedly display a variety of biochemical and morphological phenotypes. The present rather rigid description of crown gall disease will be replaced by a broader, more fundamental definition reflecting the importance of T-DNA in establishing phenotype. The strict boundary between the normal and tumor phenotype will be obscured.

C. Description of Teratomas

Tobacco is the most common host plant used in teratoma studies. *Kalanchoe* also produces excellent teratomas, but the tissue of this species is more difficult to work with *in vitro.* The descriptions presented here apply specifically to tobacco, but many features of teratoma growth in this host apply equally well to others. When tobacco is inoculated with a teratoma-producing strain of *A. tumefaciens,* an unorganized callus is produced at the site of wounding. It is only after this preliminary phase of growth that the organogenic capacity of the teratoma is observed (Levine, 1919, 1923; Levin and Levine, 1918, 1920; Braun, 1948). Zones of meristematic cells develop and begin to produce small, distorted shoots. Often the structures produced by the teratoma appear to be little more than projections of the tumor surface. In other cases very small leaves are produced. Complete shoots may be formed infrequently, or they may cover the entire surface of the tumor as a dense lawn. Teratoma shoots are usually small, but occasionally a large shoot is produced. Even these large shoots are abnormal to a certain extent. The first shoots on the teratoma are produced *de novo* from meristematic centers; thereafter they are derived largely from the axial buds of shoots already present. Growth of lateral shoots from axial buds is a common feature of teratomas, whereas apical growth is generally limited.

Teratoma stems are thick and, in tobacco at least, usually swollen abnormally at the nodes. Leaves are typically small; the proximal portion of the midrib and lateral veins may become swollen, but there is no evidence of hypertrophy or hyperplasia in the lamina. Teratoma shoots turn green in the light and become etiolated in the dark. The life span of a teratoma is usually short. After a period of active growth the tumor is subject to decay and often dies back completely. In other cases the tumor may have a prolonged life span, but portions of it and its shoots periodically become necrotic and waste away.

Teratoma tissue retains the capacity for organogenesis *in vitro* even after it is cloned (Braun, 1958). On solid medium the tissue may be only weakly organogenic. In liquid medium shoot growth is much more pronounced, and unorganized growth is inhibited. In fact, liquid-grown cultures are often composed entirely of teratoma shoots.

D. CONTROL OF TUMOR MORPHOLOGY

Plant cells of many species have a remarkable ability to regenerate. When a plant cell is transformed, however, its capacity for morphogenesis is weakened; the majority of tumors are unstructured, and even teratomas, which produce shoots, do not root. Evidence suggests that the capacity for organogenesis remains in plant tumor cells but is not normally expressed. This may well be true for all crown gall tumors, even those in which morphogenesis has not been recorded. Perhaps when our understanding of hormone physiology is more complete, we will be able to regenerate plants regularly from the most unorganized of tumor types. At present the morphology of tumors is known to be influenced by a number of factors.

1. *Bacterial Strain and the Ti Plasmid*

As discussed above, natural isolates of *A. tumefaciens* fall into three groups based on opine utilization. The majority of teratomas are induced by nopaline strains. When nopaline plasmids are introduced into plasmid-cured nonnopaline strains in plasmid transfer experiments, the recipient bacteria induce tumors which synthesize nopaline and, in appropriate hosts, induce teratomas (Bomhoff *et al.*, 1976; Gresshoff *et al.*, 1979).

Nopaline plasmids are heterogeneous, while octopine plasmids are more closely related (Currier and Nester, 1976; Sciaky *et al.*, 1978). In a study on fragments obtained by digesting plasmids with the restriction endonuclease *Sma*I, Sciaky *et al.* (1978) have identified clusters of nopaline Ti plasmids. The members of one group, including pTiT37, pTiAT181, pTiEU6, and pTiT10/73A, are closely related, while plasmids in another cluster, including pTiC58, pTi27, and pTi223, are related to one another but not to the first group. In the first group, all four plasmids confer oncogenicity and the ability to utilize nopaline, but the capacity of the tumor to synthesize nopaline is conferred only by pTiT37. Another nopaline plasmid, pTi11BV7, is apparently different from plasmids in both of the above clusters. The nopaline-utilizing strains of *A. tumefaciens* commonly contain cryptic plasmids, which may confuse analysis. The C58 strain has been particularly useful in that it contains only a single plasmid. This nopaline-utilizing strain may be cured of its plasmid by growth at elevated temperatures (Hamilton and Fall, 1971; Van Larebecke *et al.*, 1974; Watson *et al.*, 1975). The plasmid from strain 27912 [designated 1D135 by

Kado (1976)] has a SmaI fingerprint identical to that of pTiC58 (Sciaky et al., 1978) and is also eliminated by growth at elevated temperatures (Kado, 1976).

Einset and Cheng (1979a) reported that three independently isolated tobacco crown gall tumor lines incited by C58 were temperature-sensitive. At 37°C these lines irreversibly lost the ability to grow without hormones, unlike a variety of tumor tissues initiated by other strains, suggesting a correlation between temperature-sensitive pathogenicity in A. tumefaciens and a similar sensitivity for tumor expression in transformed cells. However, Meins (1975) reported that tobacco cells transformed by the temperature-insensitive strain T37 became auxin-dependent at 35°C. In contrast to the results of Einset and Cheng (1979a), Rogler (1980) found that sunflower tumors, induced by C58 at 28°C, continued to grow in situ at 37°C.

The C58 plasmid has been analyzed in detail by Holsters et al. (1980) who isolated insertion and deletion mutants in all the known Ti phenotypes. Several regions essential for pTiC58 oncogenicity and several nonessential regions were mapped. There are regions of homology between octopine and nopaline plasmids (Chilton et al., 1978; Depicker et al., 1978; Holsters et al., 1980), and it is in these regions that most, if not all, mutations that affect oncogenicity are located (Holsters et al., 1980). One of these oncogenic regions coincides in part with T-DNA (Holsters et al., 1980). Inactivation of the capacity to use nopaline as a nitrogen source occurs in mutations which map to the right of the T-DNA (Holsters et al., 1980), but these mutations do not affect nopaline synthesis in the tumor. The right end of the T-DNA determines octopine and nopaline synthesis in tumors (Holsters et al., 1980). This function is nonessential for oncogenicity and is a nonconserved region. In general, A. tumefaciens strains contain only one Ti plasmid, since other Ti plasmids belong to the same incompatibility group (Hooykaas et al., 1980). However, strains have been produced in which both octopine and nopaline plasmids are harbored, apparently as a result of cointegration (Hooykaas et al., 1980). A clone of A. tumefaciens containing a cointegrate of pAL657 (octopine type) and pTiC58 (nopaline type) induced tumors on Kalanchoe daigremontiana with the teratoma phenotype. These tumors contained both lysopine dehydrogenase and nopaline dehydrogenase activities, indicating that the information coding for the synthesis of both opine types was expressed. This A. tumefaciens strain was also able to utilize both octopine and nopaline. The cointegrated plasmid was not excluded at 37°C as would be expected of pTiC58.

Since nopaline Ti plasmids are heterogeneous, it is not surprising that they differ somewhat in their ability to initiate teratomas. Gresshoff et al. (1979) infected Nicotiana tabacum, N. langsdorffii, N. sylvestris, and N. glauca with A. tumefaciens strains T37 and C58. Both strains initiated teratomas on N. langsdorffii, but neither did so on N. glauca. Nicotiana sylvestris and N. tabacum reacted differently to the two strains; strain C58 induced teratomas on the former plant but not on the latter, and exactly the opposite result was obtained with T37. Inoculation method or plant age may have influenced these results somewhat, since C58 is known to induce teratomas on *N. tabacum* (Einset and Cheng, 1979b; Turgeon, unpublished).

Teratoma induction by octopine-utilizing strains of *A. tumefaciens* has been reported on a number of occasions, but until recently this type of teratoma had not been cloned, and the accuracy of the early reports is therefore impossible to judge. Bopp and Resende (1966) reported the induction of teratomas in certain *Kalanchoe* hybrids using strain B6. Maier and Chen (1971) used the same strain and apparently initiated teratomas in *Datura*. Stonier (1962) interrupted transformation and may have observed teratoma formation with a highly virulent strain of *A. tumefaciens*; however, this report lacks experimental details. Marton *et al.* (1979) incubated tobacco protoplasts with octopine-utilizing strains of *A. tumefaciens* and obtained tumors which were able to regenerate shoots. These tumors did not produce roots and lacked apical dominance. Neither of these criteria definitely establishes that the shoots were composed partially or entirely of transformed cells, but the authors also reported that the shoots contained lysopine dehydrogenase activity. Aerts *et al.* (1979) reported the induction of teratomas on *Arabidopsis thaliana* using the octopine strain Ach 5. These uncloned octopine-producing tumors formed normal (opine-negative) plants and teratoma (octopine-positive) shoots.

Wullems *et al.* (1980) fused protoplasts of a nonorganogenic octopine-producing tumor induced by B6S3 in *N. tabacum* with protoplasts from streptomycin-resistant shoots of the same species. Streptomycin-resistant calli produced short teratoma-like shoots that were resistant to reinfection by *A. tumefaciens*, positive for lysopine dehydrogenase activity, and unable to root.

It has also been reported that B6 can initiate teratomas on very young tobacco plants (Owens, 1980). Apparently the teratoma shoots are stable mosaics of normal and transformed cells, since clones of tumor cells isolated from these shoots are no longer organogenic (L. Owens, personal communication).

More recently an octopine-producing teratoma of *Nicotiana otophora* has been established and subsequently cloned by F. Meins, Jr. (D. Sciaky, personal communication). The *A. tumefaciens* strain used was A277 (pTiB6806). The cloned tumor was at first unorganized but, in two independent laboratories, later began to organize into well-differentiated shoots. The tumor does not root. It grows well on hormone-free medium and synthesizes octopine. This interesting teratoma is now being recloned to determine whether single-cell isolates retain the capacity for organogenesis (D. Sciaky, personal communication).

2. Influence of the Host

Not all plant species produce teratomas. In general, species that form teratomas are noted for their ability to regenerate (Braun, 1953). More than 15 genera were reported to initiate teratomas by Smith (1917) but, as indicated earlier, there is reason to believe that, in many of these instances, the structures were not true teratomas but unorganized tumors which induced shoot or root formation from untransformed tissues of the host plant.

If the morphology of the tumor is determined by hormone relationships, one would expect that the position of the tumor on the plant would affect its regenerative capacity. Both auxin and cytokinin are synthesized in specific regions of the plant, and their concentration at any point depends upon a complex physiology of synthesis, degradation, and transport. Braun (1953) cut tobacco plants through a middle internode and inoculated the two cut surfaces with strain T37. Teratomas regularly formed on the top of the basal portion of the plant, but the base of the excised top portion produced only unorganized tumor callus. Since auxin is transported basipetally in the stem (Goldsmith, 1977), the morphology of the growing tumor on this apical portion was probably influenced by transported auxin. When unorganized callus was grown *in vitro,* it produced typical teratomas. Furthermore, Stonier (1959) reported that the application of naphthaleneacetic acid to teratomas on decapitated stems resulted in the growth of unorganized callus.

Smith (1917) reported an interesting series of experiments in which he inoculated the vascular tissue and cortex of tobacco with *A. tumefaciens.* Teratomas arose from vascular tissue tumors, but only unorganized tumors grew from the cortex. It would be worthwhile to repeat these experiments and test the morphogenetic potential of the tumors in culture. Teratomas are also influenced by the flowering potential of the host. When flowering tobacco plants are inoculated at different positions on the stem, the tumors on the upper, middle, and lower levels produce flower buds, leafy shoots, and unorganized tumors, respectively (Smith, 1917; Aghion-Prat, 1965).

3. *Grafting*

It has been argued throughout this article that most, if not all, plant tumors have morphogenetic potential but that the hormonal milieu established by the host, and by the tumor itself, regulates this capability. The influence of the host is best exemplified by grafting experiments. Braun (1951, 1959), Braun and Wood (1976), and Turgeon *et al.* (1976) grafted teratoma shoots from cloned cultures to the tops of decapitated host tobacco plants. The host cultivar in these cases was morphologically different from the cultivar used to initiate the teratomas. As Braun's early experiments (1951, 1959) demonstrated, grafting encourages the teratoma shoot to grow in a more normal fashion. It seems likely that, when a teratoma shoot is grafted, it comes under the influence of the stock plant and reacts to control mechanisms which govern growth and morphogenesis in the normal shoot. Since primary teratomas seldom display the morphogenetic potential of grafted teratoma shoots, it seems reasonable to suspect that a proper vascular connection to the host is the determining factor in establishing growth control. In fact, many grafts grow in a very haphazard way, as do primary tumors, suggesting that in these cases a proper vascular connection to the host has not been established.

Grafting is usually most successful when repeated. By the third or fourth time the apex is grafted to a new stock plant the teratoma generally grows much more normally than it does when first grafted from culture. The reasons for this are not entirely clear; however, it may be that the more normal the teratoma tissue is the better chance it has to heal the graft wound and form vascular connections with the host before neoplastic growth begins. Each time the graft is made, this cycle of normalization increases chances that suitable vascularization will occur. Even when teratoma shoots grow in an almost normal manner, they typically produce tumor callus at the graft junction. This callus eventually forms distorted teratoma shoots. Grafted teratoma shoots often wilt. Presumably, neoplastic growth at the graft junction disturbs water conduction through the xylem.

Some of the abnormalities characteristic of primary tumors generally persist in successful grafts. The stem of the grafted teratoma shoot may enlarge abnormally, especially at the nodes, and lateral shoots commonly sprout, although their growth is limited. Leaves on the grafts are small, and the swellings at the nodes extend up into the proximal portion of the midrib and large lateral veins. The lamina proper typically displays no overt tumor symptoms whatsoever. In a recent study (Turgeon, in preparation) the development of grafted shoots was studied microscopically. Two major conclusions have been drawn. First, abnormalities arise *after* the organs are initiated at the apex. The appearance of the apex itself and the tissues and leaves produced from it are surprisingly normal; in fact, it is extremely difficult to distinguish the apex of a teratoma from that of a normal plant. As the leaves and stem mature, thickening of the tissues gradually becomes evident. Second, the abnormal swellings are not, despite their macroscopic appearance. the result of aberrant cell proliferation. In fact, neoplastic growth is rarely evident even in grossly distorted stems. Stem swelling is the result of cell *expansion* in the cortex. Occasionally, small proliferating tumors erupt on the stem and leaf midrib; these outgrowths originate from the cortex. Evidently all cell types are able to suppress the tumor phenotype to a considerable extent, but the cortex is particularly prone to hypertrophy. The flowers of teratoma grafts are exceptionally normal in morphology; no abnormal growths are seen, and the petals are properly pigmented. However, the petals, sepals, and filaments initiate typical teratomas when they are grown *in vitro,* as do the other somatic tissues of the plant (Turgeon *et al.,* 1976).

The growth of tumor cultures from somatic tissue of grafts demonstrates that the virtually normal growth response of these shoots is due to tumor growth *suppression*; the tissues, at least as a whole, do not revert to the untransformed condition. However, true reversion occurs in somatic tissues (Section IV). In an analysis of many grafted shoots from a cloned teratoma line, two shoots proved to be normal in phenotype (Turgeon *et al.*, 1976). These shoots originated from cultured tissue which had undergone a previous round of grafting.

It is possible that reversion accompanies, or is even a prerequisite for, the differentiation of certain cell types in teratomas. These reverted cells may not be detected when relatively large fragments of complex teratoma tissue are cultured, especially if they divide infrequently. To test the neoplastic potential of specific cell types it will be necessary to analyze each of them separately. A step toward this goal was undertaken by isolating trichomes (epidermal hairs) of teratomas and growing them *in vitro* (Huff and Turgeon, 1981). Although the trichomes did not demonstrate any capacity for neoplastic growth *in vivo* and appeared to be completely normal, they nevertheless produced typical teratoma cultures *in vitro*.

Although abnormal growth is suppressed in grafted teratomas, most evidence indicates that opine synthesis continues. Wood *et al.* (1978) studied leaves from teratoma grafts. Leaves were excised and allowed to take up a [^{14}C]arginine solution supplied to the petioles. After 24 hours of uptake [^{14}C]nopaline was detected in extracts. These authors concluded that the phenotypic characteristics of unregulated growth and opine synthesis were differentially expressed. In contrast to these studies, Einset and Cheng (1979a) reported a parallel repression of abnormal growth and nopaline concentration in regenerated teratoma shoots. Since the tissue used was uncloned, it is conceivable that the decrease in nopaline content they measured was caused by a reduction in the proportion of transformed to untransformed cells in the regenerated shoots. There is another factor which must be considered in studies on opine metabolism in teratomas. The most useful measure of this crown gall function is expression of the gene for opine dehydrogenase synthesis. Insofar as opine concentration may be controlled by a number of factors, including the *in vivo* concentration of substrates, the amount of opine in the tissue is not an entirely suitable measure of opine dehydrogenase activity.

4. Nitrogen Source

The ability of teratoma tissue to organize and therefore the degree of neoplastic growth suppression are reversibly controlled by the form in which nitrogen is supplied. When glutamine and nitrate are used as nitrogen sources, unorganized growth is stimulated by high glutamine and low nitrate, whereas low glutamine and high nitrate promote organization of shoots (Meins, 1971). The use of glutamate as the sole nitrogen source necessitates the addition of auxin to the medium; auxin autotrophy is eliminated (Meins, 1970, 1973).

IV. Recovery from Crown Gall Tumor Disease

Tumor growth *suppression* is a reversible restraint on abnormal cell proliferation. It has been recognized for some time, however, that an irreversible alteration in phenotype also occurs in crown gall disease, leading to the reversion (recovery) of transformed cells. In 1959 Braun reported experiments in which shoots from a cloned line of tobacco teratoma tissue were grafted to normal and morphologically distinct stock plants. The grafted shoots flowered and set viable seed which, when germinated, produced plants with the normal phenotype. These experiments were extended (Turgeon *et al.*, 1976) by analyzing the sexual reproduction of the shoots in more detail. The purpose of the study was to investigate the growth potential of somatic and germ cells from teratoma flowers. Fragments of the petals and filaments grew in culture in the absence of hormones; these cultures quickly took on the appearance of teratoma tissue and synthesized nopaline. The neoplastic potential of the somatic flower parts had been suppressed but not lost. Haploid plants were generated from teratoma flowers by the technique of anther culture. Haploid plants grew in culture, and later in the greenhouse, without evidence of spontaneous tumorigenicity, and when tissue from the plants was placed in culture, it was always hormone-dependent. Cultures from haploid tissue stimulated to grow by auxin and kinetin produced normal rather than distorted shoots, and the cultures were free of detectable nopaline. Diploid tissue of plants grown from the seed of teratoma flowers (self-fertilized) was also normal. Both haploid and diploid plants readily produced tumors upon inoculation with *A. tumefaciens,* unlike teratoma grafts which were resistant to reinfection. Reversion must therefore take place during or before meiosis.

Recovery takes place not only during meiosis, however. Tissue from a cloned grafted teratoma shoot was returned to culture, and a number of shoots produced by these cultures were grafted again onto normal plants (Turgeon *et al.*, 1976). Most of these shoots clearly retained neoplastic potential; they produced tumors at the graft junctions and were typically swollen and distorted. Two shoots behaved in a much more normal fashion, and in fact none of the usual symptoms of neoplasia were evident. When tissue from these shoots was analyzed, it was found to be hormone-dependent and unable to synthesize nopaline. Therefore, reversion occurs during vegetative growth as well as during sexual reproduction. The frequency of vegetative reversion is still an open question.

Recently Yang and Simpson (1981) obtained phenotypically normal shoots from BT37, a T37-induced tobacco teratoma line cloned by A. C. Braun. When BT37 was placed on a medium containing 1 mg/liter kinetin, the tissue produced shoots which rooted and were susceptible to reinfection by *A. tumefaciens.* Tissue from these shoots was hormone-dependent and nopaline-free. It may be that a somatic reversion occurred in the cloned tissue and, because of selection during transfer, the proportion of revertant cells in the culture substantially increased.

The DNA of the suppressed and reverted teratoma tissue obtained by Braun and Wood (1976) and Turgeon *et al.* (1976) was analyzed by Yang *et al.* (1980b), using the technique of solution hybridization. Tissues from the leaves and somatic flower parts of teratoma grafts were shown to retain T-DNA, although there appeared to be less of it in leaf tissue than in the parental culture line. The arrangement of the DNA also appeared to be different (Yang *et al.,* 1980b). It would be interesting to know whether grafting consistently leads to a reduction in T-DNA content or to a predictable rearrangement of the T-DNA in grafted tissue. No T-DNA could be detected in tissues derived from reverted haploid or diploid plants at the level of sensitivity of the solution hybridization technique. Therefore, at least the majority of T-DNA sequences integrated into the nuclear genome of transformed plant cells (Chilton *et al.*, 1980; Thomashow *et al.*, 1980) are specifically eliminated at meiosis.

In recent experiments conducted by the same group (Yang and Simpson, 1981) fragments of T-DNA were found, by Southern blot analysis, in reverted haploid and diploid tissues and also in spontaneous revertants of BT37. The fragments found were "end" fragments and did not include the central "oncogenic" region of the T-DNA. These results have important practical as well as theoretical implications. Although the Ti plasmid may be an ideal vehicle for the transfer of foreign DNA to higher plants, there are two major problems associated with its use. First, it is obvious that the oncogenic phenotype is not a desirable trait. Second, the loss of foreign DNA at meiosis would eliminate the possibility of permanently altering seed-propagated plants. However, if portions of T-DNA survive sexual reproduction, it may be possible, by proper placement of foreign DNA in the plasmid, to incorporate genes into the plant genome which will persist through seed formation. Furthermore, by passing these transformed cells through a sexual cycle, the central oncogenic, but not the desired region of DNA, would be eliminated. (See Note Added in Proof.)

The mechanism of reversion is not known. Three possible explanations of reversion during sexual reproduction seem reasonable at the present time. First, meiosis may select for somatic cells which have previously reverted. Perhaps the presence of T-DNA in the microspore and megaspore mother cells interferes with meiosis. This model explains the reduced fertility generally observed when teratomas flower. Second, the T-DNA may be chromosomally segregated during meiosis. The simplest, Mendelian, form of this model predicts that half the developing pollen grains contain oncogenic T-DNA and half revert. If only the oncogenic DNA-free micro- and megagametophytes were capable of continuing development, revertant plants would be produced by sexual reproduction. The third model is one of selective recognition and excision of foreign DNA during meiosis. This model predicts reversion in virtually all micro- and megagametophytes. Note, in reference to this model, that the majority of systemic viruses are eliminated at meiosis (Bawden, 1964). The persistence of small amounts of T-DNA in some revertants is not inconsistent with any of these models; recombination could protect fragments in all three cases. In fact, mitotic crossing-over is a reasonable explanation for reversion during vegetative growth. A detailed analysis of gametogenesis in crown gall teratomas is needed before we can properly understand the details of reversion.

Reversion has also been detected in unorganized octopine-type tumors (Sacristan and Melchers, 1977). In these experiments a recently established tobacco tumor, initiated by *A. tumefaciens* 542, was cloned. Three cloned cultures were grown for a further period and, when they were cloned again, nine plants were obtained which were free of neoplastic properties. Reversion took place, as it did in the study of Turgeon *et al.* (1976), during vegetative growth of the tissue. These cloned octopine-type tumors did not initiate teratoma shoots, and therefore the phenomenon of tumor growth *suppression* was not observed.

More recently Einset and Cheng (1979b) repeated the experiments described above by regenerating tobacco plants from tumors initiated with a number of *A. tumefaciens* strains. Tumors initiated by strain C58 produced teratoma shoots, whereas normal plants were obtained from other tumors. It is quite possible, however, as these authors suggested, that normal plants were regenerated by untransformed cells carried in the tumor rather than by true reverted cells. Note that in the experiments of Sacristan and Melchers (1977) the majority of clones derived from a recently established tumor were normal in phenotype.

V. Suppression and Recovery in Other Neoplastic Diseases

The phenomena of suppression and reversion are not unique to crown gall disease. A number of neoplastic cell types in animals and plants are amenable to growth control. For comparative purposes a few selected examples are discussed in this section.

A. HABITUATION

As discussed above, the majority of plant cells require an exogenous source of hormones to grow in culture. Crown gall cells, on the other hand, have no such requirement. Gautheret (1955) found that normal plant cells grown in culture often became hormone-independent spontaneously. This phenomenon is known as habituation. Cells grown for prolonged periods in culture may become autonomous for cytokinin, auxin, or both growth regulators. Habituation is genetically stable, since it persists over many mitotic generations. Cytokinins are present in tissues which have no need for this hormone in culture (Dyson and Hall, 1972; Einset and Skoog, 1973).

Gradual changes in the growth hormone requirements of habituated cells were measured by Meins and Binns (1977). Clones of habituated cells were often more cytokinin-autonomous than the parent cultures from which they were derived. In a sense this change toward greater autonomy is similar to the phenomenon of tumor progression in animals (Meins, 1977). Of more importance to the present discussion is the fact that some cells which were highly autonomous for cytokinin became *less* habituated (Meins and Binns, 1977); the need for exogenous hormone increased, presumably as a result of decreasing cytokinin production. Habituated cells also reverted to the hormone-dependent condition when shoots were grown from habituated cultures (Binns and Meins, 1973). Pith cells from these shoots grew only in the presence of cytokinin.

Cytokinin habituation is sensitive to bromodeoxyuridine (BUdR) (Section V,B) and also to temperature. Syōno and Furuya (1971) observed that cytokinin-habituated cells which grew well without cytokinin at 26°C required the hormone at 16°C. Binns and Meins (1979) confirmed this observation and demonstrated that cold-sensitive lines gave rise, after prolonged culture, to cold-resistant variants.

B. Chemical Modification of the Tumor Phenotype *in Vitro*

The difficulties encountered in modifying the cellular environment *in vivo* have encouraged the study of neoplastic potential in cultured cells. A number of substances, including naturally occurring growth factors (Braun, 1974), regulate the *in vitro* growth and differentiation of tumor cells. A few of the better known examples of growth modification will be discussed here.

Neoplastic animal cells usually differ markedly in morphology and growth habit from normal cells. Some of the properties often associated with neoplastic cells in culture are a round rather than flattened shape, higher saturation density, ability to pile up at high cell densities, decreased culture medium requirements, and synthesis of plasminogen activator.

Short-chain lipophilic acids, especially butyrate, alter the morphology of cells in culture. Wright (1973) demonstrated that Chinese hamster cells cultured with butyrate were more "fibroblast-like" in morphology than cells grown without butyrate and grew as a single layer. Similar morphological changes were noted in a variety of normal and transformed mammalian cell lines (Ginsburg *et al.*, 1973), erythroleukemic cells (Leder and Leder, 1975), HeLa cells (Ghosh *et al.*, 1975), murine sarcoma virus-transformed cells (Altenburg *et al.*, 1976), and Syrian hamster cells (Leavitt *et al.*, 1978). These same changes are brought about by analogues of adenosine 3′,5′-monophosphate (cAMP) (Wright, 1973; Ghosh *et al.*, 1975; Storrie *et al.*, 1978). Since butyrate is a hydrolytic product of dibutyryl cAMP, it has been suggested that the latter compound exerts its effect as a source of butyrate. However, Storrie et al. (1978) argued that cAMP was the real effector and that butyrate may exert its effect by increasing cAMP levels. The important point here is that the transformed state may be readily modified in vitro with a reduced expression of typical transformation characteristics.

BUdR is also a potent suppressor of the transformed phenotype in some cultured cells. BUdR preferentially suppresses cytodifferentiation, and its effects are reversible. Most investigators feel that incorporation into DNA is essential in order for BUdR to be effective. The effects of BUdR on melanoma cells have been studied by Silagi (1976). A cloned, melanotic line of mouse melanoma cells was highly tumorigenic when the cells were injected subcutaneously into mice. When the same cells were grown in the presence of BUdR (1–3 μg/ml), the growth rate was not severely affected but pigmentation levels quickly fell and reached zero within 1–2 weeks. Of particular importance is the fact that BUdR-treated cells also flattened and became contact-inhibited. The tumorigenicity of the cells was also affected by BUdR. After 2 weeks on BUdR-containing medium, melanoma cells failed to induce tumors in inoculated animals (Silagi, 1976), probably, at least in part, because of an increase in antigenicity. When BUdR-suppressed cells were grown on BUdR-free medium, they produced pigment, and when these cells were injected into immunologically compromised mice, they formed tumors in many cases. Therefore, BUdR suppression is reversible.

The reversible suppression of neoplastic growth properties in animal cells by BUdR has an interesting parallel in plant tissues. As discussed above, normal plant cells may become habituated for growth regulators when grown in culture. When cells that were habituated for cell division factors were grown in the presence of BUdR, they became cytokinin-dependent (Meins, 1976; Kandra and Maliga, 1977). This effect was prevented by thymidine and was reversible.

Murine leukemic cell lines are also susceptible to phenotypic alteration in vitro (see reviews by Harrison, 1976, 1977). These cells, known as Friend cells, arise following viral infection of erythroid stem cells, probably at the proerythroblast stage of development. Friend cells are committed to differentiate as erythroid cells, although they grow continuously in culture as transformed cells. Friend et al. (1971) discovered that dimethyl sulfoxide (DMSO) induced 80% or more of the cells to differentiate, whereas only a very small fraction of cells differentiated in the absence of inducer. Since that time, other compounds have been found to induce Friend cell differentiation (Rueben et al., 1976).

Friend et al. (1971) found that induced cells synthesized α- and β-globin. Little or no globin mRNA was detected in non-DMSO-treated cells using a sensitive cDNA probe (Ross et al., 1972). Following DMSO stimulation globin mRNA appeared in the cells within 2 days, suggesting that induction of differentiation involved the stimulation of transcription. Clearly the Friend cell, although transformed, is capable of responding to appropriate signals and differentiating along a committed pathway. It has been a matter of concern to those working in the field that Friend cells do not respond to erythropoietin, a natural inducer of erythrocyte differentiation. However, a more recently described cell type known as the Rauscher erythroleukemia cell responds both to erythropoietin and DMSO (Sytkowski *et al.*, 1980).

When considering the chemical modification of the tumor phenotype as a clinical approach to the treatment of cancer, it must be recognized that transformation properties *in vitro* and tumorigenic potential *in vivo* may not always correlate. For example, Sisskin and Weinstein (1980) isolated a mutant of the highly tumorigenic Chinese hamster ovary (CHO) cell line, which appeared by many criteria, including doubling time, culture requirements, and morphology, to be more normal than wild-type cells. However, the mutant line, as well as the wild-type culture, was tumorigenic when transplanted into athymic mice. Therefore the properties often associated with neoplasia *in vitro* can be experimentally dissociated from tumorigenicity. In fact, normal cells may often be reversibly induced to assume a typical transformed cell morphology and growth pattern *in vitro*. For example, the presence or absence of growth-stimulating polypeptides isolated from tumor cells induced normal fibroblasts to cycle back and forth between the normal and transformed phenotypes (DeLarco and Todaro, 1973; Ozanne *et al.*, 1980).

C. Temperature-Sensitive Tumor Growth

Experiments on tumor cells transformed by temperature-sensitive (*ts*) mutants of oncogenic viruses suggest that maintenance of the neoplastic state requires the continued expression of at least part of the viral genome. A number of investigators reported that *ts* mutants of RNA sarcoma viruses induced temperature-sensitive tumors; the maintenance of transformation-specific functions was temperature-dependent (Toyoshima and Vogt, 1969; Martin, 1970; Stephenson *et al.*, 1973; Boettiger *et al.*, 1977; Klarlund and Forchhammer, 1980). Temperature-sensitive expression of neoplastic properties was also reported in cells transformed by polyoma DNA tumor viruses (Dulbecco and Eckart, 1970) and SV40 (Noonan *et al.*, 1973).

Temperature sensitivity experiments have usually been conducted *in vitro* where the environment may easily be controlled. It has been more difficult to assess tumorigenicity in warm-blooded animals where normal body temperature is within the permissive range of the tumor cell. Results obtained from *in vivo* grown cells are complex and suggest the possible involvement of host control mechanisms. Poste and Flood (1980) used a *ts* mutant of avian sarcoma virus to transform cells of the chorioallantoic membrane of embryonated chick eggs.

When the eggs were raised to nonpermissive temperatures, tumor growth continued, whereas *in vitro* the cells were temperature-sensitive. Unlike the virus-transformed cells, a chemically induced *ts* line was sensitive to temperature *in vivo*. Klarlund and Forchhammer (1980) inoculated the tails of athymic mice with *ts* cells transformed by a *ts* mutant of Maloney sarcoma virus. They shifted the temperature of the tails to the nonpermissive temperature and under these conditions the tumors regressed, although temperature-insensitive secondary tumors later developed.

D. TERATOCARCINOMA

Teratocarcinomas are highly malignant tumors which occur in the gonads of mammals. Sections through testicular teratocarcinomas reveal a number of well-differentiated somatic cell types which often form primitive organs such as teeth and spinal cord. Small clusters of cells termed embryoid bodies, which resemble early embryos, are also formed. The best studied teratocarcinoma is that of the mouse. It is known that teratocarcinomas develop from pluripotent stem cells known as embryonal carcinoma cells. When embryoid bodies containing embryonal carcinoma cells were transplanted into healthy mice, the endoderm and mesenchyme in the embryoid body did not initiate neoplastic growth, but embryonal carcinoma cells proliferated and formed a mass of dividing cells which subsequently differentiated into a number of cell types (Pierce *et al.*, 1960). These results were later confirmed by cloning single embryonal carcinoma cells (Kleinsmith and Pierce, 1964).

Since the progeny of embryonal carcinoma cells are able to differentiate and thereby lose neoplastic potential, it is clear that in this case a highly malignant animal cell is capable of reverting to the nontumorous condition. Reverted cells are able to participate in morphogenesis, as demonstrated by the formation of embryoid bodies. These conclusions were dramatically substantiated by the introduction of embryonal carcinoma cells into the blastocysts of genetically distinct normal mice (Brinster, 1974; Papaioannou *et al.*, 1975; Mintz and Illmensee, 1975; Illmensee and Mintz, 1976). Mice born after the injected blastocysts were returned to the uteri of normal females were allophenic (mosaic) and composed of cells derived from both the normal parent and the injected embryonal carcinoma cells. Embryonal carcinoma-derived cells were found in tissues produced by all three germ layers of the allophenic mice and were normal in phenotype; they did not produce tumors. One male allophenic mouse subsequently had normal offspring with genetic traits of teratocarcinoma cells. Since embryonal carcinoma cells are effectively able to reproduce an entire animal, they are totipotent.

It should be emphasized that the experiments which resulted in the formation of allophenic mice and demonstrated the full potential of the embryonal carcinoma cell were based on the same reasoning that had led Braun (1951) to graft crown gall teratoma cells onto normal plants more than 20 years earlier. If a neoplastic cell has retained the capacity for normal development, this potential will best be expressed when the cell is placed in a natural environment conducive to normal morphogenesis. In the case of the crown gall cell this is achieved by grafting tissue to the tops of normal plants. In the case of the murine teratocarcinoma the neoplastic cell is transplanted into a normally developing embryo. There is apparently a major difference in response in that the crown gall cell retains the capacity for further tumor growth; i.e., the neoplastic state is suppressed, while the teratocarcinoma cell apparently reverts to the nontumorous condition. The most straightforward explanation of this difference is that the crown gall cell, while responding to controls exerted by the normal plant, retains the T-DNA responsible for its transformation.

REFERENCES

Aerts, M., Jacobs, M., Hernalsteens, J.-P., Van Montagu, M., and Schell, J. (1979). *Plant Sci. Lett.* **17,** 43–50.

Aghion-Prat, D. (1965). *Nature (London)* **207,** 1211.

Altenburg, B. C., Via, D. P., and Steiner, S. H. (1976). *Exp. Cell Res.* **102,** 223–231.

Bawden, F. C. (1964). ''Plant Viruses and Virus Diseases''. Ronald Press, New York.

Biemann, K., Lioret, C., Asselineau, J., Lederer, E., and Polonsky, J. (1960). *Bull. Soc. Chim. Biol.* **42,** 979–991.

Binns, A., and Meins, F., Jr. (1973). *Proc. Natl. Acad. Sci. U.S.A.* **70,** 2660–2662.

Binns, A. N., and Meins, F., Jr. (1979). *Planta* **145,** 365–369.

Binns, A. N., Wood, H. N., and Braun, A. C. (1981). *Differentiation* (in press).

Boettiger, D., Roby, K., Brumbaugh, J., Beihl, J., and Holtzer, H. (1977). *Cell* **11,** 881–890.

Bolande, R. P. (1977). *In* ''Handbook of Teratology'' (J. G. Wilson and F. C. Fraser, eds.), Vol. 2. Plenum, New York.

Bomhoff, G., Klapwijk, P. M., Kester, H. C. M., Schilperoort, R. A., Hernalsteens, J. P., and Schell, J. (1976). *Mol. Gen. Genet.* **145,** 177–181.

Bopp, M., and Resende, F. (1966). *Port. Acta Biol. Ser. A* **9,** 327–366.

Braun, A. C. (1943). *Am. J. Bot.* **30,** 674–677.

Braun, A. C. (1947). *Am. J. Bot.* **34,** 234–240.

Braun, A. C. (1948). *Am. J. Bot.* **35,** 511–519.

Braun, A. C. (1951). *Cancer Res.* **11,** 839–844.

Braun, A. C. (1953). *Bot. Gaz.* **114,** 363–371.

Braun, A. C. (1958). *Proc. Natl. Acad. Sci. U.S.A.* **44,** 344–349.

Braun, A. C. (1959). *Proc. Natl. Acad. Sci. U.S.A.* **45,** 932–938.

Braun, A. C. (1969). ''The Cancer Problem. A Critical Analysis and Modern Synthesis.'' Columbia Univ. Press, New York.

Braun, A. C. (1974). ''The Biology of Cancer.'' Addison-Wesley, Reading, Massachusetts.

Braun, A. C. (1977). ''The Story of Cancer.'' Addison-Wesley, Reading, Massachusetts.

Braun, A. C. (1978). *Biochim. Biophys. Acta* **516,** 167–191.

Braun, A. C., and Wood, H. N. (1976). *Proc. Natl. Acad. Sci. U.S.A.* **73,** 496–500.

Brinster, R. L. (1974). *J. Exp. Med.* **140,** 1049–1056.

Chilton, M.-D., Drummond, M. H., Merlo, D. J., Sciaky, D., Montoya, A. L., Gordon, M. P., and Nester, E. W. (1977). *Cell* 11, 263–271.

Chilton, M.-D., Drummond, M. H., Merlo, D. J., and Sciaky, D. (1978). *Nature (London)* 275, 147–149.

Chilton, M.-D., Saiki, R. K., Yadav., N., Gordon, M. P., and Quetier, F. (1980). *Proc. Natl. Acad. Sci. U.S.A.* 77, 4060–4064.

Currier, T. C., and Nester, E. W. (1976). *J. Bacteriol.* 126, 157–165.

DeLarco, J. E., and Todaro, G. J. (1978). *Proc. Natl. Acad. Sci. U.S.A.* 75, 4001–4005.

Depicker, A., Van Montagu, M., and Schell, J. (1978). *Nature (London)* 275, 150–153.

Drummond, M. H., Gordon, M. P., Nester, E. W., and Chilton, M.-D. (1977). *Nature (London)* 269, 535–536.

Dulbecco, R., and Eckart, W. (1970). *Proc. Natl. Acad. Sci. U.S.A.* 67, 1775–1781.

Dyson, W. H., and Hall, R. H. (1972). *Plant Physiol.* 50, 616–621.

Einset, J. W., and Cheng, A. (1979a). *In Vitro* 15, 917–921.

Einset, J. W., and Cheng, A. (1979b). *In Vitro* 15, 703–708.

Einset, J. W., and Skoog, F. (1973). *Proc. Natl. Acad. Sci. U.S.A.* 70, 658–660.

Firmin, J. L., and Fenwick, R. G. (1977). *Phytochemistry* 16, 761–762.

Foulds, L. (1969). "Neoplastic Development, Vol. 1." Academic Press, New York.

Friend, C., Scher, W., Holland, J. C., and Sato, T. (1971). *Proc. Natl. Acad. Sci. U.S.A.* 68, 378–382.

Gautheret, R. J. (1955). *Annu. Rev. Plant Physiol.* 6, 433–484.

Ghosh, N. K., Deutsch, S. I., Griffin, M. J., and Cox, R. P. (1975). *J. Cell. Physiol.* 86, 663–672.

Ginsburg, E., Salomon, D., Sreevalsan, T., and Freese, E. (1973). *Proc. Natl. Acad. Sci. U.S.A.* 70, 2457–2461.

Goldmann, A., Thomas, D. W., and Morel, G. (1969). *C. R. Acad. Sci. (D)* 268, 852–854.

Goldsmith, M. H. M. (1977). *Annu. Rev. Plant Physiol.* 28, 439–478.

Gresshoff, P. M., Skotnicki, M. L., and Rolfe, B. G. (1979). *J. Bacteriol.* 137, 1020–1021.

Gurley, W. B., Kemp, J. D., Albert, M. J., Sutton, D. W., and Callis, J. (1979). *Proc. Natl. Acad. Sci. U.S.A.* 76, 2828–2832.

Hamilton, R. H., and Fall, M. Z. (1971). *Experentia* 27, 229–230.

Harrison, P. R. (1976). *Nature (London)* 262, 353–356.

Harrison, P. R. (1977). *In* "MTP International Review of Biochemistry" (J. Paul, ed.), Vol. 15, pp. 227–267. Univ. Park Press, Baltimore, Maryland.

Holsters, M., Silva, B., Van Vliet, F., Genetello, C., DeBlock, M., Dhaese, P., Depicker, A., Inze, D., Engler, G., Villarroel, R., Van Montagu, M., and Schell, J. (1980). *Plasmid* 3, 212–230.

Hooykaas, P. J. J., Den Dulk-Ras, H., Ooms, G., and Schilperoort, R. A. (1980). *J. Bacteriol.* 143, 1295–1306.

Huff, M., and Turgeon, R. (1981). *Differentiation* (in press).

Huggins, C. B., and Hodges, C. V. (1941). *Cancer Res.* 1, 293–297.

Illmensee, K., and Mintz, B. (1976). *Proc. Natl. Acad. Sci. U.S.A.* 73, 549–553.

Kado, C. I. (1976). *Annu. Rev. Phytopathol.* 14, 265–308.

Kandra, G., and Maliga, P. (1977). *Planta* 133, 131–133.

Kemp, J. D. (1977). *Biochem. Biophys. Res. Commun.* 74, 862–868.

Kemp, J. D. (1978). *Plant Physiol.* 62, 26–30.

Klarlund, J. K., and Forchhammer, J. (1980). *Proc. Natl. Acad. Sci. U.S.A.* 77, 1501–1505.

Kleinsmith, L. J., and Pierce, G. B. (1964). *Cancer Res.* 24, 1544–1551.

Leavitt, J., Barrett, J. C., Crawford, B. D., and Ts'o, P. O. P. (1978). *Nature (London)* 271, 262–265.

Leder, A., and Leder, P. (1975). *Cell* 5, 319–322.

Levine, M. (1919). *Bull. Torrey Bot. Club* **46**, 447-452.

Levine, M. (1923). *Phytopathology* **13**, 107-116.

Levin, I., and Levine, M. (1918). *Proc. Soc. Exp. Biol. Med.* **16**, 21-22.

Levin, I., and Levine, M. (1920). *J. Cancer Res.* **5**, 243-260.

Maier, J. T., and Chen, P. K. (1971). *Phyton* **28**, 95-108.

McCulloch, E. A., Siminovitch, L., Till, J. E., Russell, E. S., and Bernstein, S. E. (1965). *Blood* **26**, 399.

Mak, T. W., Gamble, C. L., MacDonald, M. E., and Bernstein, A. (1979). *Cold Spring Harbor Symp. Quant. Biol.* **44**, 893-899.

Martin, G. S. (1970). *Nature (London)* **227**, 1021-1023.

Márton, L., Wullems, G. J., Molendijk, L., and Schilperoort, R. A. (1979). *Nature (London)* **277**, 129-131.

Meins, F., Jr. (1970). *Planta* **92**, 240-247.

Meins, F., Jr. (1971). *Dev. Biol.* **24**, 287-300.

Meins, F., Jr. (1973). *Planta* **112**, 57-64.

Meins, F., Jr. (1975). *Planta* **122**, 1-9.

Meins, F., Jr. (1976). *Planta* **129**, 239-244.

Meins, F., Jr. (1977). *Am. J. Pathol.* **89**, 687-702.

Meins, F., Jr., and Binns, A. (1977). *Proc. Natl. Acad. Sci. U.S.A.* **74**, 2928-2932.

Ménagé, A., and Morel, G. (1964). *C. R. Acad. Sci. Ser. D* **259**, 4795-4796.

Mintz, B., and Illmensee, K. (1975). *Proc. Natl. Acad. Sci. U.S.A.* **72**, 3585-3589.

Noonan, K. D., Renger, H. C., Basilico, C., and Burger, M. M. (1973). *Proc. Natl. Acad. Sci. U.S.A.* **70**, 347-349.

Owens, L. (1980). *Plant Physiol.* **65**, S135.

Ozanne, B., Fulton, R. J., and Kaplan, P. L. (1980). *J. Cell. Physiol.* **105**, 163-180.

Papaioannou, V. E., McBurney, M. W., and Gardner, R. L. (1975). *Nature (London)* **258**, 70-73.

Pierce, G. B. F., Dixon, F. J., and Verney, E. L. (1960). *Lab. Invest.* **9**, 583-602.

Pierce, G. B., Shikes, R., and Fink, L. M. (1978). "Cancer. A Problem of Developmental Biology." Prentice-Hall, New York.

Poste, G., and Flood, M. K. (1980). *Cell* **17**, 789-800.

Rogler, C. E. (1980). *Proc. Natl. Acad. Sci. U.S.A.* **77**, 2688-2692.

Voss, J., Ikawa, Y., and Leder, P. (1972). *Proc. Natl. Acad. Sci. U.S.A.* **69**, 3620-3623.

Rueben, R. C., Wife, R. L., Breslow, R., Rifkind, R. A., and Marks, P. A. (1976). *Proc. Natl. Acad. Sci. U.S.A.* **73**, 862-866.

Sacristan, M. D., and Melchers, G. (1977). *Mol. Gen. Genet.* **152**, 111-118.

Sciaky, D., Montoya, A. L., and Chilton, M.-D. (1978). *Plasmid* **1**, 238-253.

Silagi, S. (1976). *Int. Rev. Cytol.* **45**, 65-111.

Sisskin, E. E., and Weinstein, I. B. (1980). *J. Cell. Physiol.* **102**, 141-153.

Smith, E. F. (1917). *Bull. Johns Hopkins Hospital* **28**, 277-294.

Smith, E. F. (1921). *J. Agric. Res.* **21**, 593-598.

Stephenson, J. R., Reynolds, R. K., and Aaronson, S. A. (1973). *J. Virol.* **11**, 218-222.

Stonier, T. (1959). *Plant Physiol.* **34** (Suppl.), xvii.

Stonier, T. (1962). *Symp. Soc. Dev. Growth* **20**, 85-115.

Storrie, B., Puck, T. T., and Wenger, L. (1978). *J. Cell. Physiol.* **94**, 69-76.

Syōno, K., and Furuya, T. (1971). *Plant Cell Physiol.* **12**, 61-71.

Sytkowski, A. J., Salvado, A. J., Smith, G. M., McIntyre, C. J., and DeBoth, N. J. (1980). *Science* **210**, 74-76.

Thomashow, M. F., Nutter, R., Postle, K., Chilton, M.-D., Blattner, F. R., Powell, A., Gordon, M. P., and Nester, E. W. (1980). *Proc. Natl. Acad. Sci. U.S.A.* **77**, 6448-6452.

Toyoshima, K., and Vogt, P. K. (1969). *Virology* **39**, 930-931.

Turgeon, R. (1981). *In* "Molecular Biology of Plant Tumors" (J. Schell and G. Kahl, eds.). Academic Press, New York.

Turgeon, R., Wood, H. N., and Braun, A. C. (1976). *Proc. Natl. Acad. Sci. U.S.A.* **73**, 3562-3564.

Van Larebeke, N., Engler, G., Holsters, M., Van Den Elsacker, S., Zaenen, I., Schilperoort, R. A., and Schell, J. (1974). *Nature (London)* **252**, 169-170.

Van Larebeke, N., Genetello, C., Schell, J., Schilperoort, R. A., Hermans, A. K., Hernalsteens, J. P., and VanMontagu, M. (1975). *Nature (London)* **255**, 742-743.

Watson, B., Currier, T. C., Gordon, M. P., Chilton, M.-D., and Nester, E. W. (1975). *J. Bacteriol.* **123**, 255-264.

White, P. R., and Braun, A. C. (1942). *Cancer Res.* **2**, 597-617.

Wood, H. N., and Braun, A. C. (1961). *Proc. Natl. Acad. Sci. U.S.A.* **47**, 1907-1913.

Wood, H. N., Binns, A. N., and Braun, A. C. (1978). *Differentiation* **11**, 175-180.

Wright, J. A. (1973). *Exp. Cell Res.* **78**, 456-460.

Wullems, G. J., Molendijk, L., and Schilperoort, R. A. (1980). *Theor. Appl. Genet.* **56**, 203-208.

Wullems, G. J., Molendijk, L., Ooms, G., and Schilperoort, R. A. (1981). *Cell* **24**, 719-727.

Yang, F., and Simpson, R. B. (1981). *Proc. Natl. Acad. Sci. U.S.A.* (in press).

Yang, F., McPherson, J. C., Gordon, M. P., and Nester, E. W. (1980a). *Biochem. Biophys. Res. Commun.* **92**, 1273-1277.

Yang, F., Montoya, A. L., Merlo, D. J., Drummond, M. H., Chilton, M.-D., Nester, E. W., and Gordon, M. P. (1980b). *Mol. Gen. Genet.* **177**, 704-714.

NOTE ADDED IN PROOF. Wullems *et al.* (1981) analyzed 200 clones from protoplasts of tumorous shoots and found all to be hormone autonomous and positive for both opine and T-DNA. Binns *et al.* (1981) tested more than 500 clones derived from mesophyll cells of a previously cloned teratoma shoot and found 3 that had reverted. Both studies support the conclusion that tumor shoots do not depend upon chimerism for morphogenetic potential.

Wullems *et al.* (1981) reported that the F_1 progeny of grafted tumor shoots, initiated by somatic hybridization or by *in vitro* transformation, retained a number of markers which characterized the tumorous mother plants. The F_1 progeny grew slowly, proved insusceptible to tumor induction, were self-sterile, and the flowers exhibited heterostyly. In a limited number of cases these plants also synthesized nopaline.

INTERNATIONAL REVIEW OF CYTOLOGY, SUPPLEMENT 13

Plasmid Studies in Crown Gall Tumorigenesis

Stephen L. Dellaporta and Rick L. Pesano

Department of Life Sciences, Worcester Polytechnic Institute, Worcester, Massachusetts

I.	Introduction	83
II.	General	84
III.	Ti Plasmid-Determined Traits	85
	A. Oncogenicity and Tumor Morphology	85
	B. Opine Synthesis in Tumor Cells and Catabolism in Agrobacteria	86
	C. Arginine Catabolism	88
	D. Host Range	88
	E. Sensitivity to Agrocin 84 and Exclusion of Phage AP1	89
	F. Conjugative Transfer of the Plasmid	89
IV.	Genetic Analysis of Ti Plasmid DNA	90
V.	T-DNA	95
	A. Evidence for DNA Transfer (T-DNA)	95
	B. Experimental Approaches to the Study of T-DNA Sequences	96
	C. T-DNA Organization	97
	D. Integration of T-DNA	98
	E. Reversion from the Neoplastic State	100
	References	101
	Note Added in Proof	104

I. Introduction

The current interest in plant molecular biology focuses on the possibility of the introduction and expression of foreign DNA in plant cells and the eventual differentiation of these cells into modified plants. Before the production of genetically modified plants will be possible, the initial problems of foreign DNA integration and expression in plant cells must be understood. Natural genetic exchange from strains of the crown gall bacterium, *Agrobacterium tumefaciens*, to plant cells offers a means of studying and dissecting molecular events responsible for the stable incorporation and expression of these bacterial genes in plants. Genes carried on the tumor-inducing (Ti) plasmid of *Agrobacterium*, a portion of which is transferred to the plant cell during crown gall tumorigenesis, are responsible for profound effects on the metabolism, morphology, and physiology of the plant.

The intent of this article is to serve as a reference for Ti plasmid-encoded functions involved in crown gall tumorigenesis. Current research on transposon mutagenesis is described as it relates to Ti plasmid DNA transfer to plant cells. In addition, we present recent evidence for T-DNA transfer, integration, and its organization into the plant genome.

II. General

The neoplastic disease of dicotyledonous plants called crown gall is caused by a group of gram-negative soil bacteria belonging to the genus *Agrobacterium*. Currently, *Agrobacterium* is classified as a genus in the family Rhizobiaceae, along with the genus *Rhizobium;* and current members of the *Agrobacterium* genus are distinguished primarily by their phytopathogenicity. Four species of *Agrobacterium* are currently recognized: *A. tumefaciens* (Smith and Townsend) Conn., the causative organism of crown gall disease (Smith and Townsend, 1907); *A. rhizogenes* (Riker *et al.*) Conn., the bacterium responsible for hairy root disease (Riker *et al.*, 1930); *A. rubi* (Hildebrand; Starr and Weiss), which causes cane gall (Hildebrand, 1940); and *A. radiobacter* (Beijerinck and Van Delden) Conn., a nonpathogenic member of the genus.

Plants affected with crown gall disease are seldom killed, but gall formation generally causes decreased growth and vigor, which increases the plant's susceptibility to other agents and decreases plant productivity. Particularly susceptible areas of crown gall disease are nurseries, orchards, and vineyards which can suffer significant crop losses. Plant species susceptible to crown gall disease belong to over 61 widely separated families, including at least 142 genera of dicotyledonous plants (Elliot, 1951; DeCleene and Deley, 1976). Monocotyledonous plants do not appear to be susceptible to the disease.

Although virulent bacteria are essential for tumor induction, the continued presence of the inciting bacterium is unnecessary for maintenance of neoplastic growth of tumor cells once initial transformation has occurred. This idea that crown gall cells are "non-self-limiting" was developed with the advent of plant tissue culture techniques by the demonstration that tumors of different plant species could be cultured *in vitro* in the absence of detectable bacteria (White, 1945; Gautheret, 1947; Morel, 1948; Hildebrand and Riker, 1949; Braun, 1951). Such cultures were propagated by serial transfer for decades with no loss of the ability to form tumors when grafted onto healthy plants (Stonier, 1968). Repeated attempts to demonstrate the presence of the bacterium or altered forms of the bacterium in tumorous tissues grown *in vitro* produced consistent negative results (Smith, 1920; Riker, 1923; Robinson and Walkden, 1923; Margrou, 1927; Stonier, 1956; Hohl, 1960, 1961; Ryter and Manigrault, 1964; Rasch, 1964; Lippincott and Heberlein, 1965; Gee *et al.*, 1967; Sun, 1969).

Braun and Mandle (1948) first introduced the concept of a tumor-inducing principle as a hypothetical agent that efficiently passes from the virulent bacteria to the plant cell during the process of transformation. The development of this concept came about through experiments demonstrating that tumor plant cells could grow indefinitely on simple chemically defined medium lacking phytohormones, which could not support the continued growth of normal cells of the type from which the tumor cells were derived (Braun, 1956). Thus a crown

gall cell acquires a capacity for autonomy since, as a result of cellular transformation, it stably acquires the capacity to synthesize essential growth-regulating substances and metabolites that normal cells require for growth and division. Since the capacity for autonomous growth depends on the persistent activation of certain normally repressed biosynthetic systems, it appears that the tumor-inducing principle is responsible for deregulation of these systems.

One major discovery concerning the molecular basis of crown gall transformation was the identification of large plasmids in oncogenic strains of *Agrobacterium* (Van Larebeke *et al.*, 1974; Zaenen *et al.*, 1974). Previous to the discovery of these plasmids, Kerr (1969, 1971) demonstrated that virulence could be transferred from one *Agrobacterium* strain to a recipient strain by inoculating both strains together or in succession into a plant wound. Demonstration that oncogenicity was indeed associated with the Ti plasmid was that loss of this plasmid was concomitant with loss of virulence and that reacquisition of the plasmid rendered the bacterium virulent (Watson *et al.*, 1975; Van Larebeke *et al.*, 1974, 1975). It is now clear that the Ti plasmid of *Agrobacterium* is the tumor-inducing principle first proposed by Braun and Mandle (1948).

III. Ti Plasmid-Determined Traits

A. Oncogenicity and Tumor Morphology

The initial demonstration that certain oncogenicity traits were plasmid-encoded was that loss of virulence was concomitant with loss of Ti plasmid DNA (with restoration of oncogenicity when the plasmid was reintroduced into the strain) (Watson *et al.*, 1975; Zaenen *et al.*, 1974). Plasmid-determined oncogenicity was further confirmed with the finding that a small portion of the Ti plasmid, termed T-DNA, was found stably incorporated within cultured crown gall cells (Chilton *et al.*, 1977). The ubiquity of T-DNA insertions into the plant genome has been confirmed by many investigators (Schell *et al.*, 1979; Thomashow *et al.*, 1980a,b; Chilton *et al.*, 1980; Yang *et al.*, 1980; Merlo *et al.*, 1980; Zambryski *et al.*, 1980). It is now clear that T-DNA sequences are transferred to the plant cell during the transformation process, become integrated into the plant nuclear DNA (Thomashow *et al.*, 1980b; Chilton *et al.*, 1980), and are transcribed into mRNA and protein (Drummond *et al.*, 1977; Schell *et al.*, 1979; Gurley *et al.*, 1979; McPherson *et al.*, 1980). It is believed that proteins encoded by T-DNA sequences are directly involved in eliciting the tumorous phenotype in crown gall cells in addition to opine synthesis in these tumors (Section V).

T-DNA sequences are not the only determinants of oncogenicity in *Agrobacterium*. Certain Ti plasmid sequences which map outside T-DNA cannot be mutagenized without a loss of virulence (Garfinkel and Nester, 1980; Holsters *et al.*, 1980; Klapwijk *et al.*, 1980). These oncogenicity traits are possibly involved in bacterial attachment, plasmid transfer to plant cells, and/or T-DNA integration events (Section IV).

Ti plasmids are also determinants of the morphology of crown gall tumors. Agrobacteria carrying octopine-type plasmids usually elicit unorganized tumors, while nopaline-type plasmids are generally associated with either teratoma-like tumors or an unorganized morphology. Teratomas are characteristically super-numerary organlike structures, while unorganized tumors fail to differentiate into recognizable organlike structures. However, the morphological characteristics of tumor cells are not strictly plasmid-controlled, since both bacterial and host factors are concerned in determining tumor morphology. Factors such as host plant species and site of inoculation can influence teratoma formation (Braun, 1953; Gresshoff *et al.*, 1979). In addition, mutagenesis of specific T-DNA sequences in the octopine plasmid $pTiB_6806$ can alter the morphology of the resultant crown gall tumors (Garfinkel and Nester, 1980). This suggests the possibility that phytohormone balances can be influenced by certain Ti plasmid sequences.

B. Opine Synthesis in Tumor Cells and Catabolism in Agrobacteria

In addition to the oncogenic properties of Ti plasmids, several other charac-teristics are now known to be plasmid-coded. A major feature of crown gall cells is the capacity to synthesize opines, unusual α-N-substituted amino acid deriva-tives. Since the original demonstration that tumor cells synthesized the unusual amino acid derivative of lysine (Lioret, 1957; Biemann *et al.*, 1960), N-α-(D-1-carboxyethyl)- L- lysine or lysopine, several other basic amino acid derivatives unique to crown gall cells have been detected. These include octopine (Ménagé and Morel, 1964), octopinic acid (Ménagé and Morel, 1965), nopaline (Goldmann *et al.*, 1969), histopine (Kemp, 1977), nopalinic acid (Firmin and Fenwick, 1977; Sutton *et al.*, 1977), and agropine (Firmin and Fenwick, 1978) (Fig. 1). These compounds cannot be detected in normal plant cells *in vivo* or *in vitro* (Bomhoff, 1974; Holderbach and Beiderbeck, 1976; Kemp, 1976; Montoya *et al.*, 1977). Opine synthesis in tumor cells was shown to be bacterial strain-specific, and Ti plasmid sequences were implicated in determining the specific type of opine produced in the transformed plant cell (Bomhoff *et al.*, 1976). It is now known that specific Ti plasmid sequences are responsible for opine anabolism in tumor cells (Koekman *et al.*, 1979; Holsters *et al.*, 1980).

In addition, *Agrobacterium* strains which induce synthesis of a particular opine in tumor cells also possess the ability to utilize specifically the opine it elicits as a nutrient source and as a specific inducer of conjugative transfer of Ti plasmids to plasmidless strains (Van Larebeke *et al.*, 1974; Watson *et al.*, 1975;

1. Pyruvic opines:[a]

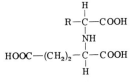

Octopine: R = $NH_2-\overset{\overset{\displaystyle NH}{\|}}{C}-NH-(CH_2)_3$

Octopinic acid: R = $NH_2-(CH_2)_3-$

Lysopine: R = $NH_2-(CH_2)_4-$

Histopine: R = [structure]

2. α-Ketoglutaric opines:[a]

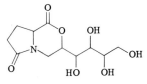

Nopaline: R = $NH_2-\overset{\overset{\displaystyle NH}{\|}}{C}-NH-(CH_2)_3-$

Nopalinic acid: R = $NH_2-(CH_2)_3-$
(ornaline)

3. Agropine:[b]

[chemical structure]

FIG. 1. Structure of opines. [a]For references see Section III,B. [b]Coxon *et al.* (1980).

Petit *et al.*, 1970; Bomhoff *et al.*, 1976; Montoya *et al.*, 1977; Ellis *et al.*, 1979). It was this correlation between the synthesis and catabolism of opines that originally led to speculation that bacterial plasmid genes were expressed in plant tumor cells. The induction of opine production by tumor cells appears to be the result of bacterial plasmid genes which redirect the metabolism of the plant in order to produce a carbon and nitrogen source which it can specifically utilize. The genetic linkage of catabolic and anabolic functions on the Ti plasmid allows the transfer of these functions as a unit to other strains during conjugation. However, the catabolic functions of the bacteria and the synthesis of opines by the tumor have been shown to be two distinct plasmid traits (Montoya *et al.*, 1977; Klapwijk *et al.*, 1976).

Ti plasmid classification is based upon the general type of opine catabolism coded for by the plasmid (e.g., octopine, nopaline, agropine) and the type of opine synthesis elicited in tumor cells. Octopine-type plasmids code for octopine and related opines such as lysopine, histopine, and agropine. These plasmids represent a group of highly homologous plasmids (approaching 100% in most cases) based on DNA reassociation kinetics (Currier and Nester, 1976; Drummond and Chilton, 1978). Many show nearly identical restriction enzyme digestion patterns (Genetello et al., 1977; Sciaky et al., 1978).

Members of the second general class of Ti plasmids, the nopaline-type, show greater internal diversification (Depicker et al., 1978; Sciaky et al., 1978) and usually exhibit less than 30% homology with octopine-type plasmids (Currier and Nester, 1976). Nopaline Ti plasmids confer nopaline and ornaline utilization on host bacterial strains and usually, but not always, induce nopaline and ornaline production in resultant tumor cells. Nopaline plasmids pTiAT-181, pTiEU-6 (Petit et al., 1970), and pTI-T10/73 (Sciaky et al., 1978) code for the ability to utilize nopaline yet do not cause nopaline to be synthesized in tumors and can be classified as "defective" nopaline-type Ti plasmids (Guyon et al., 1980).

The third class of Ti plasmids, once classified as "null type" (Petit et al., 1970; Sciaky et al., 1978), includes pTi542 and pTiAT1 which have recently been shown to induce agropine production in tumors and subsequent utilization of agropine by strains harboring these plasmids (Guyon et al., 1980). Agropine has also been identified in octopine tumors (Firmin and Fenwick, 1978), suggesting a common evolutionary ancestor of octopine- and agropine-type plasmids (Sciaky et al., 1978). This view is supported by the high degree of homology between octopine Ti plasmids and an agropine plasmid, pTiBo542 (Drummond and Chilton, 1978; Currier and Nester, 1976).

C. Arginine Catabolism

Both octopine and nopaline Ti plasmids have been shown to code for the catabolism of arginine as a carbon source (Ellis et al., 1979). In both cases the Ti plasmids must be induced or constitutive for opine catabolism for arginine to be utilized. It has been proposed that the Ti plasmids code for at least two enzymes necessary for the degradation and catabolism of the opines. The first cleaves the opine into an α-keto acid and arginine, and the second degrades an arginine derivative to a suitable carbon source.

D. Host Range

The host range of A. tumefaciens has been shown to be plasmid-determined (Loper and Kado, 1979; Thomashow et al., 1980a). Conjugative transfer of a broad-host-range Ti plasmid (pTi1D1) to a limited-host-range strain which induced tumors only on grapevine (Vitis sp.) elicited the donor plasmid phenotype (wide host range) in the transconjugant. However, a conclusive demonstration that the donor plasmid either replaced or coexisted with the recipient plasmid was not clear. Thomashow et al. (1980b), utilizing purified plasmid DNA from strains isolated from grapevine and shown to have a limited host range

(Panagopoulos and Psallidas, 1973; Panagopoulos *et al.*, 1978), transformed a plasmidless strain which had once harbored a nopaline-type plasmid. Transconjugants showed a virulence response identical to that of the donor strain with all plants tested and harbored donor strain plasmids as determined by restriction endonuclease fingerprint mapping. This suggests that the major determinant of the host range is plasmid-encoded.

E. Sensitivity to Agrocin 84 and Exclusion of Phage AP1

In addition to the above plasmid-determined traits, there is evidence suggesting that the presence of the Ti plasmid is responsible for host bacterial sensitivity to agrocin 84, a bacteriocin produced by *A. radiobacter* strain 84. When Agrocin-sensitive strains are cured of Ti plasmid DNA, the resultant bacteria are resistant to the bacteriocin (Engler *et al.*, 1975). Furthermore, all independently isolated agrocin-resistant strains were shown to be avirulent and plasmidless. Agrocin sensitivity has recently been localized by transposition mutagenesis (see below) to a specific region of the nopaline plasmid pTiC58 (Holsters *et al.*, 1980).

Exclusion of phage AP1 by *Agrobacterium* is determined by Ti plasmid sequences. Plasmidless strains are sensitive to lysis by AP1 (Schell, 1975; Genetello *et al.*, 1977). Exclusion has been correlated with the functional presence of a 9 megadalton region in pTiC58 (Holsters *et al.*, 1980).

F. Conjugative Transfer of the Plasmid

In vitro conjugative transfer of the Ti plasmid is dependent upon Ti plasmid sequences which are induced by the opine the plasmid catabolizes. Octopine-utilizing strains have been shown to be induced for conjugative transfer in the presence of octopine, octopinic acid, or lysopine, but not by the octopine analogues noroctopine, homooctopine, and desmethylhomooctopine (Hooykaas *et al.*, 1979). It has yet to be shown whether nopaline derivatives can induce the transfer of Ti plasmids from nopaline-utilizing strains.

Various studies suggest that octopine catabolism (*occ*) genes and transfer (*tra*) genes each have separate operons but are controlled by a common regulatory gene (Klapwijk *et al.*, 1978; Petit and Tempé, 1978; Tempé *et al.*, 1978). Three regulatory mutants affected in octopine utilization and/or transfer functions were isolated. The first class exhibited constitutive octopine catabolism and Ti plasmid transfer, and the second class was constitutive for octopine catabolism yet remained inducible for transfer (Tempé *et al.*, 1978). A third class isolated was inducible for octopine catabolism and constitutive for transfer (Petit and Tempé, 1978). Mutants of the first class probably contain mutations in the common regulatory gene, whereby both functions are affected. These strains have probably lost the regulatory protein completely. Mutants in the second and third class probably contain mutated regulatory proteins causing a constitutive trait in one gene (*occ/tra*) and an inducible trait in the other (*tra/occ*). According to the experimental model, the latter mutants would be considered operator constitutive mutants. Similar experiments were performed on nopaline strains. Results suggest that they contain a similar regulatory mechanism for transfer (Tempé *et al.*, 1978).

IV. Genetic Analysis of Ti Plasmid DNA

Antibiotic-resistant transposable elements [transposons (Tn)] are unique segments capable of random translocation to DNA sequences on a recipient plasmid or chromosome. Such translocation events result in direct or polar insertions and deletions which enable one to generate single random mutations (Kleckner, 1977). The site of Tn integration or deletion can be physically mapped on a plasmid by restriction endonuclease fingerprint analysis, Southern blotting techniques (Southern, 1975), and electron microscopic heteroduplex analysis (Davis *et al.*, 1971). Tn mutagenesis has proven to be a powerful tool in the genetic analysis of Ti plasmid sequences.

The general method for Tn mutagenesis of the Ti plasmid involves conjugative transfer of a plasmid harboring a Tn to *Agrobacterium* or, alternatively, conjugative transfer of the Ti plasmid to a bacterial strain (*Escherichia coli* or *Agrobacterium* sp.) harboring a Tn in the chromosome or a plasmid. Tn insertions are selectable on the basis of concomitant transfer of the appropriate antibiotic resistance with the Ti plasmid. To date, several important classes of phenotypic mutants of *A. tumefaciens* have been isolated using Tn1 (Dhaese *et al.*, 1979; Holsters *et al.*, 1980), Tn7 (Hernalsteens *et al.*, 1978; Holsters *et al.*, 1980), Tn5 (Garfinkel and Nester, 1980), and Tn904 and Tn1831 (Klapwijk *et al.*, 1980; Ooms *et al.*, 1980; Hooykaas *et al.*, 1980). Tn1, carrying resistance to carbenillin (Hedges and Jacob, 1974), and Tn7, carrying streptomycin and trimethoprim resistance (Barth *et al.*, 1976), have been successfully used to isolate insertion mutants of pTiC58 affecting oncogenicity (Onc), nopaline synthesis in tumor cells (Nos), ability to catabolize nopaline (Noc), exclusion of phage AP1 (Ape), sensitivity to agrocin 84 (Agr), and autotransfer (Tra) (Holsters *et al.*, 1980). Subsequent restriction endonuclease mapping of the Tn insertions into plasmids eliciting mutant phenotypes allowed a functional map of pTiC58 to be established. Interestingly, several regions involved in the determination of oncogenicity were found to correspond to regions conserved between octopine and nopaline plasmids, one of which maps within common T-DNA sequences. Onc⁻ insertions outside T-DNA were also confined to homologous sequences, suggesting that these sequences may play an essential role in the

TABLE I

PHENOTYPIC CHARACTERISTICS OF Ti PLASMIDS WITH MODIFIED RESTRICTION FRAGMENTS

Plasmid	Type	Tn	Modified fragment(s)	Phenotype[a]	Reference
pTiC58	Nopaline	Tn1	HindIII-9, -14a, -7, or -41	Onc⁻	Holsters et al. (1980)
pTiC58	Nopaline	Tn1	HindIII-23	Nos⁻	Holsters et al. (1980)
pTiC58	Nopaline	Tn1	HindIII-30 or -26	Nos⁻, Arc⁻, Orc⁻	Holsters et al. (1980)
pTiC58	Nopaline	Tn1	HindIII-10, -14, or -13	++	Holsters et al. (1980)
pTiC58	Nopaline	Tn7	HindIII-4, -9, -27, or -19	Onc⁻	Holsters et al. (1980)
pTiC58	Nopaline	Tn7	HindIII-4	Tra⁻	Holsters et al. (1980)
pTiC58	Nopaline	Tn7	HindIII-11	Agrʳ	Holsters et al. (1980)
pTiC58	Nopaline	Tn7	Between EcoRI-26 and -4[b]	Ape⁻	Holsters et al. (1980)
pTiC58	Nopaline	Tn7	HindIII-2, -30, -43, or -26	Nos⁻, Arc⁻, Orc⁻ᶜ	Holsters et al. (1980)
pTiC58	Nopaline	Tn7	HindIII-34	++	Holsters et al. (1980)
pAL102	Octopine	Tn904	HpaI-12	Onc⁻	Klapwijk et al. (1980)
pAL102	Octopine	Tn904	HpaI-6 and SmaI-3aᵈ	Onc⁺/⁻ᵉ	Klapwijk et al. (1980)
pAL102	Octopine	Tn904	SmaI-8 and -3aᵈ	Onc⁻/ᵉ	Klapwijk et al. (1980)
pAL102	Octopine	Tn904	HpaI-13	+	Klapwijk et al. (1980)
pAL102	Octopine	Tn904	Between HpaI-8 and SmaI-10a; SmaI-3a, -6, or -3b	++	Klapwijk et al. (1980)
pTiA6NC	Octopine	Tn5	HpaI-9	Onc⁻	Garfinkel and Nester (1980)
pTiA6NC	Octopine	Tn5	HpaI-12	Onc⁻	Garfinkel and Nester (1980)
pTiA6NC	Octopine	Tn5	HpaI-6	Onc⁻	Garfinkel and Nester (1980)
pTiA6NC	Octopine	Tn5	HpaI-3	Onc⁻	Garfinkel and Nester (1980)
pTiB₆806	Octopine	Tn5	HpaI-16	Onc⁻	Garfinkel and Nester (1980)
pTiB₆806	Octopine	Tn5	HpaI-15	Onc⁻	Garfinkel and Nester (1980)
pTiB₆806	Octopine	Tn5	HpaI-6	Onc⁻	Garfinkel and Nester (1980)
pTiB₆806	Octopine	Tn5	HpaI-7	+	Garfinkel and Nester (1980)
pTiB₆806	Octopine	Tn5	HpaI-6	+	Garfinkel and Nester (1980)
pTiB₆806	Octopine	Tn5	HpaI-14	+	Garfinkel and Nester (1980)
pTiB₆806	Octopine	Tn5	HpaI-7 or KpnI-4	Occ⁻	Garfinkel and Nester (1980)
pTiB₆806	Octopine	Tn5	KpnI-8 or -9	++	Garfinkel and Nester (1980)

[a] Noc, Nopaline catabolism; Occ, octopine catabolism; Agrʳ, agrocin 84 resistance; Orc, ornithine catabolism; Ape, phage AP1 exclusion; Arc, arginine catabolism; Onc, oncogenicity; Tra, autotransfer; Nos, nopaline synthesis; +, altered tumor morphology; ++, wild-type phenotype.

[b] Deletion strains.

[c] Strains containing these mutant plasmids cannot utilize nopaline as a nitrogen source, or arginine and nopaline as a carbon source.

[d] These mutations were due to a double Tn insertion.

[e] Weakly oncogenic.

transfer of T-DNA to the plant genome or in its integration. Since none of the Tra⁻ mutants were found to affect oncogenicity, and vice versa, it was concluded that certain Tra functions existed that were not involved in T-DNA transfer to the plant cell. However, the possibility remains that common functions do exist, yet were undetected. Also, mutations which mapped in the right, nonconserved T-DNA did not affect oncogenicity but were essential for nopaline biosynthesis in tumor cells (Holsters *et al.*, 1980). Specific insertions and deletions affecting pTi-encoded phenotypes are summarized in Table I (see Fig. 2).

Tn7 insertion into the T-DNA region of the nopaline plasmid pT37Noc[c]1 has been used as a marker in studying the potential use of Ti plasmids as vectors for the genetic manipulation of plants. An axenic tissue culture was started from a tumor induced by the Tn7 insertion mutant. With the use of Southern blot hybridization techniques, Tn7 was shown to be present in the plant tumor genome, integrated at its original site in the T-DNA region (Hernalsteens *et al.*, 1980).

Tn7 has also been used to obtain insertion mutants in the octopine plasmid pTiB6S3 (Hernalsteens *et al.*, 1978). With *A. tumefaciens* strain C58C1 as donor, the plasmid cointegrate pTiB6S3::RP4 was introduced into the *E. coli* strain J62Tn7-1. Transposition of Tn7 into the pTiB6S3::RP4 cointegrate was tested by conjugation into an *E. coli* strain, and the transconjugants selected for both Tn7 and RP4 markers. The frequency of Tn7 transposition observed was 5.5×10^{-4}. The high dissociation rate of the plasmid cointegrate pTiB6S3::RP4, with subsequent loss of the TiB6S3 plasmid, afforded a simple assay for localizing the Tn7 insertion either in the Ti or the RP4 part of the plasmid. If the Tn7 were inserted into the Ti part of the cointegrate, transconjugants selected for RP4 markers would also be trimethoprim-sensitive. Furthermore, it was shown that the majority of these clones could transfer the ability to utilize octopine upon back-conjugation into an *Agrobacterium* host. If, however, Tn7 were inserted into the RP4 of the cointegrate, transconjugants selected for RP4 markers would also be trimethoprim-resistant, and upon subsequent conjugation with *Agrobacterium*, many of these clones failed to transfer the octopine marker.

Tn7 insertion mutants were classified by AP1 phage exclusion, oncogenicity, and their ability to utilize octopine as sole nitrogen source. Although all three classes of mutants were isolated, the physical location of the insertions was not analyzed.

Recently, it has been shown that many if not all resistance (R) plasmids, which can mobilize nonconjugative plasmids, carry Tn's (Hooykaas *et al.*, 1980). When Ti-R plasmids are used as target sites for Tn insertion mutagenesis, it is essential to search for any additional Tn insertions which could have originated in the R plasmid. RP4, for example, contains the ampicillin resistance transposon Tn1. Upon dissociation of a Ti::RP4 cointegrate, and subsequent isolation of the Ti part, it was found that two out of seven isolates contained Tn1 insertions.

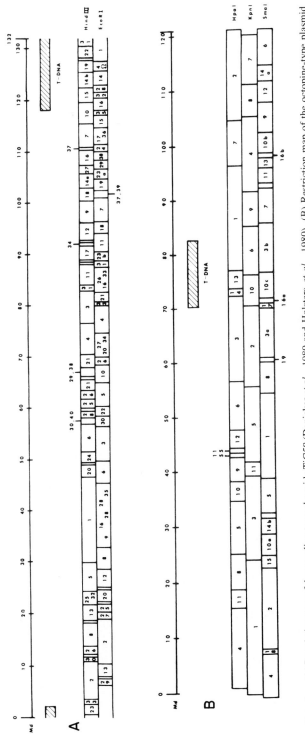

Fig. 2.   (A) Restriction map of the nopaline-type plasmid pTiC58 (Depicker *et al.*, 1980 and Holsters *et al.*, 1980).   (B) Restriction map of the octopine-type plasmid pTiB₆806 (Garfinkel and Nester, 1980).

Recently, a new transposon, designated Tn1831, has been isolated from the Ti-mobilizing plasmid R702 (Hooykaas *et al.*, 1980). Tn1831 is an 11-megadalton Tn which carries markers for resistance to spectinomycin, streptomycin, and mercury and is capable of translocating from R plasmids to Ti plasmids, and vice versa.

The streptomycin resistance transposon Tn904 has been used to isolate Tn insertion mutants of the octopine Ti plasmid pAL102 harbored by *A. tumefaciens* strain Ach5 (Klapwijk *et al.*, 1980). A transfer-deficient derivative of RP1-pMG1 was used to donate Tn904 to a constitutive self-transmissible derivative of pAL102. And Ti::Tn904 mutants were confirmed by introducing, via conjugation, an incompatible Ti plasmid, which resulted in the loss of the original resident Ti::Tn904. The transposition frequency was estimated to be 3×10^{-4} per transferred Ti plasmid. Tn904 insertion mutants were characterized with respect to virulence, octopine degradation, octopine synthesis in induced tumors, and Ti plasmid transfer. However, the only phenotypic mutants isolated were those affected in their tumor-inducing ability (Table I).

Independently isolated pTiAL102::Tn904 mutant strains (not affected in any of the known Ti-encoded traits) were analyzed for preferred Tn904 insertion sites. Eight out of 11 Tn insertions were localized in a region bordered by *Hpa*I-8 and *Sma*I-10a. The remaining three mutants showed Tn904 insertions in restriction fragments *Sma*I-3, -6, and -3b. These results suggest that, although insertions could occur at a number of sites, Tn904 appears to have a preferred site of insertion in a region close to the Ti origin of replication (Koekman *et al.*, 1979).

The transposon Tn5, which encodes resistance to kanamycin, has been used to isolate Tn insertion mutants of *A. tumefaciens* which affect virulence or the ability to catabolize octopine (Garfinkel and Nester, 1980) (Table I). Tn mutagenesis was performed by mating an *E. coli* strain carrying Tn5 in the conjugative P-type plasmid pPJB4J1 (Beringer *et al.*, 1978) with *Agrobacterium* strain A722 (pTiA6NC), A723 (pTiB$_6$806), or A6NC (pTiA6NC). Transconjugants were screened on minimal media containing kanamycin, with a resultant frequency of transposition estimated at 5×10^{-2} per recipient plated.

A number of avirulent mutants containing Tn5 insertions were not localized on the Ti plasmid. When the Ti plasmid from these mutants was transferred to a plasmid-cured strain, virulence was reestablished. Furthermore, the resulting transconjugants were kanamycin-sensitive. Presumably, Tn5 caused a mutation in a gene affecting virulence, which is not coded by the Ti plasmid. In strain A6NC, these genes are probably located on the *Agrobacterium* chromosome, as no plasmid except pTiA6NC can be detected by standard techniques. Strain A722, however, contains a cryptic plasmid which is larger than 300 megadaltons. Therefore, Tn5 insertions affecting virulence could be located in either the *Agrobacterium* chromosome or the cryptic plasmid. ^{32}P-Labeled, Tn5 hybridized to both cryptic plasmid and chromosomal DNA demonstrated that in all cases

Tn5 insertion resided in the *Agrobacterium* chromosome. Approximately one half of the chromosomal mutants were avirulent on all plants tested (sunflower, tomato, and *Kalanchoe daigremontiana*), while the remaining chromosomal mutants exhibited an altered host range.

The map location of Tn5 insertions affecting octopine catabolism (Table I) has independently confirmed the location of the *occ* gene(s) in pTiB$_6$806, previously mapped by deletion analysis (Koekman *et al.*, 1979). In all cases, Occ$^-$ mutants were virulent when assayed on *Kalanchoe*.

Studies performed to test the stability of Tn5 insertions resulted in a frequency of 5×10^{-8} for Tn5 excision. Gene function resumed in all cases, concomitant with kanamycin sensitivity, indicating loss of Tn5 from the cell.

Randomly picked, virulent, kanamycin-resistant transconjugants were assayed for preferential Tn insertion sites between the chromosome and the Ti plasmid, and 7.5% of the transconjugants screened showed Tn5 insertions in the Ti plasmid (pTiB$_6$806). If the relative target sizes of the pTiB$_6$806 (120-megadalton) plasmid are compared with the *Agrobacterium* chromosome, it can be concluded that Tn5 exhibits little or no preferential insertion site properties with reference to the plasmid and the chromosome.

As stated above, the vector used for introducing Tn5 into *A. tumefaciens* was pPJB4J1, commonly referred to as the Beringer "suicide" plasmid (Beringer *et al.*, 1978). It is a P1 group plasmid carrying both Mu and Tn5. The plasmid confers upon its host resistance to both gentamycin and kanamycin. It is also Mu-immune but is unable to produce viable Mu phage. This indicates that Tn5 is inserted either directly into Mu or beside it, causing a deletion mutation in Mu. Recent evidence has shown that, when Tn5 is transposed from pPJB4J1 in *Rhizobium*, Mu sequences are often transposed with it and are subsequently inserted with Tn5 (F. Ausubel, personal communication). Therefore, when analyzing Tn5 insertions, one must account for possible Mu sequences present. (See Note Added in Proof.)

V. T-DNA

A. EVIDENCE FOR DNA TRANSFER (T-DNA)

Indirect evidence for the transfer of Ti plasmid sequences from bacterium to plant cell was provided by the demonstration that genes controlling opine synthesis in tumor cells were localized on the Ti plasmid (Bomhoff *et al.*, 1976; Genetello *et al.*, 1977; Montoya *et al.*, 1977; Schell and VanMontagu, 1977). That a specific portion of the plasmid was responsible for plant cell transformation was initially suggested by the demonstration of reversible cointegration of the incP plasmid RP4 and pTiC58 (Holsters *et al.*, 1978; Hernalsteens *et al.*, 1978). An integration site of RP4 in pTiC58 proved to be nononcogenic and was localized in the "common DNA," a segment highly conserved between octopine- and nopaline-type Ti plasmids (Depicker et al., 1978; Chilton et al., 1978). Dissociation of the cointegrate pTiC58::RP4 restored oncogenicity.

The speculation that Ti plasmid sequences were transferred from Agrobacterium to the plant cell was confirmed when Chilton et al. (1977) found stable incorporation of a specific part of the Ti plasmid, termed T-DNA, present in an octopine-type tumor line. Indeed, this T-DNA included common DNA, previously implicated in tumorigenesis, plus adjacent Ti sequences (Chilton et al., 1977, 1978).

In addition to the original demonstration of Ti plasmid sequences in tumor cells, recent studies have confirmed the presence of T-DNA in octopine-type tumors (Merlo et al., 1980; Thomashow et al., 1980a,c) and in nopaline crown gall cells (DeBeuckeleer et al., 1978; Schell et al., 1979; Yang et al., 1980; VanMontagu and Schell, 1979; Zambryski et al., 1980; Lemmers et al., 1980). In all studies to date, the T-DNA has been shown to be derived from a similar region of the Ti plasmid (termed the T region), indicating that it is the actual homologous plasmid sequence found in transformed cells. The following sections are designed to review our present understanding of the T-DNA complement in octopine- and nopaline-type tumors and to examine such details as organization, copy number, boundaries, and integration.

B. EXPERIMENTAL APPROACHES TO THE STUDY OF T-DNA SEQUENCES

In the initial demonstration of T-DNA sequences in octopine tumor cells (Chilton et al., 1977) employed DNA solution renaturation kinetics. Each plasmid fragment was radioactively labeled in vitro to high specific activity, denatured to single-stranded DNA, and allowed to reassociate at a low concentration in the presence of high concentrations of tumor cell or normal cellular DNA ("driver" DNA). An increase in the rate of reassociation with tumor DNA versus normal DNA using a specific endonuclease fragment of Ti plasmid as probe DNA indicates the presence of DNA sequences in the tumor DNA homologous to that specific Ti fragment. Such studies are important in determining the copy number of specific plasmid sequences in a tumor line, which can be inferred from the kinetics of the reaction.

The detection of specific unique DNA sequences in a large complex eukaryotic genome has been further extended using Southern hybridization techniques (Southern, 1975). Basically, the experimental approach, which was first applied by Botcham et al. (1976) in a study on SV40 integration in mouse cell lines, consists of isolating putative transformant genomic DNA, subsequent digestion with a site-specific restriction endonuclease, and separation of the resultant fragments by agarose electrophoresis. The genomic restriction fragments are denatured in situ and transferred from the gel to a nitrocellulose filter by blotting. The filter is hybridized with a probe DNA, which has been labeled in vitro with

[32]P, and then autoradiographed. Autoradiograph bands are the result of probe DNA specifically annealed to complementary sequences on the filter, and comparison with the banding pattern of the endonuclease-restricted transformed sequences can indicate which, if any, transformed DNA sequences are present in the cellular DNA.

C. T-DNA Organization

Using DNA solution hybridization techniques, Chilton *et al.* (1977) found that the tumor $B_6806/E9$ contained T-DNA representing a contiguous region of approximately 5% of the plasmid $pTiB_6806$ at a level of about 20 copies per diploid cell. Subsequent confirmation by Merlo *et al.* (1980) extended these results to three independent octopine-type tumor lines incited by *A. tumefaciens* strains B_6806, A6S, and 15955. Results also suggested that the T-DNA complement in tumor cells could vary, especially the right-hand boundary, yet always included common DNA sequences. These findings indicated that Ti plasmid DNA incorporation was a common feature in each tumor line, yet the copy number of the right- and left-hand portion of the T-DNA could vary within a given tumor line (Merlo *et al.*, 1980).

Thomashow *et al.* (1980a) further defined T-DNA sequences in four octopine-type tumors by Southern hybridization analysis. These findings confirmed that T-DNA always included a contiguous region between *Hin*dIII fragment Y and *Eco*RI-24, which included the common DNA (Fig. 3). The ubiquity of common DNA sequences in crown gall cells was also demonstrated in a nopaline-type tumor, T37, which contained T-DNA sequences homologous to common DNA (Schell *et al.*, 1979; Yang *et al.*, 1980). Detailed mapping of the T-DNA of different nopaline-type tumor lines (Lemmers *et al.*, 1980) indicated a T-DNA complement corresponding to a pTiC58 segment extending from within *Eco*RI-35 on the left to a point within *Hin*dIII-23 on the right and also including the highly conserved common DNA (Fig. 2). In all such studies, no homology with normal, untransformed cellular DNA has been observed under high stringency conditions. The striking consistency of the common DNA found in tumor cells suggests a crucial role in the maintenance and/or induction of the transformed phenotype (Thomashow *et al.*, 1980a). This conclusion is supported by genetic analysis demonstrating that deletion of all but conserved T-DNA still results in transformation (Koekman *et al.*, 1979). However, lack of significant hybridization between limited host range plasmid DNA and the common DNA region of $pTiB_6806$ suggests these sequences are not essential for crown gall tumorigenesis on all plant species (Thomashow *et al.*, 1981).

Adjacent T-DNA sequences must also contain information for opine production. Analysis of octopine-producing and non-octopine-producing tumor lines suggests that T-DNA sequences within *Eco*RI-24 are involved in determining octopine synthesis in tumor cells (Thomashow *et al.*, 1980a). Similarly, DNA

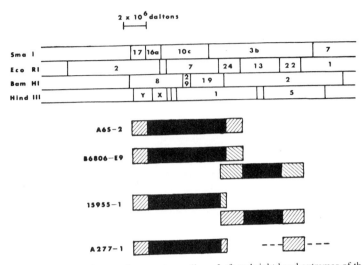

Fig. 3. T-DNA found in octopine-type tumor lines. Left and right hand extremes of the T-DNA sequences fall within the crosshatched areas of the bars. (From Thomashow *et al.*, 1980a).

sequences adjacent to the common DNA of the nopaline-type tumor T37 include *Hin*dIII-23 (Schell *et al.*, 1979; Lemmers *et al.*, 1980), which when mutagenized with a Tn1 insertion results in an Onc[+] Nos[−] phenotype (Holsters *et al.*, 1980). It will be interesting to see whether the enzyme coding for opine synthesis in tumor cells is directly coded for by T-DNA sequences. Preliminary results indicate that this may be the case.

The common feature of the organization of octopine and nopaline T-DNA is that the T region is transferred to the plant cells as a colinear core segment without major internal rearrangements (Thomashow *et al.*, 1980a; Lemmers *et al.*, 1980). This conclusion is further supported by the observation that previously established restriction maps of the T region of the corresponding Ti plasmids (Chilton *et al.*, 1979; Depicker *et al.*, 1980) are colinear with those of T-DNA found in tumor cells. Therefore, it is unlikely that significant structural changes occur in core T-DNA sequences during the course of cellular division of tumor cells. However, variation in the copy number of T-DNA sequences within tumor lines (Merlo *et al.*, 1980), especially in T-DNA sequences between *Eco*RI-22 and -24 in tumor lines $B_6806/E9$ and 15955/1, suggests the possibility of amplification of T-DNA sequences.

D. INTEGRATION OF T-DNA

T-DNA, but not entire Ti plasmid copies, is stably maintained in plant cells, which implies that T-DNA exists as free replicons and/or as T-DNA integrated into plant chromosomal or plastid DNA. However, when undigested octopine-type tumor DNA was subjected to agarose electrophoresis, blotted to nitrocellulose, and annealed with ^{32}P-labeled pTiB$_6$806, hybridization was to a broad-region high-molecular-weight DNA and not to any discrete bands, suggesting that T-DNA replicons do not exist in tumor cells (Thomashow et al., 1980a). In addition, similar filters containing BT37 and W38C58-1 undigested tumor DNA hybridized with various ^{32}P-labeled T-DNA probes derived from pTiC58 show a similar pattern of hybridization (Lemmers et al., 1980).

If T-DNA is integrated into plant DNA, then restriction endonuclease digestion of tumor DNA, which cuts at least once within integrated T-DNA sequences, should yield "composite" or "junction" fragments composed of both plasmid and plant DNA covalently joined. Such fragments should have molecular weights determined by one restriction site with the plasmid sequences and one site with the adjacent plant DNA and migrate anomalously when compared to internal plasmid fragments differing in size. The presence of anomalously migrating fragments in both octopine- (Thomashow et al., 1980a) and nopaline-type tumor cellular DNA (Schell et al., 1979; Zambryski et al., 1980; Lemmers et al., 1980) suggests T-DNA integration into plant DNA. However, the possibility of internal rearrangements of T-DNA sequences, i.e., fusion between plasmid sequences, could result in the appearance of a second type of composite fragment (type II). Therefore, demonstration that both plasmid and plant sequences are present in a composite fragment is necessary proof of covalent linkage between plasmid and plant sequences.

Molecular cloning of a putative A6S/2 junction fragment and hybridization with both plant and T-DNA sequences demonstrated recombination between Ti plasmid sequences and tobacco DNA (Thomashow et al., 1980c). In addition, composite fragments of the cloned teratoma line BT37 were shown to consist of covalently linked plant and Ti plasmid DNA (Zambryski et al., 1980).

The pattern of hybridization of the A6S/2 and T37 junction fragments is indicative of T-DNA integration into highly repeated tobacco DNA sequences (Thomashow et al., 1980c; Zambryski et al., 1980). However, the organization pattern of the Nicotiana tabacum genome indicates that at least 55% of the total DNA consists of alternating single-copy and short, repetitive sequences and 25% contains long, repetitive DNA segments (Zimmerman and Goldberg, 1977). Therefore, the insertion of T-DNA into repetitive plant DNA may not mean preferential insertion at such a site. It is apparent that more such analyses will be needed to understand T-DNA integration events.

A type II composite fragment has been isolated by molecular cloning of T37 plant DNA into the bacteriophage lambda vector charon 4A, and its structure determined by restriction mapping and hybridization analysis (Zambryski et al., 1980). When compared to the restriction pattern from the T region, the cloned fragment was found to hybridize with both ends of the T region, indicating that the left and right borders of the T-DNA were linked together. This suggests that the T-DNA in T37 is found as tandem repeats which are inserted into repetitive plant DNA sequences (Zambryski *et al.*, 1980). Preliminary sequence data suggest that several direct and one inverted repeat are present in the cloned type II composite fragment.

E. Reversion from the Neoplastic State

Under specific conditions both unorganized tumors (Sacristan and Melchers, 1977) and tissues derived from teratomas (Braun and Wood, 1976; Turgeon *et al.*, 1976) can revert to plants with normal appearances. When the tobacco teratoma BT37 was grafted onto a normal tobacco plant, some of the grafted tumors developed into plants with normal-appearing shoots, leaves, and stems which eventually flowered and set viable seed (Braun and Wood, 1976; Turgeon *et al.*, 1976). However, explants [tissues from regenerated teratoma plant leaves (TL) and from flower petals (TFP)] from these revertant shoots retained tumorous characteristics in tissue culture, since they proliferated in the absence of phytohormones, synthesized nopaline, and formed tumors when grafted onto healthy hosts. However, axenic cultures derived from the haploid portion of the anthers [tissue derived from haploid anthers (TDHA)] and diploid plant tissue lines obtained from the F_1 progeny of revertants [diploid from teratoma seed (DFTS)] did not grow without exogenously supplied hormone or synthesize nopaline.

Since nopaline synthesis was demonstrated in TL and TFP tissues (Turgeon *et al.*, 1976), it was expected that these cells should contain T-DNA sequences, which was found to be the case. Kinetic analysis suggested that most of the T-DNA present in the original teratoma was present in tissues derived from regenerated plants (Yang *et al.*, 1980). Subsequently, Southern hybridization experiments demonstrated that T-DNA present in TL and TFP tissues was co-linear and essentially the same size (except for a missing *Eco*RII-6/35 restriction site) as the T-DNA of the parental BT37 teratoma (Lemmers *et al.*, 1980). However, the type II composite border fragment detected in BT37 (Zambryski *et al.*, 1980) was missing in these tissues (Lemmers *et al.*, 1980). Neither the haploid plants derived from anther culture (TDHA) nor DFTS tissue contained measurable T-DNA sequences (Lemmers *et al.*, 1980).

The results presented to date suggest that the regenerated plants derived from the nopaline-type tumor BT37 are still transformed and contain functional T-DNA sequences. These sequences are lost by some unknown mechanism when these cells go through meiotic divisions. Whether the erasing of T-DNA is a specific excision mechanism or a common feature of foreign DNA in plant cells which go through meiosis is unknown. However, F_1 progeny of tumorous tobacco shoots, derived from callus tissues produced by *Agrobacterium*-induced transformation of protoplasts, retained morphological tumorous traits and synthesized nopaline (Wullems *et al.,* 1981). This suggests that T-DNA markers can be inherited and expressed in progeny seedlings.

REFERENCES

Barth, P. T., Datta, N., Hedges, R. W., and Grinter, N. J. (1976). *J. Bacteriol.* **125**, 800–810.

Beringer, J. E., Beynon, J. L., Buchanan-Wolaston, A. V., and Johnson, A. W. B. (1978). *Nature (London)* **276**, 633–634.

Biemann, K., Lioret, C., Asselineau, J., Lederer, E., and Polonski, J. (1960). *Biochim. Biophys. Acta* **40**, 369–370.

Bomhoff, G. H. (1974). Studies on crown-gall—a plant tumor. Investigations on protein composition and on the use of quanidines compounds as a marker for transformed cells. Thesis, University of Leyden.

Bomhoff, G. H., Klapwijk, P. M., Kester, H. C. M., Schilperoort, R. A., Hernalsteens, J. P., and Schell, J. (1976). *Mol. Gen. Genet.* **145**, 177–181.

Botcham, M., Topp, W., and Sambrook, J. (1976). *Cell* **9**, 269–287.

Braun, A. C. (1951). *Phytopathology* **41**, 963–966.

Braun, A. C. (1953). *Bot. Gaz.* **114**, 363–371.

Braun, A. C. (1956). *Cancer Res.* **16**, 53–56.

Braun, A. C., and Mandle, R. J. (1948). *Growth* **12**, 255–269.

Braun, A. C., and Wood, H. N. (1976). *Proc. Natl. Acad. Sci. U.S.A.* **73**, 496–500.

Chilton, M.-D., Drummond, M. H., Merlo, D. J., Sciaky, D., Montoya, A. L., Gordon, M. P., and Nutter, R., and Nester, E. W. (1978). *In* "Microbiology," American Society of Microbiology.

Chilton, M.-D., Drummond, M. H., Gordon, M. P., Merlo, D. J., Montoya, A. L., Sciaky, D., Nutter, R., and Nester, E. W. (1978). *In* "Microbiology."

Chilton, M.-D., Saiki, R. K., Yadav, N., Gordon, M. P., and Quetier, F. (1980). *Proc. Natl. Acad. Sci. U.S.A.* **77**, 4060–4064.

Coxon, D. T., Davies, A. M. C., Fenwick, G. R., and Self, R. (1980). *Tetrahedron Lett.* **21**, 495–498.

Currier, T. C., and Nester, E. W. (1976). *J. Bacteriol.* **126**, 157–165.

Davis, R. W., Simon, M., and Davidson, N. (1971). *In* "Methods in Enzymology" (D. L. Grossman and K. Moldave, eds.), Vol. XXID, pp. 413–428. Academic Press, New York.

DeBeuckeller, M., DeBlock, M., Dhaese, P., Depicker, A., DeVos, G., DeVos, R., DeWilde, M., Jacobs, A., Schell, J., Seurinck, J., Silva, B., VanHaute, E., VanMontagu, M., VanVliet, F., Villarroel, R., and Zaenen, I. (1978). *Proc. IVth Int. Conf. Plant Pathol. Bacteriol., 4th, I.N.R.A., Angers* pp. 153–160.

DeCleene, M., and Deley, J. (1978). *Bot. Rev.* **42**, 389–466.

Depicker, A., VanMontagu, M., and Schell, J. (1978). *Nature (London)* **275**, 150.

Dhaese, P., DeGreve, H., Decraemer, H., Schell, J., and VanMontagu, M. (1979). *Nucleic Acids Res.* **7**, 1837–1849.

Drummond, M. H., and Chilton, M.-D. (1978). *J. Bacteriol.* **136**, 1178–1183.

Drummond, M. H., Gordon, M. P., Nester, E. W., and Chilton, M.-D. (1977). *Nature (London)* **269**, 535–536.

Elliot, C. (1951). "Manual of Bacterial Plant Pathogens," 2nd ed., pp. 3–12. Chronica Botanica, Waltham, Massachusetts.

Ellis, J., Kerr, A., Tempe, J., and Petit, A. (1979). *Mol. Gen. Genet.* **173**, 263–269.

Engler, G., Holsters, M., VanMontagu, M., Schell, J., Hernalsteens, J. P., and Schilperoort, R. (1975). *Mol. Gen. Genet.* **138**, 345–349.

Firmin, J. L., and Fenwick, R. G. (1977). *Phytochemistry* **16**, 761–762.

Firmin, J. L., and Fenwick, G. R. (1978). *Nature (London)* **276**, 842–843.

Garfinkel, D. J., and Nester, E. W. (1980). *J. Bacteriol.* **144**(2), 732–743.

Gautheret, R. J. (1947). *C. R. Soc. Biol.* **141**, 598–601.

Gee, M., Sun, C. N., and Dwyer, J. D. (1967). *Protoplasma* **64**, 195–200.

Genetello, C., VanLarebeke, N., Holsters, M., Depicker, A., VanMontagu, M., and Schell, J. (1977). *Nature (London)* **265**, 561–563.

Goldmann, A., Thomas, D. W., and Morel, G. (1969). *C. R. Acad. Sci. (Paris)* **268D**, 852–854.

Gresshoff, P. M., Skotnicki, M. L., and Rolfe, B. G. (1979). *J. Bacteriol.* **137**(1), 1020–1021.

Gurley, W. B., Kemp, J. D., Albert, M. J., Sutton, D. W., and Collins, J. (1979). *Proc. Natl. Acad. Sci. U.S.A.* **76**, 2828–2832.

Guyon, P., Chilton, M.-D., Petit, A., and Tempe, J. (1980). *Proc. Natl. Acad. Sci. U.S.A.* **77**(5), 2693–2697.

Hedges, R. W., and Jacob, A. E. (1974). *Mol. Gen. Genet.* **132**, 31–40.

Hernalsteens, J. P. (1980). *Nature (London)* **287**, 654–666.

Hernalsteens, J. P., DeGreve, H., VanMontagu, M., and Schell, J. (1978). *Plasmid* **1**, 218–225.

Hildebrand, E. M. (1940). *J. Agric. Res.* **61**, 685–696.

Hildebrandt, A. C., and Riker, A. J. (1949). *Am. J. Bot.* **36**, 74–85.

Hohl, H. R. (1960). *Ber. Schweiz. Bot. Ges.* **70**, 395–439.

Hohl, H. R. (1961). *Phytopathol. Z.* **40**, 317–356.

Holderbach, E., and Beiderbeck, R. (1976). *Phytochemistry* **15**, 955–956.

Holsters, M., Silva, A., Genetello, C., Engler, G., VanVliet, F., DeBlock, M., Villarroel, R., VanMontagu, M., and Schell, J. (1978). *Plasmid* **1**, 456–467.

Holsters, M., Silva, B., VanVliet, F., Genetello, G., DeBlock, M., Dhaese, P., Depicker, A., Inze, D., Engler, G., Villarroel, R., VanMontagu, M., and Schell, J. (1980). *Plasmid* **3**, 212–230.

Hooykaas, P. J. J., Roobol, C., and Schilperoort, R. A. (1979). *J. Gen. Microl.* **110**, 99–109.

Hooykaas, P. J. J., DenDelk-Ras, H., and Schilperoort, R. A. (1980). *Plasmid* **4**, 64–75.

Kemp, J. D. (1976). *Biochem. Biophys. Res. Commun.* **69**, 816–822.

Kemp, J. D. (1977). *Biochem. Biophys. Res. Commun.* **74**, 862.

Kerr, A. (1969). *Nature (London)* **233**, 1175–1176.

Kerr, A. (1971). *Physiol. Plant Pathol.* **1**, 241–246.

Klapwijk, P. M., Hooykaas, R. J. J., Kester, H. C. M., Schilperoort, R. A., and Rörsch, A. (1976). *J. Gen. Microl.* **96**, 155–163.

Klapwijk, P. M., Scheulderman, T., and Schilperoort, R. A. (1978). *J. Bacteriol.* **136**(2), 775–785.

Klapwijk, P. M., Breukelen, J. V., Korevaar, K., Ooms, G., and Schilperoort, R. A. (1980). *J. Bacteriol.* **141**(1), 129–136.

Kleckner, N., Roth, J., and Botstein, D. (1977). *J. Mol. Biol.* **116**, 125–159.

Koekman, B. T., Ooms, G., Klapwijk, P. M., and Schilperoort, R. A. (1979). *Plasmid* **2**, 347–357.

Lemmers, M., DeBeuckeleer, M., Holsters, M., Zambryski, P., Depicker, A., Hernalsteens, J. P., VanMontagu, M., and Schell, J. (1980). *J. Mol. Biol.* **144**, 353–376.

Lioret, C. (1957). *Annee Biol.* **31**(7/8), 385–394.

Lippincott, J. A., and Heberlein, G. T. (1965). *Am. J. Bot.* **52**, 856–863.

Loper, J. E., and Kado, C. J. (1979). *J. Bacteriol.* **139**, 591–596.

McPherson, J. C., Nester, E. W., and Gordon, M. P. (1980). *PNAS* **77**(5), 2666–2670.

Margrou, J. (1927). *Ann. Inst. Pasteur.* **41**, 785–801.

Menage, T., and Morel, A. (1964). *C.R. Acad. Sci. Paris* **259D**, 4795–4796.

Menage, T., and Morel, A. (1965). *C.R. Acad. Sci. Paris* **261**, 2001–2002.

Merlo, D. J., Nutter, R. C., Montoya, A., Garfinkel, D. J., Drummond, M. H., Chilton, M.-D., Gordon, M. P., and Nester, E. W. (1980). *Mol. Gen. Genet.* **177**, 637–643.

Montoya, A. L., Chilton, M.-D., Gordon, M., Sciaky, D., and Nester, E. W. (1977). *J. Bacteriol.* **129**, 101–107.

Morel, G. (1948). *Ann. Epiphyt. N.S.* **14** (Ser. Pathol. Veg. Mem. 5), 123–234.

Ooms, G., Klapwijk, P. M., Poulis, J. A., and Schilperoort, R. A. (1980). *J. Bacteriol.* **144**(1), 82–91.

Panagopoulos, C. G., and Psallidas, P. G. (1973). *J. Appl. Bacteriol.* **36**, 233–240.

Panagopoulos, C. G., Psallidas, P. G., and Alivizatos, A. S. (1978). *Proc. Int. Conf. Plant Pathol. Bacteriol., 4th, Angers.*

Petit, A., and Tempe, J. (1978). *Mol. Gen. Genet.* **167**, 147–155.

Petit, A., Delhaye, S., Tempe, J., and Morel, G. (1970). *Physiol. Veg.* **8**, 205.

Rasch, E. M. (1964). *Exp. Cell Res.* **36**, 475–486.

Riker, A. J. (1923). *J. Agric. Res.* **25**, 119–132.

Riker, A. J., Banfield, W. M., Wright, W. H., and Keitt, G. W. (1930). *Science* **68**(1763), 357–359.

Robinson, W., and Walkden, H. (1923). *Ann. Bot.* **37**, 299–324.

Ryter, A., and Manigault, P. (1964). *Bull. Soc. Franc. Physiol. Veg.* **10**, 44–56.

Sacristan, M. D., and Melchers, G. (1977). *Mol. Gen. Genet.* **152**, 111–117.

Schell, J. (1975). *In* "NATO Adv. Study Institute on Genetic Manipulation with Plant Material, Liege" (L. Ledoux, ed.), Vol. A3, pp. 163–181. Plenum, New York.

Schell, J., and VanMontagu, M. (1977). *Brookhaven Symp. Biol.* **29**, 36–49.

Schell, J., VanMontagu, M., DeBeuckeleer, M., DeBlock, M., Depicker, A., DeWilde, M., Engler, G., Genetello, C., Hernalsteens, J. P., Holsters, M., Seurinck, J., Silva, B., VanVliet, F., and Villarroel, R. (1979). *Proc. R. Soc. London Ser. B* **204**, 251–266.

Sciaky, D., Montoya, A. L., and Chilton, M.-D. (1978). *Plasmid* **1**, 238–253.

Smith, E. F. (1920). "An Introduction to Bacterial Diseases of Plants." Saunders, Philadelphia, Pennsylvania.

Smith, E. F., and Townsend, C. O. (1907). *Science* **25**, 671–673.

Southern, E. M. (1975). *J. Mol. Biol.* **98**, 503–517.

Stonier, T. (1956). *Am. J. Bot.* **43**, 647–655.

Stonier, T. (1968). *In Vitro* **3**, 185.

Sun, C. N. (1969). *Adv. Frontiers Plant Sci.* **23**, 119–135.

Sutton, D., Kemp, J., and Hack, E. (1977). *Plant Physiol.* **59**, 108.

Tempe, J., Estrade, C., and Petit, A. (1978). *Proc. Int. Conf. Plant Pathol. Bacteriol., 4th, Angers* pp. 153–160.

Thomashow, M. F., Nutter, R., Montoya, A. L., Gordon, M. P., and Nester, E. W. (1980a). *Cell* **19**, 729–739.

Thomashow, M. F., Panagopoulos, C. G., Gordon, M. P., and Nester, E. W. (1980b). *Nature (London)* **283**, 794–796.

Thomashow, M. F., Nutter, R., Poste, K., Chilton, M.-D., Blattner, F. R., Powell, A., Gordon, M. P., and Nester, E. W. (1980c). *Proc. Natl. Acad. Sci. U.S.A.* **77**(11), 6448–6452.

Thomashow, M. F., Knauf, V. C., and Nester, E. W. (1981). *J. Bacteriol.* **146**, 484–493.

Turgeon, R., Wood, H. N., and Braun, A. C. (1976). *Proc. Natl. Acad. Sci. U.S.A.* **73**, 3562–3564.

Van Larebeke, N., Engler, G., Holsters, M., Van den Elesecher, S., Zaenen, I., Schilperoort, R. A., and Schell, J. (1974). *Nature (London)* **252**, 169–170.

Van Larebeke, N., Genetello, C., Schell, J., Schilperoort, R. A., Hermans, A. K., Hernalsteens, J. P., and Van Montagu, M. (1975). *Nature (London)* **255**, 242–243.

VanMontagu, M., and Schell, J. (1979). *In* "Plasmids of Medical, Environmental and Commercial Importance" (K. Timmis and A. Puhler, eds.), pp. 71–96. Elsevier, Amsterdam.

Watson, B., Currier, T. C., Gordon, M. P., Chilton, M.-D., and Nester, E. W. (1975). *J. Bacteriol.* **123**, 255–264.

White, P. R. (1945). *Science* **94**, 239–241.

Yang, F. M., Montoya, A. L., Merlo, D. J., Drummond, M. H., Chilton, M.-D., Nester, E. W., and Gordon, M. P. (1980). *Mol. Gen. Genet.* **177**, 707–714.

Zaenen, I., VanLarebeke, N., Teuchy, H., VanMontagu, M., and Schell, J. (1974). *J. Mol. Biol.* **86**, 109–127.

Zambryski, P., Holsters, M., Kruger, K., Depicker, A., Schell, J., VanMontagu, M., and Goodman, H. M. (1980). *Science* **209**, 1385–1391.

Zimmerman, J. L., and Goldberg, R. B. (1977). *Chromosoma* **59**, 227–252.

NOTE ADDED IN PROOF. These experiments, based on random site insertion of Tn, enable localization of genetic functions to sites or restriction fragments on pTi. Recently, however, techniques have been developed which will allow fine structure mapping by fragment-specific Tn mutagenesis (Ruvkun and Ausubel, 1981). A defined restriction fragment, first cloned in an appropriate *E. coli* cloning vector, is mutagenized with Tn and screened for Tn insertion in the cloned fragment. Subsequent recloning of the mutagenized fragment into a wide host range conjugative plasmid (e.g., pRK290), transfer to parental strain, and homologous recombination with wild-type parental DNA sequences allow site-directed mutagenesis and fine structure genetic analysis. This procedure, suitable for a wide variety of gram-negative bacteria, has recently been applied for site-directed mutagenesis of pTi sequences (D. Garfinkel *et al.*, submitted).

INTERNATIONAL REVIEW OF CYTOLOGY, SUPPLEMENT 13

The Position of *Agrobacterium rhizogenes*

Jesse M. Jaynes and Gary A. Strobel

Department of Plant Pathology, Montana State University, Bozeman, Montana

I.	Introduction	105
II.	Biology of the Organism	106
	A. Relationship to *Agrobacterium tumefaciens*	106
	B. Physiology of Root Induction	107
	C. Taxonomy	110
	D. Host Range	111
III.	Molecular Biology of the Hairy Root Plasmid	111
	A. Large Plasmids and Their Role in Crown Gall	111
	B. Large Plasmids and Their Role in Hairy Root	114
	C. Isolation and Restriction Endonuclease Digestion of the Hairy Root Plasmid	115
	D. Hairy Root Plasmid Promoters	118
	E. Conjugal Transfer of the Hairy Root Plasmid	122
IV.	Future Prospects	122
	A. Biology of *Agrobacterium rhizogenes*	122
	B. Molecular Biology of *Agrobacterium rhizogenes*	122
	References	124

I. Introduction

Agrobacterium rhizogenes causes a proliferation of roots on the plants it infects, and this disease has been termed hairy root. The etiology of hairy root is quite distinct from that of crown gall, which is caused by *Agrobacterium tumefaciens*. The factor(s) in *A. tumefaciens* responsible for causing plant neoplasms has been shown, without equivocation, to lie on a large plasmid (pTi). Furthermore, a discrete portion of the plasmid DNA is found in the DNA of the infected plant tissue. This represents a classic example of bacterial transformation of a eukaryotic organism. It is tempting to extend this evidence in order to explain the causality of hairy root. However, to date, there are no published data demonstrating that hairy root is caused by bacterial transformation of plant cells.

Nothing is known about the stability and expression of foreign genes in plants. Also, very little is known about the mechanism of gene transfer mediated by agrobacterium plasmids. However, it is clear that agrobacterium plasmids offer a great deal of potential for altering the genetic composition of plants in a predictable and desirable way.

105

II. Biology of the Organism

A. RELATIONSHIP TO *Agrobacterium tumefaciens*

Agrobacterium rhizogenes (Fig. 1) is the causal organism of the hairy root disease of numerous species of higher plants (Elliot, 1951). In the carrot, for example, *A. rhizogenes* invades the primary root via wounds. Subsequently, there is a proliferation of literally hundreds of secondary roots, hence the designation hairy root.

Argobacterium rhizogenes is closely related to *A. tumefaciens,* a well-known bacterium which is the etiological agent of crown gall (Smith and Townsend, 1907). In contrast to *A. rhizogenes, A. tumefaciens* causes unorganized neoplastic growth on numerous host plants (Brown and Wood, 1961). Galls most commonly are produced on the roots, but occasionally above-ground neoplasms are

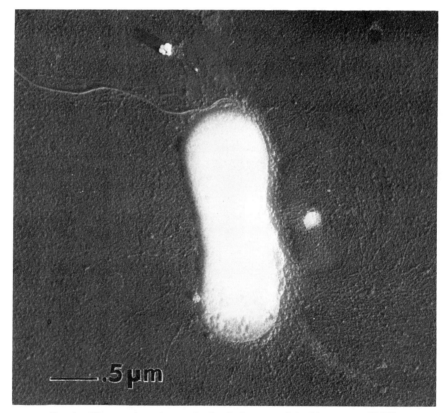

FIG. 1.   Electron photomicrograph of a shadow-cast of *A. rhizogenes.* ×22,500.

also observed. Typically, no root proliferations appear on the galls on plants infected with *A. tumefaciens*.

The factor in *A. tumefaciens* responsible for causing plant neoplasms was named the tumor-inducing principle (Braun and Mandle, 1958). Van Larebeke *et al*. (1974) and Watson *et al*. (1975) described it as a plasmid of approximately 10^8 daltons. This plasmid causes a regulatory and metabolic transformation in plant tissue that is a classic example of bacterial transformation of a eukaryotic organism (Kado, 1976).

Argobacterium rhizogenes strains also harbor large plasmids (Currier and Nester, 1976). Albinger and Beiderbeck (1977) have shown the "in planta" transfer of the plasmid of the infectious strain of agrobacteria. Recently, Moore *et al*. (1979) and White and Nester (1980) have provided convincing evidence that also associates the pathogenicity of *A. rhizogenes* with a large plasmid.

B. Physiology of Root Induction

Seventy-three years ago it was demonstrated that the etiological agent of the plant disease known as crown gall was a bacterium, *A. tumefaciens* (Smith and Townsend, 1907). Since that time, other agrobacteria species have been shown to cause cell-proliferative diseases. The most notable, *A. rhizogenes*, induces cell differentiation and growth in discrete and functional root tissue, quite unlike *A. tumefaciens* which directs a dedifferentiation with concomitant neoplastic development (Fig. 2). However, little is known regarding the physiology of tumor induction caused by *A. tumefaciens*, and even less is known about *A. rhizogenes*-induced root proliferation. In both cases virulence is correlated with the possession of a large plasmid and, with respect to *A. tumefaciens*, a portion of this plasmid DNA has been found within the nuclear DNA of the plant (Chilton *et al*., 1980).

It has been suggested that a plant cell can respond to the tumorigenic signal only during a particular stage in cellular development (Bopp, 1966). Likewise, it is not unreasonable to assume that root induction caused by *A. rhizogenes* follows a similar pathway. For root induction, *A. rhizogenes* must be in contact with wounded plant tissue. Moore *et al*. (1979) demonstrated the need for cell-to-cell interaction by inducing root proliferation in a novel way. Carrot root disks were overlaid with a 0.2-μm-pore-size Nucleopore membrane. The upper surface of the membrane was inoculated with a suspension of *A. rhizogenes* sufficient to initiate a rooting response. The bacteria grew prolifically on the membrane surface, but no roots developed on the carrot disks even after 4–6 weeks. However, inoculated carrot disks without a filter sprouted roots in approximately 2 weeks. It was reasoned that the *A. rhizogenes*–host cell interaction required cell-to-cell contact.

Presumably, bacterial attachment to the plant cell wall is the necessary first

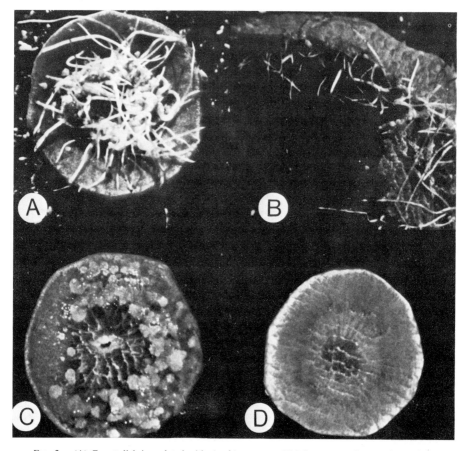

Fig. 2.   (A) Carrot disk inoculated with *A. rhizogenes,* which has sprouted roots about 14 days after inoculation. (B) Carrot disk showing root formation originating primarily in the pericycle region of the root. (C) Gall proliferation on a carrot disk inoculated with *A. tumefaciens* B_6. (D) Control disk.

step in the infection process. It has been demonstrated that avirulent agrobacteria are capable of competing with virulent bacteria for the presumed attachment sites and thus can inhibit tumor formation (Lippincott and Lippincott, 1969). These data suggest that the bacteria bind to plant cells and probably transform them by an unknown mechanism of plasmid DNA transferral.

Moore *et al.* (1979) have shown that the pericycle region of the primary root probably is the essential tissue in the induction of rooting by *A. rhizogenes.* The pericycle normally functions as the tissue from which secondary roots arise. Furthermore, undifferentiated callus tissue or cell cultures, when inoculated with *A. rhizogenes,* do not differentiate into roots. In addition, stem cuttings of a number of plants (juniper, elm), because they do not have a preformed pericycle, do not demonstrate root proliferation when treated with *A. rhizogenes* (unpublished observations).

We have also examined the ability of *A. rhizogenes* to infect the seedlings of a limited number of plants. Typically, the seed is inoculated with sufficient bacteria and allowed to germinate for approximately 2 weeks. We have been able to induce hairy root disease in a number of different plants in this fashion (Fig. 3).

Unusual amino acid derivatives have been found in tumors induced by *A.*

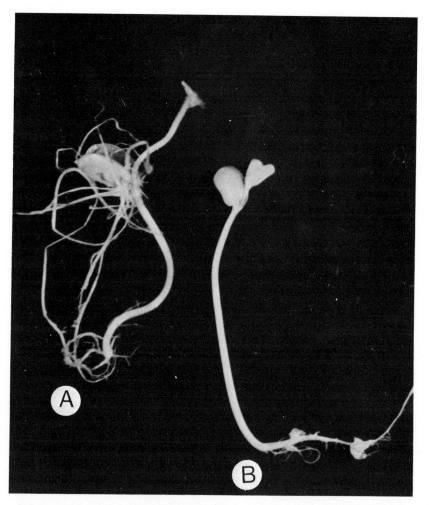

FIG. 3.   (A) Soybean seedling inoculated with *A. rhizogenes* by seed treatment. (B) Control soybean seedling receiving no inoculum.

tumefaciens (Hooykass *et al.*, 1977). It has been amply demonstrated that the genes coding for the synthesis of these novel compounds reside within the large plasmid DNA of *A. tumefaciens* (Kemp *et al.*, 1980). The majority of these tumor-inducing plasmids can be assigned to one of two groups, rough tumors and smooth tumors, depending upon whether the tumors they incite synthesize octopine or nopaline (Montoya *et al.*, 1977).

Presumably, *A. tumefaciens* can ensure its niche by transforming plant cells and thus causing these modified cells to produce a specified metabolite which only *A. tumefaciens* can utilize for growth (the genes coding for octopine and nopaline degradation also reside on the plasmid). In contrast, no plasmid-borne trait other than virulence has been ascribed to the root-inducing plasmid of *A. rhizogenes* (Schilperoot *et al.*, 1978).

C. Taxonomy

Agrobacteria are very common soil inhabitants. They are motile, gram-negative rods with one to four peritrichous flagella usually laterally attached. *Bergey's Manual* (1974) divides the genus into four major species according to the type of disease incited which, together with rhizobia, are listed in the family of Rhizobiaceae. These species were characterized as such on the basis of their phytopathogenicity and symbiotic properties.

A number of taxonomic studies and DNA compositional analyses have shown that agrobacteria and fast-growing rhizobia are rather closely related and more distantly related to slow-growing rhizobia (Moffet and Colwell, 1968).

Strains of agrobacteria have been generally catalogued into two groups, biovars 1 and 2. These two categories are based on the ability of the organism to grow on separate selective media (Panagopolus *et al.*, 1978). Recently, a third

TABLE I

Comparison of Agrobacteria Biovars 1 and 2[a]

Characteristic	Biovar 1	Biovar 2
Utilization of erythritol	−	+
Utilization of citrate	−	+
Utilization of malonate	−	+
Utilization of ammonium citrate	+	−
Utilization of glycerophosphate	+	−
Sodium chloride tolerance	+ (4%)	− (1%)
Utilization of L-tyrosine	+	−
Oxidase reaction	+	−
3-Ketolactose	+	−
Litmus milk	Alkaline	Acid

[a] From Keane *et al.* (1970).

biovar group was discovered which cannot grow on either selective medium (Kersters *et al.*, 1973). These biovar-3 strains have been isolated only from tumorigenic grape tissue and are intermediate in characteristics between biovars 1 and 2. In both biovar 1 and biovar 2 pathogenic and nonpathogenic strains exist, and therefore division of the genus *Agrobacterium* into species according to pathogenicity is unsound, as genes responsible for the phytopathogenic properties reside on transmissible plasmids (Table I).

D. Host Range

Various isolates of *A. tumefaciens* are pathogenic (i.e., tumor-producting) for 643 separate species of 1193 host plants tested (De Cleene and De Ley, 1976). This is by far the most extensive list available and provides convincing evidence of the wide host range of *A. tumefaciens*. It is significant that only 5 of 643 species susceptible are monocots, found only in the Liliaceae and Araceae families.

It is obvious that there is a great deal of information regarding the host range of *A. tumefaciens* (Anderson and Moore, 1979). In contrast, very little is known about the host range of *A. rhizogenes*. We have tested a limited number of plants by inoculating root disks with *A. rhizogenes* (Fig. 4). It is interesting to note that *A. rhizogenes* strain A4 can induce rooting in carrot, sweet potato, soybean, alfalfa, birdsfoot trefoil, kalenchoe, ginseng, and safflower.

Loper and Kado (1979) have shown, by transfer of different Ti plasmids into a strain of *A. tumefaciens* of narrow host range, that it is possible to increase the number of hosts susceptible to tumor induction by this strain of agrobacterium. This is evidence that the host range determinants lie within the Ti plasmid rather than within the host agrobacterium chromosome. It is logical to assume that the root-inducing plasmid has its own host range determinants. This can be illustrated by obtaining a plasmid-cured strain of agrobacterium which is totally nonpathogenic. If the Ti plasmid is transferred to the nonpathogenic strain, the bacterium becomes capable of producing crown gall in a number of different plants. Likewise, if the root-inducing plasmid is transferred to the same nonpathogenic strain, the bacteria become capable of inducing hairy root disease in a number of plants (not necessarily the same plants, unpublished results).

III. Molecular Biology of the Hairy Root Plasmid

A. Large Plasmids and Their Role in Crown Gall

Ti plasmids have been found in all oncogenic strains of *A. tumefaciens*. Strains cured of pTi lose their oncogenic properties but, when pTi is introduced

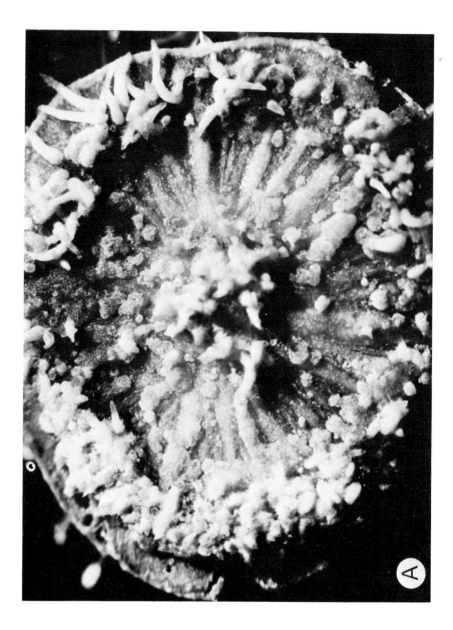

FIG. 4. (A) Alfalfa root disk inoculated with *A. rhizogenes* about 10 days after inoculation. (B) Control disk receiving no inoculum.

back into the strain, the ability to induce tumors is restored (Schell and Van Montagu, 1980).

The Ti plasmids of tumor-inducing *A. tumefaciens* are large, ranging in size from 90 to 160 × 10⁶ daltons. The Ti plasmid encodes a number of traits including oncogenicity, interbacteria conjugal transfer, octopine production and catabolism, nopaline production and catabolism, host specificity, and agrocin sensitivity (Kado, 1980).

Convincing experimental evidence indicates that part of the Ti plasmid is found within the plant DNA (Depicker *et al.*, 1978; Chilton *et al.*, 1978). This transferred DNA has been termed the T-DNA and is approximately 7 × 10⁶ daltons in size. The T-DNA is responsible for the production of opines by the infected plant tissue, and recently it has been shown that the T-DNA is found within the nuclear DNA of the plant (Chilton *et al.*, 1980).

It is clear that *A. tumefaciens* possesses a unique quality—the ability to integrate a selected piece of its own DNA within the nuclear genome of the plant. It remains to be seen whether or not *A. tumefaciens* T-DNA can be utilized to alter predictably the traits of plants.

B. LARGE PLASMIDS AND THEIR ROLE IN HAIRY ROOT

Large plasmids have been isolated from *A. rhizogenes* and have been given the designation pHr (hairy root) (Moore *et al.*, 1979). Several lines of evidence correlated pathogenicity with the presence of a large plasmid. The ability of these bacteria to confer hairy root disease was lost when they were grown in the presence of intercalative agents known to eliminate plasmids in bacteria. Furthermore, by conjugal transfer, it was possible to demonstrate restoration of virulence in these cured strains. It has been shown that the replication of pTi is restricted at 37°C, resulting in loss of the plasmid and a concomitant loss of oncogenic properties (Drummond, 1979). Likewise, Moore *et al.* (1979) were able to correlate temperature sensitivity with apparent loss of the Hr plasmid.

White and Nester (1980) have also shown that pHr encodes virulence traits in *A. rhizogenes*. They found that three large plasmids, pAr15834a (107 × 10⁶ daltons), pAr15834b (154 × 10⁶ daltons), and pAr15834c (258 × 10⁶ daltons), were harbored by their pathogenic *A. rhizogenes* strain. Apparently, pAr15834c is actually a cointegrate of the two smaller plasmids. Cointegrates of R factors and pTi have been constructed (Holsters *et al.*, 1978a), but this was the first report (White and Nester, 1980) of the cointegration of endogenous agrobacterium plasmids. The significance of this cointegration is not apparent. However, they were isolated within a transconjugant, pAr15834b, which was virulent. It appears that pAr15834a is mobilized only by pAr15834b, since the smaller plasmid has never been found alone (F. F. White, personal communication).

While Ti plasmid sequences have been found within the nuclear DNA of infected plant tissue, no data have been published on the presence of pHr DNA sequences within the DNA of plants exhibiting hairy root disease. It is possible that the induction of roots by *A. rhizogenes* is simply the plant's response to the external stimuli provided by the bacteria. However, if one considers the many similarities between the induction of crown gall and the inciting of hairy root, it is reasonable to assume that pHr fragments will ultimately be found within the DNA of infected plant tissue. The *A. rhizogenes*-caused transformation, with the resultant production of discrete plant tissue, i.e., roots, may ultimately be the more important disease to study.

C. Isolation and Restriction Endonuclease Digestion of the Hairy Root Plasmid

Plasmids exist as extrachromosomal genetic elements in a wide variety of gram-positive and gram-negative bacteria. These elements show a great deal of variation in size as well as copy number and function within the cell. Relatively large plasmids usually possess a set of genes capable of promoting bacterial mating and subsequent transfer of the plasmid to a recipient bacterial cell. Presumably, agrobacterial plasmids also contain a set of genes capable of transferring the plasmid to a receptive plant cell. A general but distinctive feature of plasmid elements is that they exist in a covalently closed, circular, duplex form physically separated from the chromosome of the host cell. It is precisely this property (circularity) which allows one to isolate plasmid DNA physically and purify it. This is based on the fact that covalently closed circular DNA binds less of an intercalative dye (usually ethidium bromide) and thus can be separated from nicked circles and linear pieces of DNA by ultracentrifugation of the preparation in a gradient of cesium chloride (Bauer and Vinograd, 1968). Usually, the amount of plasmid DNA represents a small fraction of the total bacterial DNA in the cell, but methods have been developed which attempt to enrich for plasmid DNA (Clewell and Helinski, 1969; Guerry *et al.*, 1973). However, these techniques have proven to be inadequate for the isolation of *A. rhizogenes* plasmids (unpublished observation).

We have found the general method of plasmid isolation described by Currier and Nester (1976) to be satisfactory in allowing us to isolate adequate numbers of *A. rhizogenes* plasmids. This procedure utilizes a shearing of the bacterial lysate followed by an alkaline denaturation treatment which can be an enrichment step for the plasmid if the shearing treatment is not too vigorous. Following this treatment, plasmid DNA is concentrated by precipitation and further purified by dye–buoyant density gradient centrifugation (Bolivar *et al.*, 1977).

Figure 5 illustrates the mobility of purified *A. rhizogenes* plasmids exposed to agarose gel electrophoresis. It is readily apparent from the gel that a large plasmid is under extreme shear pressure evinced by the great number of linear

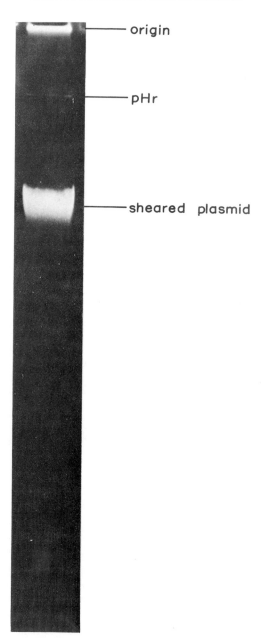

FIG. 5.   The behavior of the Hr plasmid in agarose gel electrophoresis. The Hr plasmid is approximately 150×10^6 daltons in size.

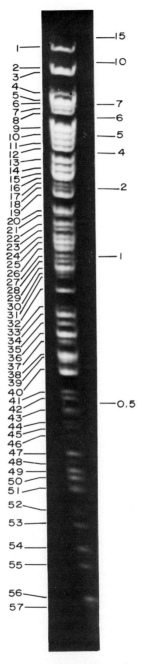

FIG. 6.   *Eco*RI digest pattern of the Hr plasmid. The numbers on the left are fragment designations, while the numbers on the right are size calibrations in megadaltons.

plasmid DNA fragments present. Contrary to the published reports of White and Nester (1980), we have never observed the cointegration of *A. rhizogenes* plasmids. However, these authors report the presence of the cointegrate in our strain of *A. rhizogenes* (F. F. White and E. W. Nester, personal communication). Undoubtedly, our inability to isolate the cointegrate can be explained by the utilization of different techniques and illustrates that the cointegration of *A. rhizogenes* plasmids may be a generalized phenomenon.

The necessary first step toward physically characterizing a plasmid is to determine the effect that various restriction endonucleases have on it. Figure 6 illustrates a restriction endonuclease fragmentation pattern of pHr. It is possible to obtain a low estimate of thé size by agarose gel electrophoresis of the restriction endonuclease fragments, but it is difficult to obtain more than just an estimate because of the comigration of fragments that are distinct but have the same size. Also, large DNA fragments (above 7×10^6 daltons) do not migrate linearly in agarose. However, agarose gel electrophoresis provides a good first estimate of size. Restriction endonuclease digestion fragmentation patterns are quite useful in fingerprinting the plasmid, as well as in locating certain genes within the plasmid by Southern blotting techniques.

Experiments have been conducted which allow us to isolate and characterize, by restriction endonuclease digestion, the hairy root plasmids. These are necessary experiments for eventual realization of the goal of localizating the putative R-DNA and construction of pHr plant-cloning vectors.

D. Hairy Root Plasmid Promoters

A fundamental and necessary property of all living cells is their ability to modulate gene function in response to extracellular stimuli. The translation from gene to protein begins when the enzyme RNA polymerase copies the DNA base sequence into a complementary sequence of RNA. This RNA in turn is translated into a protein product on the ribosome. The accessibility of RNA polymerase to the attachment sites of genes is the hallmark of genetic regulation in cells. These initiating attachment sites on the DNA are called promoters.

It is clear that the study of gene promoters can lead to greater understanding of the complexities of genetic regulation and perhaps can be useful in constructing new gene combinations by fusing the promoter and controlling elements of one gene to the structural part of another well-characterized gene (Casadaban and Cohen, 1979).

A number of techniques have been utilized in an attempt to locate the promoter fragments of genes (Seeburg and Schaller, 1975). Perhaps one of the more useful methods involves the cloning of DNA restriction fragments of the altered promoters of antibiotic resistance genes located on small multicopy plasmids (Neve *et al.*, 1979).

Derivatives of the *Escherichia coli* drug resistance plasmid pBR316 have been constructed, which act as molecular probes for promoter-containing DNA restriction fragments (Rodriguez *et al.*, 1979). We have utilized one such plasmid, designated pBRH4, in an attempt to locate putative pHr promoters. The plasmid pBRH4 contains a unique *Eco*RI restriction endonuclease site located within the promoter region for the tetracycline resistance (TC^r) gene. This site was created by inserting a chemically synthesized octanucleotide, containing the *Eco*RI cleavage sequence, into the *Hin*d III site of pBR316. Subsequent base pair alterations within the Tc^r promoter, caused by this insertion, resulted in elimination of the ability of the plasmid to confer tetracycline resistance on the recipient bacterial strain. *Eco*RI-cleaved foreign DNA fragments cloned into the *Eco*RI site of these plasmids allow the isolation of recombinant transformants with tetracycline resistance levels greater than that of the original plasmid vector. Thus, utilizing these promoter–probe plasmids, one can select for cloned fragments containing promoter-like sequences by insertional activation of the Tc^r gene (Fig. 7) (West *et al.*, 1980).

When pHr is cleaved with the restriction endonuclease *Eco*RI, a complex pattern of fragments is developed (Fig. 6). If these fragments are cloned into the promoter–probe plasmid pBRH4, two of these fragments confer increased levels

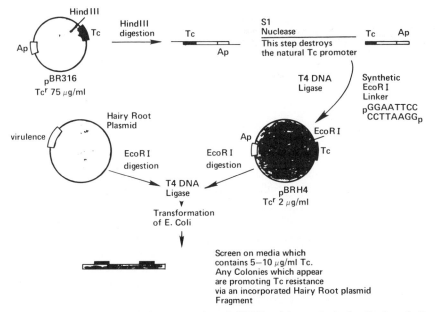

FIG. 7.   Methods utilized in the construction of pBRH4 and its use in the localization of pHr promoters. Tc, Tetracycline resistance gene; Ap, ampicillin resistance gene.

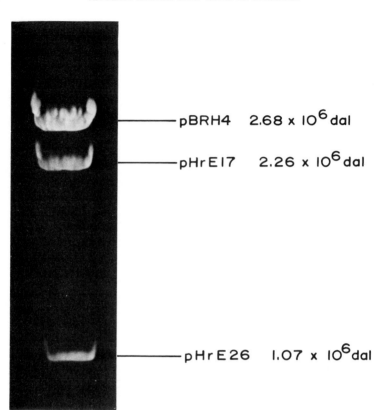

Fig. 8.   Agarose gel electrophoresis of the cloned promoter fragments. The size of the fragment is shown on the right.

of tetracycline resistance (Fig. 8). Presumably, this occurs because promoter-like sequences, located on the pHr fragments, allow transcription into the Tc^r gene. The fact that these fragments are derived from pHr is confirmed when the cloned pieces are radiolabeled and allowed to hybridize with a Southern blot of pHr fragments generated by EcoRI (Fig. 9). EcoRI-generated fragments E17 and E26 appear to be the fragments with promoter-like activity. RNA polymerase binding studies need to be carried out in order to prove that these fragments harbor promoter sequences (Seeburg and Schaller, 1975). Also, in terms of importance, localization of eukaryotic RNA polymerase-accessible promoters within the putative R-DNA (or T-DNA, for that matter) seems to be a necessary prerequisite for the construction of a useful plant cloning vector. Experiments have been designed with these contingencies in mind.

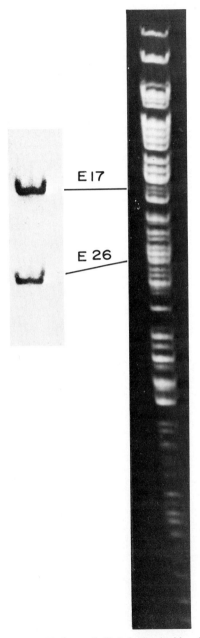

FIG. 9.   Southern blot of an *Eco*RI digest of pHr hybridized with radiolabeled cloned promoter fragments.

E. Conjugal Transfer of the Hairy Root Plasmid

It is known that Ti plasmids cannot be maintained in *E. coli* (Holsters *et al.*, 1978a). Also, R-factor-pTi cointegrates are unstable and yield *E. coli* strains that retain R-factor plasmids that have lost the Ti portion of the cointegrate. Conversely, it is possible, by standard genetic means, to introduce Ti plasmids into rhizobial bacteria. Hooykaas *et al.* (1977) introduced pTi into a strain of *Rhizobium trifolii*. Not only was the Ti plasmid stable, but they also found the Ti plasmid to be fully expressed in *R. trifolii*. Also, depending upon which plasmid was received, *R. trifolii* induced either smooth tumors containing nopaline or rough tumors containing octopine. It is interesting that *R. trifolii* strains harboring pTi were still able to nodulate their normal hosts effectively. Neither octopine or nopaline was ever detected in nodules, while tumors induced by these pTi-harboring strains of *R. trifolii* were never capable of nitrogen fixation. This is significant, since it indicates that the processes of tumor induction and nodulation are completely different but can be induced by the same organism. The ability to transfer pHr to other bacteria may be of some importance.

IV. Future Prospects

A. Biology of *Agrobacterium rhizogenes*

For many years, *A. rhizogenes* and the hairy root syndrome it produces on plants were considered pathogenically unimportant. Primarily, because there was little or no observed crop loss among infected plants, the relationship was considered just another biological curiosity. With the advent of impressive breakthroughs in work on *A. tumefaciens*, it has become clear that many of these phenomena, including the involvement of a plasmid and its role in the transformation of a plant cell, are similar to those reported for *A. rhizogenes*.

Biologically, the controlled induction of rooting in plants appears to have numerous practical applications. Desert plants commonly have a root/shoot ratio greater than 1. Because of the wide host range of *A. rhizogenes*, many agronomically important crops could be stimulated to produce more roots, thus rendering them more tolerant of drought conditions. Moore and Millikan (unpublished observations) demonstrated that apple trees exhibiting the hairy root syndrome survived the 1975–1976 drought period much better than control counterparts (93 versus 30%, respectively). Still other interesting and potentially useful rhizosphere relationships may be discovered in plants infected by *A. rhizogenes*.

B. Molecular Biology of *Agrobacterium rhizogenes*

A great deal is known about the role of the Ti plasmids of *A. tumefaciens* in crown gall tumor formation. It is also very clear that practically nothing is known about *A. rhizogenes* and the hairy root syndrome. Certainly, the most immediate goal should be the isolation and characterization of the putative R-DNA. The worthwhile objectives of physical mapping, with the aid of transposon mutagenesis and restriction endonuclease digestion, should not be overlooked. This will allow the localization of other potentially useful traits on the Hr plasmid. The Hr plasmid has a great deal of potential as far as construction of cloning vectors for plants is concerned, if the R-DNA can be found (Fig. 10). Necessarily, promoters must be found within the R-DNA of the Hr plasmid. This is the critical first step toward the eventual goal of transforming plant cells with an exogenous DNA of choice. That such a feat is possible has been demonstrated by Schell and colleagues (Schell and Van Montagu, 1980). By *in vivo* recombination, a bacterial transposon, Tn7, was incorporated into the T-DNA. Subsequently, they demonstrated that tobacco tumor lines induced with such a Tn7-harboring Ti plasmid contained the complete T-DNA as well as all the Tn7 DNA still inserted at its original site. The transcription of this T-DNA has not been established.

With continued experimentation, it will be possible to utilize agrobacterium plasmids to their fullest potential as natural genetic engineers in introducing entirely new and useful traits into plants.

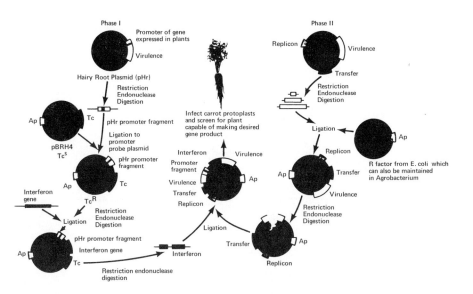

FIG. 10. Hypothetical scheme whereby a foreign gene can be transplanted into the genome of a plant protoplast.

ACKNOWLEDGMENTS

The authors wish to thank T. Carroll and S. Zaske for the electron micrograph of *A. rhizogenes*. Also, the help of A. Hartwigsen in the seedling tests was greatly appreciated. We also thank R. Rodriguez for the gift of pBRH4 and his guidance in conducting the promoter experiments. We thank the NSF, the Herman Frasch Foundation, and the Dow Chemical Co. for their financial assistance on aspects of this project.

REFERENCES

Albinger, G., and Beiderbeck, R. (1977). *Phytopathol. Z.* **90,** 306–310.
Allen, O. N., and Holding, A. J. (1974). *In* "Bergey's Manual" (R. E. Buchanan and N. E. Gibbons, eds.) Ed. VIII, pp. 264–267. William & Wilkins, Baltimore, Maryland.
Anderson, A. R., and Moore, L. W. (1979). *Phytopathology* **69,** 320–323.
Bauer, W., and Vinograd, J. (1968). *J. Mol. Biol.* **33,** 141–171.
Bolivar, F., Rodriguez, R. L., Betlach, M. C., and Boyer, H. W. (1977). *Gene* **2,** 75–93.
Bopp, M. (1966). *Biol. Rundsch.* **4,** 25–37.
Braun, A. C., and Mandle, R. J. (1958). *Growth* **12,** 255–269.
Braun, A. C., and Wood, H. N. (1961). *Adv. Cancer Res.* **6,** 81–109.
Casadaban, M. J., and Cohen, S. N. (1979). *Proc. Natl. Acad. Sci. U.S.A.* **76,** 4530–4533.
Chilton, M. D., Drummond, M. H., Merlo, D. J., and Sciaky, D. (1978). *Nature (London)* **275,** 147–149.
Chilton, M. D., Saiki, R. K., Yadav, N., Gordon, M. P., and Quetier, F. (1980). *Proc. Natl. Acad. Sci. U.S.A.* **77,** 4060–4064.
Clewell, D. B., and Helinski, D. R. (1969). *Proc. Natl. Acad. Sci. U.S.A.* **62,** 1159–1166.
Currier, T. C., and Nester, E. W. (1976). *Anal. Biochem.* **76,** 431–441.
De Cleene, M., and DeLey, J. (1976). *Bot. Rev.* **42,** 389–466.
Depicker, A., VanMontagu, M., and Schell, J. (1978). *Nature (London)* **275,** 150–153.
Drummond, M. (1979). *Nature (London)* **281,** 343–347.
Elliot, C. (1951). *In* "Manual of Bacterial Plant Pathogens," 2nd rev. ed. Chronica Botanica, Waltham, Massachusetts.
Guerry, P., LeBlanc, D. J., and Falkow, S. (1973). *J. Bacteriol.* **116,** 1064–1066.
Holsters, M., Silva, B., Genetello, C., Engler, G., VanVliet, F., deBlock, M., Villarroel, R., Van Montagu, M., and Schell, J. (1978a). *Plasmid* **1,** 456–467.
Holsters, M., Silva, B., VanVliet, F., Hermalsteens, J. P., Genetello, C., Van Montagu, M., and Schell, J. (1978b). *Mol. Gen. Genet.* **163,** 335–338.
Hooykaas, P. J. J., Klapwijk, P. M., Nuti, M. P., Schilperoort, R. A., and Rörsch, A. (1977). *J. Gen. Microbiol.* **98,** 477–484.
Kado, C. I. (1976). *Annu. Rev. Phytopathol.* **14,** 265–308.
Kado, C. I. (1980). Recombinant DNA Tech. Bull. No. 2, 145–153.
Keane, P. J., Kerr, A., and New, P. B. (1970). *Aust. J. Biol. Sci.* **23,** 585–595.
Kemp, J. D., Hack, E., and Sutton, D. W. (1980). *In* "Genome Organization and Expression in Plants" (C. J. Leaver, ed.), pp. 489–496. Plenum, New York.
Kersters, K., DeLey, J., Sneath, P. H. A., and Suckin, M. (1973). *J. Gen. Microbiol.* **78,** 227–239.
Lippincott, B. B., and Lippincott, J. A. (1969). *J. Bacteriol.* **97,** 620–626.
Loper, J. E., and Kado, C. I. (1979). *J. Bacteriol.* **139,** 591–596.
Moffet, M. L., and Colwell, R. R. (1968). *J. Gen. Microbiol.* **51,** 245–266.
Montoya, A. L., Chilton, M.-D., Gordon, M. P., Sciaky, D., and Nester, E. W. (1977). *J. Bacteriol.* **129,** 101–107.
Moore, L., Warren, G., and Strobel, G. (1979). *Plasmid* **2,** 617–626.

Neve, R. L., West, W. W., and Rodriguez, R. L. (1979). *Nature (London)* **277**, 324–325.

Panagopolus, C. G., Psallidas, P. G., and Alivizatos, A. S. (1978). *Proc. Int. Conf. Plant Pathol. Bacteriol., 4th, Angers* **1**, 221–228.

Rodriguez, R. L., West, R. W., Heynecker, H., Bolivar, F., and Boyer, H. W. (1979). *Nucleic Acids Res.* **6**, 2167–2187.

Schell, J., and Van Montagu, M. (1980). *In* "Genome Organization and Expression in Plants" (C. J. Leaver, ed.), pp. 453–470. Plenum, New York.

Schilperoort, R. A., Klapwijk, P. M., Hooykaas, P. J. J., Koekman, B. P., Ooms, G., Otten, L. A. B. M., Wurzer-Figurelli, E. M., Wullems, G. J., and Rörsch, A. (1978). *Proc. Int. Congr., 4th, Plant Tissue Cell Culture* pp. 85–94.

Seeburg, P. H., and Schaller, H. (1975). *J. Mol. Biol.* **92**, 261–277.

Smith, E. F., and Townsend, C. O. (1907). *Science* **25**, 671–673.

Van Larebeke, N., Engler, G., Holsters, M., Van de Elsacker, S., Zaenen, I., Schilperoort, R. A., and Schell, J. (1974). *Nature (London)* **252**, 169–170.

Watson, B., Currier, T. C., Gordon, M. P., Chilton, M. D., and Nester, E. W. (1975). *J. Bacteriol.* **123**, 255–264.

West, R. W., Neve, R. L., and Rodriguez, R. L. (1980). *Gene* **7**, 271–288.

White, F. F., and Nester, E. W. (1980). *J. Bacteriol.* **141**, 1134–1141.

INTERNATIONAL REVIEW OF CYTOLOGY, SUPPLEMENT 13

Recognition in *Rhizobium*-Legume Symbioses

Terrence L. Graham

Monsanto Agricultural Products Company, St. Louis, Missouri

I. Introduction 127
II. Recognition in the Soybean–*Rhizobium japonicum* System . . . 129
   A. *In Vitro* Lectin Binding to *Rhizobium japonicum* 129
   B. Tissue Distribution of Soybean Lectin 132
   C. Other Aspects of Recognition in Soybean 134
   D. Summary 135
III. Recognition in the Clover–*Rhizobium trifolii* System 136
   A. Serological Studies and *in Vitro* Lectin Binding 137
   B. Root Adsorption Studies and Localization of Clover Lectin . 139
   C. Factors Affecting Expression of Clover and *Rhizobium trifolii*
      Receptors 141
   D. Summary 142
IV. The *Rhizobium* Cell Surface 142
V. Concluding Remarks 145
   References 145

I. Introduction

Recently a number of reviews have appeared on "recognition" in plant–microbe interactions. These include reviews on initial molecular events in pathogenicity and induced disease resistance (Albersheim and Anderson-Prouty, 1975; Callow, 1975, 1977; DeVay and Alder, 1976; Sequeira, 1978) and in *Rhizobium* symbiosis (Bauer, 1977, 1980a; Broughton, 1978; Dazzo, 1980a,b,c; Napoli *et al.*, 1980; Solheim and Paxton, 1980). Several other reviews on *Rhizobium* recognition are in press (Bauer, 1980b; Bhuvaneswari, 1980; Dazzo and Hubbell, 1980; Carlson, 1980). These areas are in extensive review not because they are thoroughly understood, but because of the complexity of the results and theories that have been presented. There is a real need for critical analysis of the problems, progress, and future approaches of the research efforts. It is hoped that this review will complement others in the area of *Rhizobium*–legume recognition in achieving this end.

Most of the work discussed in this chapter deals with experimental evidence for the involvement of plant lectins in the recognition of *Rhizobium* by the roots of their leguminous hosts. The lectin hypothesis had its experimental origin in the early work of Hamblin and Kent (1973) and Bohlool and Schmidt (1974). It simply states that plant lectins, specific carbohydrate-binding proteins, are in-

128   TERRENCE L. GRAHAM

volved in the binding and/or bridging of *Rhizobium* bacteria to the host root surface. Genetically, the event could be considered to correspond to the phenotypic event termed Roa (*r*oot *a*dhesion) and described by Vincent (1980) and Lepidi *et al.* (1979).

The lectin hypothesis filled a vacuum; it was an attractive hypothesis, and it represented the first clear-cut hypothesis on the molecular basis of recognition between host and *Rhizobium*. For similar reasons, lectins have also been suggested as playing a possible role in plant host–pathogen interactions (Anderson and Jasalavich, 1979; Doke *et al.*, 1975; Goodman *et al.*, 1976; Graham and Sequeira, 1977; Graham *et al.*, 1977; Kojima and Uritani, 1974; Mirelman *et al.*, 1975; Sequeira and Graham, 1977). However, other possible mechanisms for the specific molecular interactions of *Rhizobium* with root epidermal cells can be envisioned. For example, simple polysaccharide–polysaccharide interactions have been hypothesized as the basis of recognition in the very closely related *Agrobacterium* infection process (Whatley *et al.*, 1976) and for *Xanthomonas* infection (Morris *et al.*, 1977).

Little work has been done on the mechanisms of recognition between *Rhizobium* species and hosts other than clover and soybean, even though the first hypothesis on the involvement of lectin in *Rhizobium* recognition came from the *Rhizobium phaseoli–Phaseolus vulgaris* system (Hamblin and Kent, 1973). It is of obvious importance that work on these other systems be pursued; systematic and careful studies on other systems could help immensely in clearing up many of the ambiguities emerging from the soybean and clover systems. I have limited discussion largely to the soybean and clover symbioses. The reader is referred to the following pertinent research papers on the possible involvement of lectins in other individual systems: pea–*Rhizobium leguminosarum* (Chen and Phillips, 1976; Kamberger, 1979a,b; Kijne, 1978; Kijne *et al.*, 1977, 1980; Planque and Kijne, 1977; Rouge and Labroue, 1977; Schaal and Kijne, 1979; Wolpert and Albersheim, 1976); bean–*R. phaseoli* (Chen and Phillips, 1976; Hamblin and Kent, 1973; Kamberger, 1979a; Wolpert and Albersheim, 1976); lupine–*R. lupini* (Chen and Phillips, 1976; Kamberger, 1979a); alfalfa–*R. meliloti* (Chen and Phillips, 1976; Kamberger, 1979a); *Rhizobium* spp. (cowpea miscellany) (Chen and Phillips, 1976; Dazzo and Hubbell, 1975c; Kamberger, 1979a; Wolpert and Albersheim, 1976). The last system is unique in that it represents a heterogeneous mixture of host legumes and *Rhizobium* species for which recognition appears to be "promiscuous" or relatively nonspecific.

Another rapidly developing area of research on *Rhizobium*–legume symbioses involves study of both the chromosomal and the extrachromosomal genetics of symbiosis. This area is still young as it relates to recognition per se, and the reader is thus referred to several recent general reviews (Denarie and Truchet, 1979; Nuti *et al.*, 1980; Postgate, 1979; Subba Rao, 1980) and to Kondorosi and Johnson, Denarie *et al.*, and Newcomb in this volume.

II. Recognition in the Soybean-*Rhizobium japonicum* System

Research on the mechanisms of recognition in the soybean-*R. japonicum* symbiosis has been characterized by recurring controversy and complexity. Although this was one of the first systems for which a molecular basis was proposed (Bohlool and Schmidt, 1974), work progressed slowly and sporadically until recently and was complicated by numerous inconsistencies and contradictions in data from various laboratories. Although some major questions still need to be answered, the soybean system is now emerging as one of the better understood systems.

The work of Bohlool and Schmidt (1974) established the hypothesis that soybean lectin is involved in recognition of the soybean symbiont *R. japonicum*. In an examination of 48 *Rhizobium* strains (25 isolates of *R. japonicum* and 23 isolates which do not nodulate soybean), Bohlool and Schmidt microscopically observed binding of fluorescein isothiocyanate (FITC)-labeled soybean seed lectin to 22 of the 25 *R. japonicum* strains and to none of the 23 nonsymbionts. Bacterial cells were heat-fixed prior to incubation with the lectin. Binding did not occur on all bacteria within a population of cells (lower than 1% of cells responded with 7 of the "positive" strains), nor was it homogeneous on the bacterial cell surface. Unfortunately only one lectin concentration was used in the binding studies. The obvious questions that emerged from this work were: (1) What is the reason for the lack of binding of soybean lectin *in vitro* to 3 of the *R. japonicum* strains examined and for the heterogeneous distribution and low levels of lectin binding within bacterial populations and on the bacterial cell surface? (2) Is the soybean lectin present in seeds also present on the root surface in a form accessible as a recognition molecule? (3) What is the actual cell surface structure of *R. japonicum* involved in binding the soybean lectin? These questions are addressed separately below.

A. *In Vitro* LECTIN BINDING TO *Rhizobium japonicum*

As research in other laboratories progressed, what appeared to be serious problems involving the original hypothesis of Bohlool and Schmidt (1974) began to emerge. The first of these involved the reproducibility of the lectin-binding results. Chen and Phillips (1976) reported that soybean seed lectin did not bind to the single strain of *R. japonicum* examined (a strain not examined by Bohlool and Schmidt) but did bind with higher frequency and affinity to five *Rhizobium* strains which were soybean nonsymbionts. Once again, however, only a single lectin concentration was examined. Also, consistently low levels of lectin binding (less than 8% of the cells responded) were seen for any of the strains examined. In an attempt to reduce background binding of the FITC-labeled lectin to free extracellular polysaccharides on the slides used for fluorescent microscopy, Chen and Phillips (1976) incubated the bacteria with lectin and washed the cells extensively prior to fixation. Since the extracellular polysaccharides of gram-negative bacteria often differ greatly in their structural integrity, this may have been partially responsible for the low levels of lectin binding they observed and the apparently conflicting results. Work in several other laboratories (Brethhauer, 1977; Law and Strijdom, 1977), however, also demonstrated a failure of soybean seed lectin to bind to certain *R. japonicum* strains and a disturbing ability of the lectin to bind to nonsymbiotic strains.

A series of critical studies from the laboratory of W. D. Bauer have helped somewhat to clear the anomalies and apparent inconsistencies in the *in vitro* lectin–*R. japonicum* binding studies or at least to define their possible sources. In the first of these investigations, Bhuvaneswari *et al.* (1977) repeated the FITC-labeled lectin-binding studies (at saturating lectin concentrations) and developed an additional highly quantitative binding assay using tritium-labeled lectin. Using both assays they demonstrated that the soybean lectin receptors on the cell surface of *R. japonicum* changed markedly with the age of the culture in synthetic media. Three classes of *R. japonicum* strains were identified; those (the majority) for which soybean lectin receptors were at a maximum in early and midlog phases of growth, one strain (3Ilb 123) for which lectin receptors were present in stationary, lag, and early log phases but absent in midlog and late log phases, and a third class for which no lectin receptors could be detected at any stage of growth. The results, moreover, were in close agreement with the original results of Bohlool and Schmidt. FITC-labeled lectin was observed to bind to 15 of 22 *R. japonicum* strains examined and to none of 9 strains which do not nodulate soybean. Most importantly, with positive strains at their binding maxima, greater than 30% of the cells responded. Binding was hapten-reversible, and the tritium-labeled lectin demonstrated an affinity constant at optimal culture conditions of 4×10^{-7} M for *R. japonicum* 3Ilb 138. Several strains showed biphasic binding curves, but these were not interpreted. Besides advancing considerably the quantitation of soybean lectin-binding studies, these results provided a possible explanation of the conflicting results from other laboratories and suggested that the lectin receptors on *R. japonicum* were transient and that their appearance may be regulated differently for different strains of the organism. Similar phenomena have been reported for the lectin receptors on *Rhizobium trifolii* (Dazzo *et al.*, 1979) and *R. leguminosarum* (J. W. Kijne, personal communication).

This important suggestion of a dynamic and transient nature of the bacterial receptors was further confirmed by Bhuvaneswari and Bauer (1978) in a second study. They demonstrated that, when the strains were grown in root exudate media or in association with intact roots, specific soybean lectin receptors appeared to develop on all six strains examined which normally did not possess them regardless of growth phase in synthetic media. Although the percentage of cells of these strains binding FITC-labeled soybean lectin was low (1–5% of the cells) in comparison to that for strains which normally possess the receptor (> 50% of cells), the results suggested that the environment of the plant root may somehow affect the development of lectin receptors on the symbiont bacterium. Unfortunately, Bhuvaneswari and Bauer did not determine the nature of the induced receptors nor did they show that they were actually of bacterial origin. An interesting possibility that cannot be dismissed on the basis of this study is that the anomalous strains of rhizobia adsorbed molecules of plant origin from the root media and that these molecules were responsible for the very limited binding of soybean lectin. This possibility is not unreasonable, especially if soybean tissue contains cross-reactive lectin receptors as proposed for the clover system (Dazzo and Hubbell, 1975b). On the positive side, however, of the six heterologous strains of rhizobia examined, only two cowpea strains (known to be promiscuous) and none of the single strains of *Rhizobium trifolii, R. meliloti, R. lupini,* or *R. phaseoli* examined bound the soybean lectin after soybean root culture. Obviously, more heterologous *Rhizobium* strains should be examined, and the data would be more convincing if a higher percentage of the cells were influenced by the root culture. Studies by Fjellheim and Solheim (personal communication) may also be relevant here; these workers report that root extracts of clover and pea cause specific but limited breakdown of polysaccharide preparations from their respective symbionts *R. trifolii* and *R. leguminosarum.* Do the suggested host enzymes play a role in modification of *Rhizobium* polysaccharides as a preliminary to bacterial binding?

Other recent studies from the laboratories of Bauer, Schmidt, and Bal (Bohlool and Schmidt, 1976; Calvert *et al.,* 1978; Mort and Bauer, 1978, 1980; Shantharam *et al.,* 1980; Tsien and Schmidt, 1977) have examined the nature, heterogeneity, and distribution of the actual cell surface components of *R. japonicum* in synthetic media; the results of these investigations generally confirm that extracellular polysaccharide is the likely receptor for soybean lectin on the *R. japonicum* surface. These studies also provide further molecular explanations for the heterogeneity or variation of lectin binding within *R. japonicum* in terms of both the intensity of lectin binding and the percentage of cells responding to lectin. It is quite obvious that regulated changes in the cell surface of *R. japonicum* do occur and could account for the earlier contradictory results from various laboratories.

Bhuvaneswari *et al.* (1977) earlier had noted the presence of an inhibitor of soybean lectin binding in culture filtrates of *R. japonicum* strains. Freund and Pueppke (1978) further described the release of these inhibitors from *R. japonicum* in culture. Regardless of whether these inhibitors represent released lectin receptors from cultured cells, this finding adds yet another dimension of complexity to the *in vitro* lectin-binding studies. Thus, in addition to changes in the surface of rhizobia with culture conditions and the simple stability and/or integrity of the receptor on the bacterial surface, the presence of soluble inhibitors in bacterial suspensions may influence *apparent* binding of the lectin in the assay used. It is obvious that much still needs to be resolved and, although there is obviously an entire array of complexities that one could evoke to explain apparent anomalies, it is not yet clear which explanations are actually related to which difficulties in *in vitro* lectin-binding experiments. The *in vitro* binding of lectins to cell surfaces is just one research approach and because of its complexity perhaps should not be overemphasized in research on host–microorganism interactions.

B. Tissue Distribution of Soybean Lectin

A second major problem of the lectin recognition hypothesis in soybean was the finding that the seed lectin used in the original lectin recognition studies did not appear to be present at detectable levels in older plant tissues, especially root tissues (Pueppke *et al.*, 1978). Also, it was found that the seed lectin might not even be present at all in certain soybean cultivars that are nonetheless nodulated by *R. japonicum* (Pull *et al.*, 1978).

Pueppke *et al.* (1978), using three different quantitative assays for the seed lectin, examined its presence in soybean seeds, seed coats, embryo axes, and cotyledons, and also in the primary roots, secondary roots, leaves, and stems of Beeson soybean seedlings as a function of plant age. In ungerminated seeds, these workers found the highest concentrations of the lectin in cotyledons (9.14 mg/gm defatted flour), although some lectin was also found in the seed coat and embryo axis (1.81 and 1.39 mg/gm defatted flour, respectively). After germination, the seed lectin concentration fell off sharply in the cotyledons ($>$ 5000 μg/gm fresh weight at planting), reaching levels of only 0.1–1.0 μg/gm fresh weight just before abscission of the cotyledons. In the growing seedling, the seed lectin was detected in all other plant tissues examined, although at much lower levels than in the cotyledons (approximately 0.5–5.0 μg/gm fresh weight in tissues of 4-day-old seedlings). Most importantly, in all tissues examined, including primary and secondary roots, the seed lectin was undetectable in plants after 14–16 days (limit of detection $\sim$0.1 μg/gm fresh weight). At this time soybeans are reportedly still readily nodulated (Skrdleta, 1971). Unfortunately, other soybean cultivars have not been examined in such detail. Also, the plants used in these studies were grown in the presence of nitrate, a condition proposed to affect lectin levels on clover roots (Dazzo and Brill, 1978).

In another study, Pull *et al.* (1978) examined the seeds of 102 lines of soybean for the presence of the classic 120,000-dalton seed lectin. In 97 lines, the lectin concentration varied from 2.5 to 12.2 mg/gm defatted flour. However, in 5 lines the lectin was not detected (detection limits apparently were similar to those of the former study) by four criteria, even though these soybean cultivars were readily nodulated by four strains of *R. japonicum*. Three of these soybean lines were derived from lines known to contain the seed lectin. Unfortunately, root tissues of the soybean lines were not examined for the seed lectin or other lectin activity.

Although neither of these studies disprove the proposed role of soybean lectins in the recognition of *R. japonicum,* they raise very serious problems which must be resolved. Several possible explanations for the negative results have to be considered. One possibility is that the classic 120,000-dalton soybean seed lectin is indeed not involved in recognition but that another receptor (lectin or otherwise) with a similar molecular binding specificity is. In addition to the 120,000-dalton soybean lectin, several minor isolectins of similar specificity have been reported in several cultivars (Lis *et al.,* 1966). Moreover, preliminary studies by Keegstra (Pueppke *et al.,* 1978) have suggested the presence of another lectin in soybean roots extractable with detergent and having a binding specificity similar to but distinguishable from that of the soybean seed lectin. Very recently an abstract from the laboratory of Schmidt (Jack *et al.,* 1979) described the isolation of a root lectin from soybean, which differs from the seed lectin only in that it contains ornithine; otherwise the lectins appear to have a similar binding specificity. No bacterial binding studies have been presented, however. Finally, Fett (1980a,b), observed a lectin activity in soybean extracts that agglutinated *Xanthomonas phasedi* var. *sojensis* but not red blood cells. Although this lectin obviously had a specificity different from that of the classic soybean lectin, this observation underlines the possible danger in looking only at classic hemagglutinating lectins.

Another unanswered question is whether the levels of soybean lectin extracted from plant tissues truly reflect its concentration *in vivo*. The techniques used to extract soybean lectin may not have solubilized the lectin in older plant tissues where it could have become immobilized in membrane or wall fractions. Such soybean lectins, solubilized only by detergent treatment, are suggested both by the preliminary work of Keegstra (Pueppke *et al.,* 1978) and by detailed studies from the laboratory of Sharon (Bowles and Sharon, 1979). In fact, Bowles and Sharon have detected lectins in all soybean tissues up to 7 weeks old.

A related question is whether the lectin is concentrated in highly localized tissue compartments (e.g., growing root hair tips) actually involved in recognition and whether its apparent absence is due to subsequent decompartmentalization and dilution during extraction. This possibility actually gains some support from recent studies (described in Section II,C) on the heterogeneous and highly transient infectibility of soybean root tissues (Bhuvaneswari *et al.,* 1980) and from work with clover root lectin (Dazzo and Brill, 1977, 1978).

A last possibility which certainly should not be overlooked, but which has not yet received detailed attention, is that the presence of soybean lectin in root tissues either is dependent on growth conditions (e.g., soil nutrients) or is even actively induced by the presence of *Rhizobium* or other microorganisms in the rhizosphere. The presence of nitrate or ammonium ions has been shown to decrease dramatically both detectable levels of clover lectin on roots and binding of *R. trifolii* to clover root hairs (Dazzo and Brill, 1978). Critically, the above studies of Pueppke *et al.* (1978) were performed on soybean roots from seedlings grown in the presence of nitrate. Also, it is not unreasonable that in their coevolution a mechanism may have developed in which factors from *R. japonicum* induce soybean seed lectin in root tissues; this situation would certainly benefit the energy balance in the plant, since the plant might expectedly have only restricted and highly localized needs for the lectin. As discussed above, Bhuwaneswari and Bauer (1978), in defining complexities in another area of the soybean recognition system, have demonstrated the apparent inducibility of the *R. japonicum* receptors for soybean lectin when anomalous strains which normally do not bind the lectin are grown in soybean root exudate or in association with intact soybean roots. Thus the host may produce factors which influence the presence of receptors on the symbiont; it does not seem unreasonable to assume that the opposite could occur. Preliminary evidence from Bauer's laboratory in fact suggests that both extracellular polysaccharides from *R. japonicum* and N-acetylgalactosamine (the specific hapten of soybean lectin) specifically induce infectivity of emerging soybean root hair cells by *R. japonicum* (Bauer *et al.*, 1979). It will be of great interest to see if these changes involve localized changes in soybean lectin levels within the host.

C. OTHER ASPECTS OF RECOGNITION IN SOYBEAN

As one might expect, the complexity of events leading to active recognition and successful symbiosis does not end (or even necessarily begin) at cell surface interactions. For example, Bhuvaneswari *et al.* (1980) have demonstrated that the soybean cells infectible by *R. japonicum* are located in a very specifically defined area of soybean root immediately above the area of root elongation and below the area of root hair emergence. The infectibility (i.e., the *ability* to become nodulated) of these cells is a highly transient property which is gained and lost within a few hours by a process of acropetal development. Thus infectibility does not appear to develop before axial growth of the epidermal cells ceases and is totally lost when the cells are approximately 4–5 hours old. Beyond this developmental regulation, the critical host responses to recognition of the symbiont leading to nodulation per se appear to be externally *induced* only in these developmentally "ready" host cells within 1 or 2 hours after contact with the symbiont. Induction may be mediated by substances of bacterial origin (possibly *R. japonicum* exopolysaccharide and notably also by the hapten of soybean lectin, N-acetylgalactosamine) [Bauer *et al.*, 1979]. We thus obviously again have several layers of dynamic host–symbiont developmental conditions and molecular interactions possibly influencing both bacterial attachment (Section II,A and B) and its outcome. Preliminary and somewhat related work with *R. trifolii* suggests an induction of infection thread formation in clover by substances highly specific to *R. trifolii* (Higashi and Abe, 1980). Moreover, the involvement of *R. trifolii* extracellular polysaccharide in root hair deformation has been suggested by Dazzo and Hubbell (1975b). Although neither of these later studies examined developmental changes or measured nodulation per se, they too indicated that factors from the symbiont may affect events well beyond simple bacterial attachment.

It is clear that we are just beginning to understand the total recognition system between plant and microorganism and that events such as root hair curling, host cell penetration, and infection thread formation represent additional *levels* of recognition which may be subject to highly specific control whether based on overlapping or distinct molecular mechanisms. A separation of the molecular mechanisms for attachment per se and for further events in the total recognition system is suggested by the preliminary work of Dazzo and Brill (1979) on the clover system. These workers showed that *Azotobacter vinelandii* transformants could be obtained which carried the *R. trifolii* receptor for clover lectin and anti-clover root antiserum. These transformants bound to clover roots but were noninfective. It should not be ruled out, however, that the same molecular partners involved in the specific attachment of rhizobia to root epidermal cells (putatively in soybean the bacterial extracellular polysaccharide and the host lectin) may play a partial or overlapping role in the induction of additional changes in the host which play a role in total infectibility. An interesting parallel to this exists in host–pathogen interactions where it has been proposed that one of the molecules possibly responsible for simple binding recognition of *P. solanacearum* by its host plants (the bacterial lipopolysaccharide) may also be responsible for the later induction of host resistance mechanisms *against* the invading microorganism (Graham and Sequeira, 1977; Graham *et al.*, 1977). Interestingly this latter recognition system may also involve the host lectin as a receptor (Sequeira and Graham, 1977).

D. SUMMARY

Although the molecular basis for specificity in soybean–*R. japonicum* recognition is far from fully understood, great strides have been made in defining some of the complexities which may have contributed to earlier ambiguities and inconsistencies in the data from various laboratories. Generally, an excellent correlation is now seen in the *in vitro* interaction between *R. japonicum* and soybean seed lectin. Further studies are needed, however, to strengthen the suggestions made in explaining past anomalies. Problems related to the presence and distribution of soybean lectin still need to be resolved. However, the soybean system has provided some very valuable lessons. It is quite obvious that recognition between two living organisms is not simple. That which in the absence of experience we would normally consider a fairly simple lock-and-key phenomenon is obviously a complex array of dynamic and active developmental conditions, influences, and responses involving the interacting partners. The evidence for lectin involvement in soybean–*R. japonicum* recognition is still ambivalent, but the basic hypothesis is certainly still viable. Our understanding of the total system is far from final.

III. Recognition in the Clover–*Rhizobium trifolii* System

Research on mechanisms of recognition in the clover–*R. trifolii* symbiosis initially progressed more rapidly than research on the other systems. Although at first sight the information on the clover system may seem more unified than that on the soybean system, it must be remembered that the research is largely based on the work of a limited research group. The work of Dazzo and co-workers (Dazzo, 1980a), however, has served not only to establish the lectin hypothesis in the clover system but has also presented a detailed hypothesis that other workers hopefully will evaluate.

The basic hypothesis for the clover system differs from that for other systems in that it developed largely out of an initial search for serological differences between infective and noninfective species of *Rhizobium*. This emphasis on antigenic characteristics in mammalian systems as probes of the cell surface of both plant host and symbiont has persisted even though the unique antigens involved are demonstratably weak antigenic determinants. The proposed hypothesis for the clover system (Dazzo and Hubbell, 1975b) is in fact based largely on the existence of "cross-reactive antigens" on the surface of both host and symbiont. In this model, clover lectin is proposed to bridge these cross-reactive antigens to form an initial preinfective recognition system. Although this difference may seem slight from a mechanistic point of view, it dramatically affected initial experimental approaches to research on the clover system. Although cross-reactive antigens have been implicated in recognition in mammalian host–parasite interactions where their development is readily understandable (Zabriskie, 1967; Pearlman *et al.*, 1965; Markowitz *et al.*, 1960; Cahill *et al.*, 1971), their true meaning in plant host–parasite interactions is less clear. No doubt antigenic comparisons represent useful cell surface probes, but it is not likely that plants contain recognition molecules with the complexity and specificity of mammalian antibodies. Nonetheless, the initial work of Dazzo and his co-workers generally suggests the possibility that cross-reactive *lectin* receptors represent a somewhat parallel system in the plant kingdom and, although plant lectins are generally *assumed* to have very simple sugar-binding specificities, only a very few lectins have been studied in sufficient detail in this regard.

A. SEROLOGICAL STUDIES AND *in Vitro* LECTIN BINDING

Although several previous attempts to demonstrate antigenic differences between infective and noninfective *Rhizobium* strains had been largely unsuccessful (Graham, 1969; MacGregor and Alexander, 1972; Marshall, 1967; Rensburg *et al.*, 1968), Dazzo and Hubbell (1975a) used a repeated and prolonged immunization schedule in rabbits to demonstrate the presence in *R. trifolii* strains of both host nodulation-specific and nonspecific antigens. With the use of four infective strains and homologous noninfective mutants derived from the infective strains, several antigens were found to be common to all strains, but an additional unique antigen was found to be common only to infective strains. Importantly, the results suggested that the previously elusive *R. trifolii* antigens were either weak antigens or were not fully accessible. Although this certainly justly prompted the reexamination of other related serological phenomena (see below), it also strengthened the importance of alternative research techniques.

Previous studies from the laboratory of Hubbell (Charudattan and Hubbell, 1973) had detected cross-reactive antigens from eight legumes and three species of *Rhizobium* but had failed to demonstrate a correlation with nodulation specificity. However, Dazzo and Hubbell (1975b) reexamined this phenomenon in the clover system, using prolonged immunization schedules. Several assay methods and eight strains of *R. trifolii* (four infective and mutant noninfective pairs) were used. This preliminary work suggested the presence of specific antigenic cross-reactivity between infective strains and the host clover *Trifolium repens* (white clover). In tube agglutination tests, end titers were 8 to 16-fold higher for the interaction between infective strains and crude *T. repens* root antiserum than for corresponding noninfective mutant strains with one exception. These results suggested that, although cross-reactive antigens existed on both infective and noninfective cells, infective cells were generally more highly reactive with *T. repens* root antiserum.

Immunofluorescent studies described in the same paper confirmed that both infective and noninfective strains bound FITC-labeled *T. repens* root antiserum. However, when adsorbed previously with whole cells of noninfective strains, without exception the antiserum reacted detectably only with infective cells. Conversely, when preadsorbed on infective cells, no antiserum reaction with noninfective or infective cells could be detected. Generally these studies supported the original serological studies (Dazzo and Hubbell, 1975a) in suggesting the presence of unique determinants common to infective strains but not present on noninfective strains. Unfortunately, the results were presented as simple positive or negative reactions, and no information was given on the quantitative sensitivity of the immunofluorescence assay. Other *Rhizobium* species, including *R. japonicum, R. leguminosarum, R. phaseoli, R. meliloti,* and *Rhizobium* spp. (of the cowpea miscellany), generally had negative reactions with FITC-labeled *T. repens* root antiserum. However, one strain (*Rhizobium* spp.

HR1) was equally reactive with *normal* rabbit serum and *T. repens* root antiserum. This latter result strengthens one's concern about the overall specificity and usefulness of the immunological comparisons, especially those involving crude antiserum.

Conversely, FITC-labeled antiserum against several infective strains was observed to bind to the surface of both white and strawberry clover roots, especially at the growing root hair tip. Unfortunately, no information was presented on whether antiserum against the noninfective strains bound to clover roots or whether antiserum against infective strains bound to roots of other plants.

Because of the semiquantitative methods employed, the use of crude antiserum, and the lack of several key controls, these studies and those above generally suggested the possibility of, but did not demonstrate, specific cross-reactive antigens on clover roots and infective *R. trifolii* cells. Bohlool *et al.* (1979) recently reported an inability to detect cross-reactive antigens on infective and noninfective *R. trifolii* and roots of red clover (*Trifolium pratense*). Their immunization schedule was not reported.

The isolation and partial characterization of the cross-reactive capsular antigens from strain 403 (infective) and the homologous Bart A (noninfective) mutant were reported in the same paper (Dazzo and Hubbell, 1975b). The capsular polysaccharide of strain 403 reacted with antiserum against *T. repens* roots by immunoprecipitation and immunofluorescence; however, no data on the reaction of the antiserum with the polysaccharide from strain Bart A were presented.

Also reported (Dazzo and Hubbell, 1975b) was the demonstration in crude clover seed extracts of lectin activity which caused the agglutination of all infective strains of *R. trifolii* examined (a total of 9 including several infective revertants). None of the 4 noninfective *R. trifolii* strains or the 11 other *Rhizobium* species examined were agglutinated. The agglutination results were not quantified, however, and the autoagglutination of several strains prevented their examination. Preincubation of crude lectin with 30 mM 2-deoxyglucose or N-acetylgalactosamine inhibited the agglutination of strains 403 and 2S-2 (infective); unfortunately these were the only two strains examined for hapten inhibition, and inhibition was only examined at a single lectin concentration (the most *dilute* titer giving agglutination).

Also, in the same paper it was reported that the crude clover lectin could "agglutinate" rabbit erythrocytes coated with the polysaccharide antigen from strain 403, but curiously only when the lectin was followed by antiserum against crude lectin. Although this could suggest that the lectin had indeed bound to the coated erythrocytes, it could simply be a reflection of nonspecific adsorption of other proteins from the crude lectin preparation.

It was on the basis of these extensive but preliminary studies that Dazzo and Hubbel (1975b) proposed the recognition model in which clover lectin cross-bridges common antigens on the clover and *Rhizobium* cell surfaces.

Further studies by Dazzo and Brill (1979) have somewhat strengthened the original *in vitro* serological observations, but again the work is limited by the number of strains examined and the use of semiquantitative methods. Binding of anticlover root antiserum to *R. trifolii* was confirmed using monovalent Fab fragments from a purified IgG fraction. Although binding was not demonstrated directly, the Fab fragments inhibited both lectin-mediated agglutination of *R. trifolii* and *R. trifolii* adsorption to roots. Unfortunately only one strain of *R. trifolii* was examined. The work suggested, however, that anti-clover root antibody and lectin could possibly bind to the same or interacting sites on *R. trifolii*. Binding of the extracellular polysaccharide antigen from *R. trifolii* to anti-clover root antiserum was also further studied (Dazzo and Brill, 1979) with more highly purified polysaccharide fractions (but unfortunately again with crude antiserum). Two fractions (again from a single strain) were found to precipitate with both the clover lectin and anti-clover root antiserum. Results were negative with single strains of *R. meliloti* and *R. japonicum*. Precipitation by the antiserum of one of the polysaccharide fractions from a single *R. trifolii* strain was partially inhibited by 2-deoxyglucose. This was taken as partial evidence to suggest again that the lectin and antibody receptors on the *Rhizobium* polysaccharide are similar.

B. Root Adsorption Studies and Localization of Clover Lectin

In another type of study (Dazzo *et al.*, 1976) the adsorption of *R. trifolii* to root hairs of *T. repens* was examined. After a 12-hour incubation of clover roots with 1.9×10^8 *Rhizobium* cells, the roots were washed with saline and examined by phase-contrast microscopy to quantify single-cell bacterial binding for a given root hair length. As reported by previous workers (Salman and Fahraeus, 1963; Menzel *et al.*, 1972; Marshall *et al.*, 1975; Napoli *et al.*, 1975), attachment of the rod-shaped bacteria was polar. Six combinations of infective *R. trifolii* strains and their respective noninfective mutants and three *R. meliloti* strains were examined. The mean number of cells of infective *R. trifolii* cells which bound to clover root hairs was four to five times greater than that of noninfective *R. trifolii* or *R. meliloti* cells. Unfortunately, no other heterologous *Rhizobium* strains or other unrelated bacteria were examined, although in a subsequent study Dazzo extended this examination to all *Rhizobium* species (Dazzo, 1980b). Binding of *R. trifolii* to clover root hairs was inhibited 10-fold by the clover lectin hapten (2-deoxyglucose) and not by two related sugars; again, however, only a single strain was examined. The preincubation of one infective *R. trifolii* strain with crude clover lectin caused massive coating of the clover root surface with *R. trifolii* cells. Although this massive binding was reported to be far greater than that of infective *R. trifolii* cells without lectin, the experiment was ambiguous since simple lectin-mediated agglutination or cross-bridging of *R. trifolii* cells to *R. trifolii* already bound to the root was not excluded (i.e., actual binding to the root cell surface may not have been greater) and since only a single infective strain of *R. trifolii* and a single noninfective *R. japonicum* strain (no noninfective *R. trifolii* strains) were examined.

In a related study Dazzo and Brill (1977) examined the binding of partially purified FITC-labeled capsular polysaccharides from two *R. trifolii* and a single *R. meliloti* strain to clover and alfalfa root hairs. In this interesting work the FITC-labeled capsular polysaccharides of *R. trifolii* were found to bind preferentially to clover root hair tips. The binding could be inhibited by unlabeled *R. trifolii* polysaccharide and by 2-deoxyglucose, but not by unlabeled *R. meliloti* polysaccharide. Labeled polysaccharide from the single *R. meliloti* strain examined bound to alfalfa root hair tips but not to clover root hairs. When washed with 30 mM 2-deoxyglucose, a highly active proteinaceous fraction was obtained from clover roots which agglutinated *R. trifolii* but not *R. meliloti* or *R. japonicum*. The agglutination was inhibited by 2-deoxyglucose. Unfortunately only single isolates of these three *Rhizobium* species were examined. These studies further supported the role of lectin and *Rhizobium* capsular polysaccharide in bacterial adsorption to the root surface. Perhaps more importantly, they represented the first evidence for the presence of the clover lectin in root tissue at a site readily accessible to the bacterium.

Further evidence for localization of clover lectin on the root surface was provided in a study dealing with purification of clover lectin (Dazzo *et al.*, 1978). Clover lectin was purified to apparent homogeneity from seeds of white clover and also from washings of clover roots with 2-deoxyglucose. The two proteins were shown to be similar if not identical by electrophoretic and immunological criteria and by their 2-deoxyglucose-inhibited specificity for *R. trifolii*. The purified seed lectin was tested for agglutination against 34 strains of *Rhizobium*. These results strengthened previously reported data on binding specificities (Dazzo and Hubbel, 1975b; Dazzo and Brill, 1977). Unfortunately, apparently because of limited amounts of the protein, the lectin from root washings was again not tested against a wide range of *Rhizobium* species or strains. Antibody against the purified seed lectin was prepared. Quantitative immunocytofluorimetry showed that antibody against clover lectin bound selectively to epidermal cells in the root hair region of young clover seedlings. Only background levels were found to bind to roots of alfalfa, joint vetch, and birdsfoot trefoil. The importance of this work was that it suggested that the seed and root lectins were apparently similar if not identical and that clover lectin may be localized on the root hair tips where maximum binding of *R. trifolii* and capsular polysaccharide had been reported (Dazzo and Brill, 1977; and Dazzo *et al.*, 1976).

Further root adsorption studies (Dazzo and Brill, 1979) indirectly demonstrated the binding of antiserum against partially purified *R. trifolii* polysaccharide fractions to white clover roots. Binding was uniform on the root surface and not localized on root hair tips. The results further suggest the possible cross-reactivity of the clover root and *R. trifolii* antigens and suggest that these antigens, unlike the lectin itself, may be uniformly distributed on the root surface.

C. Factors Affecting Expression of Clover and *Rhizobium trifolii* Receptors

As is true in the soybean system (Section II), important parameters influencing the presence and nature of the receptors on white clover and on *R. trifolii* have begun to be examined. The data again suggest that recognition is not a simple, nonregulated event.

The influence of fixed nitrogen levels on lectin expression on the clover root surface has been studied (Dazzo and Brill, 1978). Both nitrate (16 mM) and ammonium ion (1 mM) were found to reduce dramatically the binding of *R. trifolii* to clover root hairs to background levels. The fixed nitrogen sources also dramatically reduced immunologically detectable levels of clover lectin on the root hair surface. The effects of either fixed nitrogen source were concentration-dependent, and both bacterial binding and detectable lectin concentrations showed similar slopes of decay. It will be of interest to see if these results can be extrapolated to additional *R. trifolii* strains (only one strain was examined). Neither nitrogen source directly affected clover seed lectin activity. The root lectin appears to be similar in this regard (F. B. Dazzo, personal communication). Direct ligand-binding studies using radioactive N^{13}O$_3$ have shown that nitrate does not interact directly with the lectin (E. M. Hraback and F. B. Dazzo, personal communication). Moreover, recent studies (E. M. Hraback and F. B. Dazzo, personal communication) have shown that the accessibility of lectin receptor sites on purified clover root walls is significantly reduced if walls are isolated from seedlings grown in 15 mM nitrate.

Factors affecting the occurrence of lectin receptors on *R. trifolii* have also been examined (Dazzo *et al.*, 1979), but again only with a single strain. The results suggest that both lectin receptors and crude anti-clover root antibody receptors change with time in synthetic media. On agar-grown cells, lectin receptors (measured by lectin-mediated agglutination and indirect immunofluorescence) maximized at 5 days. This correlated with the presence of a distinct extracellular matrix and maximum root hair binding. On broth-grown cells, anti-clover root antiserum binding showed two distinct maxima at early log and early stationary phases of growth. Unfortunately, root hair binding of broth-grown *R. trifolii* was only examined for the early stationary-phase peak and not the early log-phase peak. Experiments with agar-grown cells were not performed with anti-clover root antiserum and, vice versa, experiments with broth grown cells were not performed with the lectin, although the lectin and antiserum may recognize the same *R. trifolii* determinants (Dazzo and Brill, 1979). Nonetheless, the results suggest possible variations in receptors with culture age. To be less ambiguous, the experiments must be extended to include more exhaustive lectin and/or anti-root antiserum studies and more *R. trifolii* (and other *Rhizobium* species) strains. This work is now under way (Urbano and Dazzo, 1980).

Recently, Dazzo (personal communication) obtained results which suggest that growth of *R. trifolii* in the presence of clover root exudate may cause polar arrangement of the *R. trifolii* capsule and lectin receptor sites. If this is true, it may help explain the preferential polar attachment of *R. trifolii*. Similar polarities for capsular polysaccharide and lectin binding have been reported for *R. japonicum* (Bohlool and Schmidt, 1976; Tsien and Schmidt, 1977). It will be of interest to see if the very diverse lectin receptor responses of *R. japonicum* strains to synthetic and root exudate media (Bhuvaneswari *et al.*, 1977; Bhuvaneswari and Bauer, 1978) involve similar morphological changes.

D. Summary

Taken in its entirety, the work on the clover–*R. trifolii* system presents a picture of a possible recognition mechanism involving clover lectin as a cross-bridging protein for similar surface carbohydrate determinants on clover root hairs and *R. trifolii*. Unfortunately, studies to date provide a patchwork of somewhat ambiguous information, involving crude or partially purified lectin, antibody, and *Rhizobium* polysaccharide preparations. Although some studies have been extended (Dazzo *et al.*, 1978; Dazzo and Brill, 1979), much work still needs to be done in more depth with the purified molecules. Moreover, several studies lack important controls and/or extensive examination of strains, often making definitive conclusions difficult to reach. Nonetheless, in the clover system, unlike the soybean system, the lectin has been identified as an apparently readily accessible molecule on the root surface (Dazzo and Brill, 1977; Dazzo *et al.*, 1978), although more complete studies on the *Rhizobium*-binding specificity of this lectin are needed. It has also been shown that the *in situ* levels of the clover root lectin may be regulated by the levels of fixed nitrogen in the rhizosphere (Dazzo and Brill, 1978). Moreover, problems dealing with the transience of receptors on *R. trifolii* have been examined (Dazzo *et al.*, 1979) and, as in the soybean system, should eventually lead to a more valid and specific chemical understanding of the *Rhizobium* cell surface (Urbano and Dazzo, 1980).

IV. The *Rhizobium* Cell Surface

An additional controversy accompanying the lectin recognition hypothesis for *Rhizobium*-legume interactions involves the chemical and morphological character of the *Rhizobium* cell surface molecules involved in specific interactions with the host plant. This is of obvious importance in any recognition mechanism whether it involves plant lectins or not. Most work has focused on the extracellular polysaccharides (EPS) and lipopolysaccharides (LPS) of *Rhizobium,* since there is no evidence that the flagella play a mandatory role in recognition, at least at the level of bacterial attachment and infection (Napoli and Albersheim, 1980). The relative roles of LPS and EPS in recognition are not yet fully clear. In the end, both molecules may be found to play some role in the infection process with rhizobia, as they appear to in plant pathogenesis (Whatley *et al.,* 1976; Graham and Sequeira, 1977; Graham *et al.,* 1977; Sequeira and Graham, 1977; Morris *et al.,* 1977; Ayers *et al.,* 1979).

Chemical elucidation of the sugar composition and primary structures of the cell surface polysaccharides of rhizobia is progressing rapidly and may advance at an even more rapid rate as a result of recent dramatic developments in methods for sequencing complex polysaccharide molecules (Valent *et al.,* 1980; Lindbergh and Lönngren, 1978). It must be stressed, however, that past studies have been limited to a small number of strains. More importantly, these studies have not always sufficiently taken into account such factors as purity of preparations or new information on the critical nature of bacterial culture conditions (Sections II,A and III,C). Also, most of the earlier studies were purely compositional and not structural, an extremely important factor since it is certainly the structural position of sugar residues within a molecule that dictates the specificity of its reaction with other molecules. Rhizobia EPS have been studied more carefully in this context than their LPS.

It is beyond the scope of this article to discuss individual chemical studies on *Rhizobium* surface molecules in detail. A detailed review on this topic will soon appear (Carlson, 1980). Although a considerable amount of information is available on the chemistry of *Rhizobium* surface molecules, generally information available to date is so limited by the above considerations as to allow at best only qualitative and gross correlations for most systems. Discussion here will be limited to very recent examinations of *R. japonicum* and *R. trifolii* which have taken these complications into more critical account. These studies are presented simply to stress again the complexity of the plant–microbe interactions and the caution that must be exercised in chemical analyses and their interpretation.

In the soybean–*R. japonicum* system biological evidence generally suggests that the EPS are involved in specific recognition (Section II). In very recent work, Mort and Bauer (1980) have shown that, in two *R. japonicum* strains which normally express lectin-binding sites only during log growth phase in synthetic media (Bhuvaneswari *et al.,* 1977), simultaneous changes occur both in the percentage of cells showing encapsulation and in the composition of the EPS. The EPS of both strains contain mannose, glucose, galacturonic acid

(1:2:1), and variable amounts of either 4-O-methylgalactose or free galactose (the total galactose remains the same). These results generally agree with those of Dudman (1976, 1978) for *R. japonicum*. However, the ratio of free galactose to methylated galactose is considerably higher in cultures with a high proportion of cells that bind soybean lectin. Acetyl groups are also present, and Mort and Bauer suggest that the lectin-binding residues are possibly 2-O-acetylgalactose, closely related to the assumed sugar hapten of soybean lectin, 2-deoxy-N-acetylgalactosamine.

In the clover–*R. trifolii* system evidence on the relative roles of EPS and LPS is still confusing. Generally the work of Dazzo and co-workers (Section III) suggests involvement of the capsular EPS in recognition. A preliminary report by Urbano and Dazzo (1980) on the capsular EPS of *R. trifolii* suggests that changes in the ratios of four monosaccharides (all reportedly deoxyhexoses or substituted deoxyhexoses), as well as a decrease in O-acetylation and an increase in O-pyruvylation, may correlate with an increased ability of clover lectin to bind to *R. trifolii* during early stationary-phase growth in synthetic media. However, this system is not a simple one, and much still needs to be resolved. Most workers (Jansson *et al.,* 1979; P. Albersheim, personal communication) have studied the EPS isolated from culture media or cell washings and have characterized a distinct heteropolysaccharide made up of repeating oligosaccharide (eight-residue) units containing glucose, galactose, glucuronic acid, and pyruvyl substitution. Dazzo's results, although involving more crude chemical analyses, are generally in agreement with these (Dazzo and Hubbell, 1975b; Dazzo and Brill, 1979), except that deoxy sugars have been suggested to be present in the EPS. Other workers have reported deoxy sugars in the LPS of *R. trifolii* (Carlson, 1980). It is important to point out that Dazzo worked with capsular EPS (isolated from whole cells). Two obvious possibilities exist: Either the EPS preparation of Dazzo was contaminated by cell wall LPS constituents or the intact capsular EPS indeed contained covalently linked LPS constituents (Dazzo and Brill, 1979). The later possibility is not without precedence, since certain *E. coli* strains are known to produce an extracellular polysaccharide composed of polymerized O antigen (Jann and Westphal, 1975). On the other hand, in this case, the EPS would have to lose deoxy sugars specifically when released into the culture media to explain the results of other workers. These possibilities will not be resolved until further more careful analyses of the surface of *R. trifolii* are made. If Dazzo's suggestions are right, the O antigen containing capsular EPS, produced only at specified culture ages, is involved in lectin-mediated recognition. Isolated cell wall LPS would also be expected to bind to clover lectin. The data of Dazzo and Brill (1979) lend preliminary support to this theory. Whether or not both the EPS and LPS are involved in actual root attachment (e.g., at different cultural ages) still needs to be resolved.

V. Concluding Remarks

We are certainly still at the frontier of our understanding both of cell surfaces and of early recognition events at the molecular level. From the studies presented in this chapter only one thing is clear—that even the earliest interaction between *Rhizobium* and its host plants is a highly complex and dynamic *matrix* of events. Beyond this, we need to know far more about events involved in other stages of the infection process and *their* relative contributions to the overall specificity of recognition. Thus future research on one hand should emphasize the chemical nature and organization of the plant and bacterial cell surfaces. This information should allow us to understand more clearly their initial interaction. On the other hand, more research is needed on events such as chemotaxis, root hair deformation, root hair curling, and infection thread formation which are also critical to successful symbiosis. These processes certainly complement initial cell surface-binding interactions and may involve related and molecularly inseparable events. Conceptually we need to stop thinking of models for particular events and to begin to examine each biological interaction as a dynamic system of interactions which have coevolved in a unique way between any two interacting partners.

ACKNOWLEDGMENTS

The author wishes to acknowledge Drs. W. D. Bauer, R. W. Carlson, F. B. Dazzo, and L. Sequeira who offered a critical review of the manuscript. In addition, I wish to acknowledge personal communications from Drs. P. Albersheim, W. D. Bauer, W. J. Brill, F. B. Dazzo, D. H. Hubbell, J. W. Kijne, and N. P. Nuti.

REFERENCES

Albersheim, P., and Anderson-Prouty, A. J. (1975). *Annu. Rev. Plant Physiol.* **26**, 31–52.
Anderson, A. J., and Jasalavich, C. (1979). *Physiol. Plant Pathol.* **15**, 149–159.
Ayers, A. R., Ayers, S. B., and Goodman, R. N. (1979). *Appl. Environ. Microbiol.* **38**, 659–666.
Bauer, W. D. (1977). *Basic Life Sci.* **9**, 283–297.
Bauer, W. D. (1980a). *In* "Nitrogen Fixation" (W. E. Newton and W. H. Orme-Johnson, eds.), Vol. II, pp. 205–214. Univ. Park Press, Baltimore, Maryland.
Bauer, W. D. (1980b). *Annu. Rev. Plant Physiol.* **32**, 407–449.
Bauer, W. D., Bhuvaneswari, T. V., Most, A. J., and Turgeon, G. (1979). *Plant Physiol.* **63**, 745. (Abstr.)
Bhuvaneswari, T. V. (1980). *Econ. Bot.* **35**, 204–223.
Bhuvaneswari, T. V., and Bauer, W. D. (1978). *Plant Physiol.* **62**, 71–74.
Bhuvaneswari, T. V., Pueppke, S. G., and Bauer, W. D. (1977). *Plant Physiol.* **60**, 486–491.
Bhuvaneswari, T. V., Turgeon, B. G., and Bauer, W. D. (1980). *Plant Physiol.* **66**, 1027–1031.
Bohlool, B. B., and Schmidt, E. L. (1974). *Science* **185**, 269–271.

Bohlool, B. B., and Schmidt, E. L. (1976). *J. Bacteriol.* **125**, 1188–1194.

Bohlool, B., Schmidt, E. L., and Jack, M. A. (1979). *Proc. N. Am. Rhizobium Conf., 7th, June 17–21, College Sta., Texas.*

Bowles, D. J., and Sharon, N. (1979). *Planta* **145**, 193–198.

Brethauer, T. S. (1977). "Soybean Lectin Binds to *Rhizobia* Unable to Nodulate Soybean. Lectins may not Determine Host Specificity." Master's thesis, University of Illinois.

Broughton, W. J. (1978). *J. Appl. Bacteriol.* **45**, 165–194.

Cahill, J. F., Cole, B. C., Wiley, B. B., and Ward, J. R. (1971). *Infect. Immun.* **3**, 24–35.

Callow, J. A. (1975). *Curr. Adv. Plant Sci.* **18**, 181–193.

Callow, J. A. (1977). *Adv. Bot. Res.* **4**, 1–49.

Calvert, H. E., Lalonde, M., Bhuvaneswari, T. V., and Bauer, W. D. (1978). *Can. J. Microbiol.* **24**, 785–793.

Carlson, R. W. (1981). *In* "Ecology of Nitrogen Fixation" (W. J. Broughton, ed.), Vol. II. Oxford Univ. Press, London and New York, in press.

Charudattan, R., and Hubbel, D. H. (1973). *Antonie van Leeuwenhoek J. Microbiol. Serol.* **39**, 619–627.

Chen, A. P., and Phillips, D. A. (1976). *Physiol. Plant.* **38**, 83–88.

Dazzo, F. B. (1980a). *In* "Advances in Legume Science" (A. J. Bunting and R. Summerfield, eds.), University of Reading Press, England.

Dazzo, F. B. (1980b). *In* "Adsorption of Microorganisms to Surfaces" (G. Britton and K. C. Marshall, eds.), pp. 253–316. Wiley, New York.

Dazzo, F. B. (1980c). *In* "Nitrogen Fixation" (W. E. Newton and W. H. Orme-Johnson, eds.), Vol. II, pp. 165–187. Univ. Park Press, Baltimore, Maryland.

Dazzo, F. B., and Brill, W. J. (1977). *Appl. Environ. Microbiol.* **33**, 132–136.

Dazzo, F. B., and Brill, W. J. (1978). *Plant Physiol.* **62**, 18–21.

Dazzo, F. B., and Brill, W. J. (1979). *J. Bacteriol.* **137**, 1362–1373.

Dazzo, F. B., and Hubbell, D. H. (1975a). *Appl. Microbiol.* **30**, 172–177.

Dazzo, F. B., and Hubbell, D. H. (1975b). *Appl. Microbiol.* **30**, 1017–1033.

Dazzo, F. B., and Hubbell, D. H. (1975c). *Plant Soil* **43**, 713–717.

Dazzo, F. B., and Hubbell, D. H. (1981). *In* "Ecology of Nitrogen Fixation" (W. J. Broughton, ed.), Vol. II. Oxford Univ. Press, London and New York, in press.

Dazzo, F. B., Napoli, C. A., and Hubbell, D. H. (1976). *Appl. Environ. Microbiol.* **32**, 166–171.

Dazzo, F. B., Yanke, W. E., and Brill, W. J. (1978). *Biochim. Biophys. Acta* **539**, 276–286.

Dazzo, F. B., Urbano, M. R., and Brill, W. J. (1979). *Curr. Microbiol.* **2**, 15–20.

Deinema, M. H., and Zevenhuizen, L. P. T. M. (1971). *Arch. Microbiol.* **78**, 42–57.

Dénarié, J., and Truchet, G. (1979). *Physiol. Vég.* **17**, 643–667.

DeVay, J. E., and Alder, H. E. (1976). *Annu. Rev. Microbiol.* **30**, 147–168.

Doke, N., Tomiyama, K., Nishimura, N., and Lee, H. S. (1975). *Ann. Phytopathol. Soc. Jpn.* **41**, 425–433.

Dudman, W. F. (1976). *Carbohydr. Res.* **46**, 97–110.

Dudman, W. F. (1978). *Carbohydr. Res.* **66**, 9–23.

Fett, W. F., and Sequeira, L. (1980a). *Plant Physiol.* **66**, 847–852.

Fett, W. F., and Sequeira, L. (1980b). *Plant Physiol.* **66**, 853–858.

Freund, T. G., and Pueppke, S. G. (1978). *Plant Physiol.* **61**, 59.

Goodman, R. N., Huang, P. Y., and White, J. A. (1976). *Phytopathology* **66**, 754–764.

Graham, P. (1969). *In* "Analytical Serology of Microorganisms" (J. Kwapinski, ed.), Vol. II, pp. 353–378. Wiley (Interscience), New York.

Graham, T. L., and Sequeira, L. (1977). *In* "Cell Wall Biochemistry Related to Specificity in Host-Plant Pathogen Interactions" (B. Solheim and J. Raa, eds.), pp. 417–422. Columbia Univ. Press, New York.

Graham, T. L., Sequeira, L., and Huang, T.-S.R. (1977). *Appl. Environ. Microbiol.* **34**, 424-432.
Hamblin, J., and Kent, S. P. (1973). *Nature (London)* **245**, 28-30.
Higashi, S., and Abe, M. (1980). *Appl. Environ. Microbiol.* **39**, 297-301.
Jack, M. A., Schmidt, E. L., and Wold, F. (1979). *Fed. Proc. Fed. Am. Soc. Exp. Biol.* **38**, 411.
Jann, K., and Westphal, O. (1975). *In* "The Antigens" (M. Sela, ed.), Vol. 3, pp. 1-125. Academic Press, New York.
Jansson, E., Lindberg, B., and Ljunggren, H. (1979). *Carbohydr. Res.* **75**, 207-220.
Kamberger, W. (1979a). *Arch. Microbiol.* **121**, 83-90.
Kamberger, W. (1979b). *FEMS Microbiol. Lett.* **6**, 361-365.
Kijne, J. W. (1978). *Acta Bot. Neerl.* **27**, 239-240.
Kijne, J. W., Planque, K., and Swinkels, P. P. H. (1977). *In* "Cell Wall Biochemistry Related to Specificity in Host-Plant Pathogen Interactions" (B. Solheim and J. Raa, eds.). Columbia Univ. Press, New York.
Kijne, J. W., van der Schaal, I. A. M., and Gert, E. V. (1980). *Plant Sci. Lett.* **18**, 65-74.
Kojima, M., and Uritani, I. (1974). *Plant Cell Physiol.* **15**, 733-737.
Law, I. J., and Strijdom, B. W. (1977). *Soil Biol. Biochem.* **9**, 79-84.
Lepidi, A. A., Nuti, M. P., Bagnoli, G., Filippi, C., and Galluzzi, R. (1979). *In* "Some Current Research on *Vicia faba* in Western Europe" (D. A. Bond, G. T. Scarascia-Mugnozza, and M. H. Poulsen, eds.), pp. 436-460. Off. Publ. Eur. Community.
Lindberg, B., and Lönngren, J. (1978). *Methods Enzymol.* **L**, 3-33.
Lis, H., Fridman, C., Sharon, N., and Katchalski, E. (1966). *Arch. Biochem. Biophys.* **117**, 301-309.
MacGregor, A., and Alexander, M. (1972). *Plant Soil* **36**, 129-139.
Markowitz, A. S., Armstrong, S. H., and Kushner, D. S. (1960). *Nature (London) New Biol.* **187**, 1095-1097.
Marshall, K. (1967). *Aust. J. Biol. Sci.* **20**, 429-438.
Marshall, K. C., Cruickshank, R. H., and Bushby, H. V. A. (1975). *J. Gen. Microbiol.* **91**, 198-200.
Menzel, G. H., Uhig, H., and Weischsel, G. (1972). *Zentralbl. Bakteriol. Parasitenkd. Infektionski. Hyg. Abt. 2* **127**, 348-358.
Mirelman, D., Galun, E., Sharon, N., and Lotan, R. (1975). *Nature (London)* **256**, 414-416.
Morris, E. R., Rees, D. A., Young, G., Walkinshaw, M. D., and Darke, A. (1977). *J. Mol. Biol.* **110**, 1-16.
Mort, A. J., and Bauer, W. D. (1980). *Plant Physiol.* **66**, 158-163.
Mort, A. J., and Bauer, W. D. (1980). *Plant Physiol.*, in press.
Napoli, C. A., and Albersheim, P. (1980). *J. Bacteriol.* **141**, 1454-1456.
Napoli, C., Dazzo, F., and Hubbell, D. (1975). *Appl. Microbiol.* **30**, 123-132.
Napoli, C., Sanders, R., Carlson, R., and Albersheim, P. (1980). *In* "Nitrogen Fixation" (W. E. Newton and W. H. Orme-Johnson, eds.), Vol. II, pp. 189-203. Univ. Park Press, Baltimore, Maryland.
Nuti, M. P., Lepidi, A. A., Prakash, R. K., Hooykaas, P. J. J., and Schilperoort, R. A. (1980). *In* "Molecular Biology of Plant Tumors" (G. Kohl and J. Schell, eds.). Academic Press, New York.
Pearlman, P., Hammarstrom, S., Lagercrantz, R., and Gustafson, B. E. (1965). *Ann. N. Y. Acad. Sci.* **124**, 337-394.
Planque, K., and Kijne, J. W. (1977). *FEBS Lett.* **73**, 64-66.
Postgate, J. R. (1979). *Agric. Ital.* **108**, 142-146.
Pueppke, S. G., Bauer, W. D., Keegstra, K., and Ferguson, A. L. (1978). *Plant Physiol.* **61**, 779-784.
Pull, S. P., Pueppke, S. G., Hymowitz, T., and Orf, J. H. (1978). *Science* **200**, 1277-1279.
Rensburg, H. V., Strijdom, B., and Rabie, C. (1968). *S. Afr. J. Agric. Sci.* **11**, 623-626.

Rouge, P., and Labroue, L. (1977). *C. R. Acad. Sci. Ser. D.* **284,** 2423-2426.
Salman, K., and Fahraeus, G. (1963). *J. Gen. Microbiol.* **33,** 425-427.
Schaal, I., and Kijne, J. (1979). *Proc. Am. Rhizobium Conf., 7th.*
Sequeira, L. (1978). *Annu. Rev. Phytopathol.* **16,** 453-481.
Sequeira, L., and Graham, T. L. (1977). *Physiol. Plant Pathol.* **11,** 43-54.
Shantharam, S., Gow, J. A., and Bal, A. K. (1980). *Can. J. Microbiol.* **26,** 107-114.
Skrdleta, V. (1971). *Rost. Výruba (Praque)* **17,** 309-360.
Solheim, B., and Paxton, J. (1981). *In* "Plant Disease Control: Resistance and Susceptibility."
    Wiley (Interscience), New York, in press.
Subba Rao, N. S. (1981). "Recent Advances in Biological Nitrogen Fixation." Oxford Sci. IBH
    Publ., Janpath, New Delhi, in press.
Tsien, H. C., and Schmidt, E. L. (1977). *Can. J. Microbiol.* **23,** 1274-1284.
Turgeon, G., Bhuvaneswari, T. V., and Bauer, W. D. (1979). *Plant Physiol.* **63,** 744. (Abstr.)
Urbano, M. R., and Dazzo, F. B. (1980). Abstracts of the Annual Meeting of the American Society
    for Microbiology, p. 163.
Valent, B. S., Darvill, A. G., McNeil, M., Robertsen, B. K., and Albersheim, P. (1980). *Car-
    bohyd. Res.* **79,** 165-192.
Vincent, J. M. (1980). *In* "Nitrogen Fixation" (W. E. Newton and W. H. Orme-Johnson, eds.),
    Vol. II, pp. 103-129. Univ. Park Press, Baltimore, Maryland.
Whatley, M. H., Bodwin, J. S., Lippincott, B. B., and Lippincott, J. A. (1976). *Infect. Immun.* **13,**
    1080-1083.
Wolpert, J. S., and Albersheim, P. (1976). *Biochem. Biophys. Res. Commun.* **70,** 729-737.
Zabriskie, J.B. (1967). *Adv. Immunol.* **7,** 147-188.

INTERNATIONAL REVIEW OF CYTOLOGY, SUPPLEMENT 13

The *Rhizobium* Bacteroid State

W. D. SUTTON, C. E. PANKHURST, AND A. S. CRAIG

*Plant Physiology and Applied Biochemistry Divisions, D.S.I.R.,
Palmerston North, New Zealand*

I.	Introduction	149
II.	Structural Aspects of the Bacteroid State	150
	A. Infection and Nodule Development in Legumes	150
	B. Nodule Development in *Parasponia* spp.	151
	C. The Structure of Mature Bacteroids	152
III.	Genetic Aspects of the Bacteroid State	156
	A. The Concepts of Infectiveness and Effectiveness	156
	B. Effects of Plant Genotype	157
	C. Effects of *Rhizobium* Genotype	158
	D. *Rhizobium* Plasmids	159
IV.	Bacteroid Biochemistry and Physiology	159
	A. Stages of Bacteroid Development	159
	B. The Bacteroid Environment	160
	C. Viability	161
	D. The Cell Wall–Cell Membrane Complex	162
	E. Nucleic Acids	163
	F. Protein Synthesis	164
	G. Nitrogenase	165
	H. Nitrogen Metabolism	166
	I. Heme Synthesis	167
	J. Electron Transport Systems	168
	K. Respiration and Carbohydrate Metabolism	169
	L. Lipid Metabolism	170
	M. Bacteroid Senescence	171
	References	171

I. Introduction

The term "bacteroid," introduced by Brunchorst (1885), has been used to describe a variety of structures, including swollen forms of *Rhizobium* in cultures and some or all of the *Rhizobium* cells present in legume nodules. Bergersen (1974) suggested that this term be reserved for "the symbiotic forms of the bacteria which are found within the host cells of N_2-fixing legume root nodule tissue, without regard to size or shape." This definition places rather more emphasis on the nitrogen fixation process than currently appears warranted, so in this chapter we have used "bacteroid" to refer to all *Rhizobium* cells found

149

within the central tissue cells of legume root nodules, without regard to morphology or physiology. *Rhizobium* cells located in infection threads or nodule intercellular spaces are regarded as extracellular and are therefore referred to as bacteria.

The historical development of ideas relating to bacteroids was described by Jordan (1962) and Dart (1977). There have been three significant new developments in the last 7 years. First, it was shown in 1973 that a nonlegume, subsequently identified as *Parasponia rugosa* Bl. of the family Ulmaceae, was able to form nitrogen-fixing nodules with *Rhizobium* (Trinick, 1973; Akkermans *et al.*, 1978). Second, it was shown by Keister and others in 1975 that at least some *Rhizobium* strains were capable of synthesizing nitrogenase in pure cultures. Third, the long-standing view that bacteroids were nonviable was challenged in 1977, when three different groups reported that bacteroids could have high viability when isolated and cultured in suitable media.

The effect of these developments has been to remove the previous major distinctions between *Rhizobium* bacteroids and bacteria. It now appears likely that most if not all bacteroid characteristics may also be manifested by *Rhizobium* bacteria cultured in appropriate media.

II. Structural Aspects of the Bacteroid State

A. Infection and Nodule Development in Legumes

The mechanisms whereby *Rhizobium* bacteria infect the roots of legumes will not be dealt with in this chapter. The classic literature relating to infection has been reviewed by Dart (1974, 1977), while recent developments have been surveyed by Dazzo (1980), Robertson and Farnden (1980), Graham, this volume, and Newcomb, this volume.

The subsequent development of infected tissue follows different patterns in different legume species. Many workers from Bieberdorf (1938) onward have argued for a distinction between two nodule types that may be labeled "meristematic" and "spherical." The meristematic type of nodule, as seen in well-studied examples such as *Trifolium subterraneum, Medicago sativa,* and *Pisum sativum,* is characterized by one or more persistent apical or peripheral meristems. The meristem is initiated in the root cortex, adjacent to the protoxylem point closest to the infected root hair, before the infection thread has penetrated into the cortex (Newcomb *et al.,* 1979). The invading rhizobia present in infection threads grow toward the meristematic cells; as they penetrate them, the plant cells cease dividing, but new cell divisions are initiated in the next outermost layer of the middle cortex. The infection threads then branch back and grow toward the outside of the root, while the infected plant cells expand and push the meristem away from the central stele. As this process continues, it results in a nodule with several distinct zones; in sequence from the growing tip these are the nodule cortex (derived from the root cortex), nodule meristem, zone of infection, immature bacteroid-containing tissue, mature bacteroid-containing tissue, and senescent tissue. The nodule vascular tissue is contained in the cortex and connects with tissue in the central stele of the root. Uneven activity of the nodule meristem may result in a wide variety of nodule shapes.

The spherical type of nodule, as seen for example in *Glycine max* and *Arachis hypogaea*, has no persistent meristem. The initial zone of meristematic activity, which may occur either deep in the cortex or close to the root surface, grows into a roughly spherical mass of dividing infected cells; the rhizobia are disseminated partly via infection threads and/or intercellular spaces and partly via division of infected plant cells. Within about 2 weeks nodule meristematic activity is largely completed (Bieberdorf, 1938; Newcomb *et al.*, 1979), and the subsequent growth of the bacteroid-containing tissue is by cell expansion. Mature nodules are surrounded by a cortex containing vascular strands; in some legumes the cortex may develop lenticels and/or a distinct schlerenchyma layer.

The intracellular release of rhizobia involves different infection structures in different legumes; however, in all cases the rhizobia appear to enter the plant cytoplasm by becoming attached to pockets in the plant cell or infection thread membrane and then budding off into individual membrane vesicles referred to as peribacteroid membranes. The polysaccharide capsules previously seen near the rhizobia rapidly disappear (Brenchley and Thornton, 1925; Dixon, 1964), and the newly formed bacteroids, together with their associated peribacteroid membranes, undergo numerous division cycles in the plant cytoplasm. At the nodule maturation stage the bacteroids cease dividing and undergo a variety of host-specific ultrastructural changes.

Detailed descriptions of legume nodule development may be found in the reviews by Dart (1977), Goodchild (1978), Robertson and Farnden (1980), Sutton (1981), and Newcomb, this volume.

B. Nodule Development in *Parasponia* Spp.

The pattern of nodule development in *Parasponia* spp. is markedly different from that occurring in legumes. The nodule vascular system is central rather than peripheral (Trinick and Galbraith, 1976), and each bundle is surrounded by several layers of noninfected cortical cells, then a zone of bacteria-containing tissue, and then several further layers of noninfected cells. The rhizobia are disseminated throughout the infected tissue in highly branched threadlike structures reminiscent of the infection threads in the nodules of some legumes. The thread walls generally appear thinner than the plant cell walls, and in some places the wall is entirely absent, leaving only a membrane separating the rhizobia in the thread from the plant cytoplasm (Becking, 1979; Trinick, 1979). In *Parasponia parviflorum* Miq. and *P. andersonii* Planch. no individualized bacteroid-like forms were observed; however, in squash preparations from *P. rugosa* Bl. about one-third of the infected cells contained individualized rhizobia (Trinick and Galbraith, 1976). Adjacent cells in mature nodules of *P. andersonii* contained threads with quite different morphologies (Trinick, 1979). *Parasponia* spp. nodules do not contain leghemoglobin (Coventry *et al.*, 1976).

C. THE STRUCTURE OF MATURE BACTEROIDS

1. *Size and Shape*

The wide diversity of bacteroid forms has been recognized for many years (Lechtova-Trnka, 1931; Fred *et al.*, 1932) and has been amply confirmed by electron microscopy. The forms observed by reliable methods include branched rods and large pear-shaped or rounded bacteroids in *Medicago* spp. (Jordan, 1962; Dart and Mercer, 1963a), *P. sativum* (Dixon, 1964; Newcomb, 1976), and *Trifolium* spp. (Mosse, 1964; Gourret and Fernandez-Arias, 1974); moderately swollen rods in *Lupinus* spp. (Jordan and Grinyer, 1965; Dart and Mercer, 1966) and *Ornithopus sativus* (Kidby and Goodchild, 1966); small rod-shaped bacteroids in *Lotus* spp. (Craig and Williamson, 1972), *Phaseolus* spp. (Prasade and De, 1971), *Acacia, Viminaria,* and *Vigna* spp. (Dart and Mercer, 1966), and *G. max* (Goodchild and Bergersen, 1966; Basset *et al.*, 1977); and large, perfectly spherical bacteroids in *A. hypogaea* (Aufeuvre, 1973). The volume of the larger, more swollen forms has been estimated as up to 40-fold greater than that of the small rods found in infection threads (Dixon, 1964; Gourret and Fernandez-Arias, 1974). Transitional forms during bacteroid maturation have been described for *Arachis* spp. by Van Rensburg *et al.* (1977) and for *T. subterraneum* and *M. sativa* by Gourret and Fernandez-Arias (1974) and Tu (1977), respectively.

The size and shape of bacteroids and the numbers enclosed in each peribacteroid membrane are largely determined by the plant, since bacteroids of different *Rhizobium* strains in effective nodules of a given host nearly always take on the same morphology. In some cases it has been shown that a single *Rhizobium* strain forms bacteroids of differing morphology in different hosts (Kidby and Goodchild, 1966; Dart, 1977), the most dramatic example being rod-shaped bacteroids contained multiply in peribacteroid membranes in *Vigna mungo* as contrasted with large, spherical bacteroids enclosed singly in peribacteroid membranes in *A. hypogaea* (Dart, 1977). A similar example is illustrated in Figs. 1 and 2.

2. *Ultrastructure*

The cell wall of *Rhizobium*, like that of other gram-negative bacteria, consists of two bilayered membranes. In cultured rhizobia the outer membrane or cell

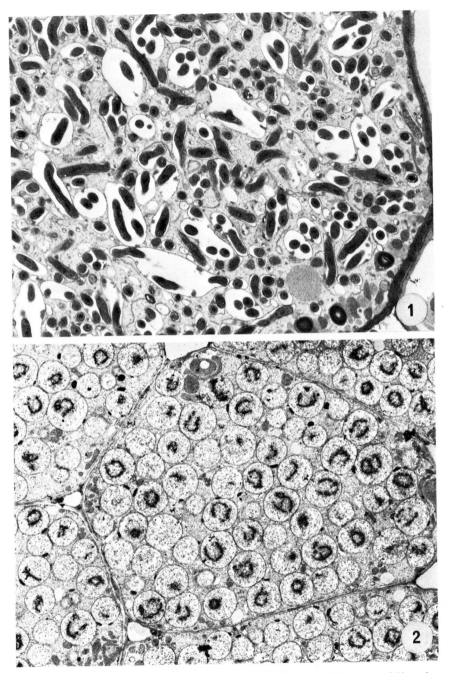

FIG. 1.   Rod-shaped bacteroids of *Rhizobium* strain CB2364 in mature Eff+ nodules of *Phaseolus angularis*. ×5000.

FIG. 2.   Spherical bacteroids of *Rhizobium* strain CB2364 in mature Eff+ nodules of *A. hypogaea*. ×5000.

envelope can be seen to consist of a thin outer layer and a more electron-dense inner layer presumed to contain peptidoglycan (murein) (Tsein and Schmidt, 1977). The inner or cytoplasmic membrane surrounding the bacterial cytoplasm is often found to be separated from the outer membrane, forming a periplasmic space.

In *M. sativa* nodules the formation of bacteroids is accompanied by a reduction in wall thickness and rigidity and by a more particulate appearance of the outer membrane (MacKenzie *et al.*, 1973). The results suggest a partial loss of structural macromolecules, possibly peptidoglycan. A reduced thickness of the presumed peptidoglycan layer has been observed in *Lupinus angustifolius* bacteroids (Robertson *et al.*, 1978) and in *Rhizobium* strain CB756 bacteroids in *Stylosanthes hamata* but not *Macroptilium atropurpureum* (A. S. Craig, unpublished observations). In *Vigna unguiculata* nodules, the thickness and appearance of the bacteroid wall are similar to those seen in *Rhizobium* bacteria (Van Brussel *et al.*, 1979).

In mature bacteroids of *Trifolium* spp. (Dart and Mercer, 1963b; Mosse, 1964), *M. sativa* (Hornez *et al.*, 1974; Tu, 1977), and *P. sativum* (Gourret and Fernandez-Arias, 1974), deep invaginations of the bacteroid cytoplasmic membrane give rise to complex, vesiculate, mesosome-like structures. Similar structures have not been observed in small, rod-shaped bacteroids (e.g., Dart and Mercer, 1966).

Rhizobium bacteria and bacteroids contain a variety of inclusion bodies, the most commonly observed being poly-3-hydroxybutyrate (PHB). In cultured soybean rhizobia, PHB and associated glycogen granules occupy one pole (approximately half) of each cell (Tsien and Schmidt, 1977). The polarity of the cell is further marked by an accumulation of extracellular polysaccharide at the cytoplasmic end and an extracellular polar body at the PHB end (Tsien and Schmidt, 1977). Little or no PHB occurs in the rhizobia within *G. max* infection threads (e.g., Basset *et al.*, 1977), but PHB rapidly accumulates at both poles of the developing bacteroids and eventually accounts for most of the bacteroid volume (Marks and Sprent, 1974; Werner and Morschel, 1978). In *Medicago* and *Trifolium* spp., PHB-like bodies may sometimes be observed in the rhizobia within infection threads, but they rapidly disappear from the developing bacteroids (Dart and Mercer, 1963a; Paau *et al.*, 1978) and only reappear during nodule senescence (Gourret and Fernandez-Arias, 1974). Both the *Rhizobium* genotype and the plant host probably influence PHB formation; in a survey of bacteroids from *Lotus* spp. nodules, Craig *et al.* (1973) found PHB to be absent in 18 out of 83 *Rhizobium* strains. A single *Rhizobium* strain studied by Kidby and Goodchild (1966) formed PHB-like bodies in *O. sativus* but not *Lupinus luteus* nodules, although biochemical tests indicated that the bodies may not have been PHB.

Glycogen granules are commonly present in cultured rhizobia (e.g., Tsien and

Schmidt, 1977; Pankhurst and Craig, 1978) and have been identified by cytochemical methods in bacteroids (Dixon, 1967; Craig and Williamson, 1972; Craig, 1974). Bergersen (1955) reported that *T. subterraneum* bacteroids contained glycogen only in ineffective nodules, and Dixon (1967) observed a rapid disappearance of glycogen granules in the early stages of *Trifolium repens* bacteroid multiplication; however, Craig *et al.* (1973) found glycogen in mature bacteroids of 27 strains of *Rhizobium* that nodulated *Lotus* spp., 15 of these associations being fully effective.

The bacteroid nuclear material or nucleoid takes on two distinct forms depending on the *Rhizobium* strain (Craig *et al.*, 1973). Fast-growing strains, which usually produce acid in culture, have finely fibrillar nuclear material dispersed throughout the cell. Slow-growing strains, which do not produce acid in culture, have a prominent, centrally located nucleoid which frequently appears highly condensed and has been shown by Gourret (1978), in bacteroids from *Sarothamnus scoparius,* to have morphological characteristics reminiscent of the chromosomes of dinoflagellates. The condensed type of nucleoid is invariably associated with spherical, densely staining polyphosphate bodies. These characteristic inclusions, which occur in a wide range of bacterial and algal cells, are thought to consist of polyphosphate, protein, and lipid, with smaller amounts of carbohydrate, RNA, and polyvalent cations (Widra, 1959; Harold, 1966; Freidburg and Avigad, 1968). The exact correlation observed among nucleoid morphology, the presence of polyphosphate bodies, and the culture growth rate led Craig *et al.* (1973) to suggest that a major taxonomic division of the rhizobia could be based on the morphological features of bacteroids. This concept has since been elaborated by Gourret (1975) and Paau (1978) and has been supported by measurements of enzyme levels in cultured *Rhizobium* bacteria (Kennedy, 1981).

Several other *Rhizobium* strain-specific inclusions have been observed in *Lotus* spp. bacteroids, including lipid polar bodies, nonlipid polar bodies, crystals, rods, and dense periplasmic inclusions (Craig *et al.*, 1973).

3. *Bacteroid Forms in Vitro*

A method for studying the *Rhizobium* cell morphology associated with *in vitro* nitrogen fixation was developed by Pankhurst and Craig (1978) using soft agar cultures. Cells in the uppermost layers of agar were rod-shaped, produced large amounts of extracellular polysaccharide, and had a moderately dense cytoplasm containing PHB, glycogen, and polyphosphate. Cells in the bottom layers were also rod-shaped but had an electron-translucent cytoplasm, a highly condensed nucleoid, and a heavily staining cell wall. Between these layers was a narrow zone of highly pleiomorphic bacteroid-like cells with moderately condensed nuclear material and polyphosphate bodies, little extracellular polysaccharide, and a polar distribution of PHB and glycogen. It was suggested that the latter type were responsible for the nitrogen-fixing activity in the cultures. Similar studies were reported by Van Brussel *et al.* (1979) for rhizobia from agar plate cultures. The nitrogen-fixing cells resembled bacteroids of the same strain from *V. unguiculata* nodules in their pleiomorphism and polyphosphate granule content but had a higher PHB content and a greater proportion of highly branched forms.

Swollen bacteroid-like forms of *Rhizobium* can be produced in bacterial cultures by a variety of media supplements including high concentrations of thiamine, alkaloids, or amino acids (Jordan, 1962; Jordan and Coulter, 1965; Skinner *et al.*, 1977). The swollen cells are enlarged and distorted and contain numerous PHB-like bodies and whorls of intracytoplasmic membrane; however, they do not exhibit any obvious changes in their cell wall structure (Skinner *et al.*, 1977).

III. Genetic Aspects of the Bacteroid State

A. THE CONCEPTS OF INFECTIVENESS AND EFFECTIVENESS

Interactions between the legume host and strains of *Rhizobium* present a wide variety of responses ranging from complete failure of the rhizobia to invade the host roots and form nodules to the production of nodules that may or may not fix nitrogen to varying degrees. The description and interpretation of these various responses has always presented difficulties, and these have become aggravated in recent years with the increasing number of plant and *Rhizobium* mutants presenting new and unusual nodulation responses.

Infectiveness (Inf), as the term implies, defines the ability of a *Rhizobium* strain to infect the roots of a given legume, but with the added proviso that this infection results in the formation of a nodule. Strains classified as Inf$^-$ are nonnodulating. Effectiveness (Eff) defines the type of nodulation and the level of nitrogen fixation. *Rhizobium* strains classified as Inf$^+$ Eff$^+$ form nodules that fix significant quantities of nitrogen, although these may range from partially effective to fully effective; strains classified as Inf$^+$ Eff$^-$ form nodules that fix little or no nitrogen.

Vincent (1980) has analyzed the sequence of events leading to the development of an effective nitrogen-fixing nodule and recognizes 11 distinct steps. He has proposed a phenotypic code for describing each step. Such a code should prove useful in describing various plant and *Rhizobium* mutants and in categorizing *Rhizobium* strains previously referred to as invasive but nonnodulating (e.g., Rolfe *et al.*, 1980), parasitic (Jordan and Garrad, 1951), producing pseudonodules or tumorlike growths (Pankhurst, 1974), or producing apparently normal nodules containing leghemoglobin but incapable of fixing nitrogen (Caldwell and Vest, 1977).

B. EFFECTS OF PLANT GENOTYPE

Many host factors under genetic control can influence nodulation and bacteroid development (Nutman, 1969; Holl and LaRue, 1976; Caldwell and Vest, 1977; Gibson, 1979). In general, research on this subject has had two basic aims: (1) to use plant genes to define factors responsible for the development and maintenance of nodules, and (2) to improve the productivity of legumes by selection and breeding for appropriate symbiotic attributes. Only the former will be considered in this section, with particular emphasis on plant genes directly involved in nodule development and function.

Genes responsible for nodulation failure have been described in *Trifolium pratense* (Nutman, 1949), *G. max* (Williams and Lynch, 1954), *P. sativum* (Lie, 1971; Holl and LaRue, 1976), and *T. subterraneum* (Gibson, 1968). In all but the last of these species the nonnodulating character appears to be controlled by a single recessive gene. In the case of *T. pratense,* nodulation failure was apparently due to failure of the bacteria to infect the root hairs (Nutman, 1949), while in *P. sativum* 'Afghanistan' infection threads were formed in the root hairs but did not give rise to nodules (Degenhardt *et al.,* 1976). With *G. max* the *rjl rjl* genotype confers resistance to nodulation by the majority of Inf$^+$ *Rhizobium* strains (Devine and Weber, 1977). Strains that nodulate *rjl rjl* plants also produce rhizobiotoxine, an enol ether amino acid which causes plant foliar chlorosis (Devine and Weber, 1977), but the reason for this association is unknown.

Plant genes that condition an ineffective nodulation response have also been described in several legume species. For example, Nutman (1969) described 4 single recessive genes (namely, *i, n,* and *d*) that confer ineffectiveness in *T. pratense.* In all cases the Eff$^-$ nodules were characterized by the absence of leghemoglobin and the presence of abundant starch. The *i i* genotype resulted in failure of rhizobia released into the host cells to develop into bacteroids. The *ie ie* genotype resulted in abnormal cell divisions at the point where bacteria were released from the infection threads, producing tumorlike growths within the nodule and preventing bacteroid development. Both *ie ie* and *i i* plants, however, could be restored to normal nodulation by modifying genes (Nutman, 1969). Nodules on *n n* plants were small and contained few infection threads or infected cells: any rhizobia that were released from the threads degenerated rapidly. Nodules formed on *d d* plants contained mainly parenchyma-like cells with few infection threads or infected cells (Nutman, 1969). An ineffective genotype of *M. sativa* similar to the *i i* genotype of *T. pratense* was reported by Viands *et al.* (1979). In this case also the nodule cells contained no leghemoglobin, a few short-lived bacteroids, and abundant starch granules.

In *P. sativum* 'Afghanistan,' two recessive genes affecting nodulation have been described (Holl and LaRue, 1976); of these, *sym 2* results in nodulation failure, while *sym 3* results in Eff$^-$ nodules, possibly by restricting the supply of carbohydrate to the bacteroids (Holl and LaRue, 1976). In *G. max* three dominant genes resulting in Eff⁻ nodules have been reported (Caldwell and Vest, 1977). There is some evidence that two of these, Rj_2 and Rj_3, are linked (Caldwell and Vest, 1977); however, their effects can be distinguished by the plant response to different strains of *Rhizobium*. Structural observations on the development of ineffective nodules have not yet been published.

Other aspects of nodule development have been shown to be under the control of the plant genotype by experiments in which different plant species or cultivars were inoculated with the same *Rhizobium* strains. Thus it is known that the plant determines the structure of the globin chains of leghemoglobin (Dilworth, 1969; Broughton and Dilworth, 1971), the time required for the nodules to commence nitrogen fixation (Caldwell and Vest, 1977), the size and shape of the bacteroids (e.g., Dart, 1977), bacteroid viability (Sutton and Paterson, 1980), hydrogenase activity (Dixon, 1972), and the number of bacteroids enclosed per peribacteroid membrane envelope (Kidby and Goodchild, 1966). The host also has an important influence on the actual nitrogen-fixing capacity of the nodules, a characteristic which can be very variable and is complicated by a strong host–*Rhizobium* strain interaction (Caldwell and Vest, 1968; Gibson and Brockwell, 1968; Jones and Hardarson, 1979). In *T. subterraneum* there are indications that effectiveness is under polygenic control (Nutman, 1961, cited by Gibson, 1979).

Exactly how many plant genes are involved in controlling nodule development is still a matter for conjecture. Holl and LaRue (1976) have conservatively estimated the involvement of 10 plant genes, 3 of which control nodule differentiation (bacterial release from infection threads, bacteroid maturation, and nodule structure). Unquestionably, the system is much more complex than this, and it is likely that the ancillary effects of many plant genes not directly involved in nodule development per se will also have an influence. The genes involved may not necessarily all be chromosomal genes, as recent evidence from studies on *Lotus* spp. hybrids has indicated a degree of maternal inheritance of nodule-forming ability (R. M. Greenwood, unpublished observations).

C. EFFECTS OF *Rhizobium* GENOTYPE

Mutational loss of symbiotic ability is not uncommonly encountered in *Rhizobium*. It may occur naturally or after mutagenesis and may be associated with other changed characteristics such as colony form (large "gummy" to small "nongummy") (Wilson *et al.*, 1975), additional nutritional requirements (auxotrophy) (Schwinghamer, 1969), and resistance to bacteriophages (Patel, 1978), excess metabolites (Staphorst and Strijdom, 1971), antimetabolites (Schwinghamer, 1968), or antibiotics (Schwinghamer, 1967). Many such mutants have been characterized phenotypically, and the block in nodule development leading to loss of symbiotic ability identified (see reviews by Vincent, 1979; Kuykendall, this volume). Mutations resulting in the failure of rhizobia to be released from infection threads into the host cells, failure of released rhizobia to multiply, failure to develop normal bacteroid morphology, and failure to develop nitrogenase activity have all been described.

Studies with such mutants have provided useful information concerning *Rhizobium*-plant interactions at different stages of nodule development. In addition to this, mutants may prove valuable in studies on possible cooperative interactions between different *Rhizobium* strains. For example, in the experiments reported by Rolfe and Gresshoff (1980), combinations of an Inf⁺ Eff⁺ wild-type strain and an Inf⁻ mutant derivative often yielded nodules containing both strains. Furthermore, the Inf⁻ mutant was occasionally found as bacteroids in protoplasts isolated from such nodules. In one instance, an Inf⁻ *Rhizobium* mutant when mixed with an Inf⁺ Eff⁻ mutant gave rise to Eff⁺ nodules (Rolfe *et al.*, 1980). When the bacteria isolated from such nodules were retested for their nodulating characteristics, they were shown to have retained the initial mutant phenotypes.

D. *Rhizobium* PLASMIDS

Many *Rhizobium* strains contain plasmid DNAs of high molecular weight (e.g., Sutton, 1974; Zurkowski and Lorkiewicz, 1976; Nuti *et al.*, 1977; Spitzbarth *et al.*, 1979; Schwinghamer and Dennis, 1979). Although there has been only a single report on the presence of plasmid DNA in bacteroids (Sutton, 1974), there are clear indications that a number of genes essential for plant infection and bacteroid development are located on extrachromosomal genetic elements (reviewed by Dénarié *et al.*, this volume).

Plasmid genes can determine *Rhizobium* host range (Higashi, 1967; Johnston *et al.*, 1978), effectiveness (Brewin *et al.*, 1980), and nitrogenase structure (Nuti *et al.*, 1979). The formation of unusual amino acids may perhaps be another plasmid-determined character. Although many of the *Rhizobium* strain-specific, ninhydrin-positive compounds formed in nodules have not yet been identified (Greenwood and Bathurst, 1968, 1978), several appear to be structurally related to 4-aminobutyrate (G. J. Shaw and L. Nixon, unpublished observations), indicating that they may form a family of related compounds reminiscent of the opines whose production is controlled by plasmid-borne genes in *Agrobacterium tumefaciens* (Watson *et al.*, 1975; Guyon *et al.*, 1980).

IV. Bacteroid Biochemistry and Physiology

A. STAGES OF BACTEROID DEVELOPMENT

When discussing bacteroids, it is convenient to recognize three developmental stages:

1. Immature bacteroids lack nitrogenase activity and are present in nodule tissue that has not yet produced significant quantities of leghemoglobin. The bacteroids, and in some cases the host cells, are capable of active cell division. Immature bacteroids depend on the plant cytoplasm for both energy and combined nitrogen, but they should not necessarily be regarded as parasitic, since it is possible that they may already be excreting vitamins, hormones, or other substances useful in the metabolism of the infected cells.

2. Mature bacteroids are characterized by high nitrogenase activity and are normally found in tissue with a high leghemoglobin content. In many plant species, the mature bacteroids are grossly enlarged (Section II,C,1) and do not appear to be capable of cell division *in situ*. In plant species with small, rod-shaped bacteroids, most bacteroids probably remain capable of cell division. Mature bacteroids depend on the plant cytoplasm for energy, but they excrete substantial quantities of combined nitrogen in the form of NH_4^+.

3. Senescent bacteroids represent the terminal stage of the nodule symbiosis, when nitrogenase activity and leghemoglobin content decline and the peribacteroid membranes disintegrate.

B. THE BACTEROID ENVIRONMENT

Many features of bacteroid metabolism can be regarded as adaptations to the unusual environment in which they function. This environment is largely under plant control, since all substances entering or leaving the bacteroid-containing tissue must pass through either the nodule cortex or the plant vascular system, and all substances entering the bacteroids must pass through at least one layer of peribacteroid membrane.

The bacteroid-containing tissue of mature nodules contains high concentrations of ammonia (e.g., Boland *et al.*, 1978; Peterson and Evans, 1978), amino acids [especially asparagine, glutamine, and 4-aminobutyrate (e.g., Butler and Bathurst, 1958; Werner *et al.*, 1980)], and organic acids, especially malonate and malate (Stumpf and Burris, 1979). Very low oxygen tensions occur at the center of mature nodules, as shown by spectroscopic measurements of leghemoglobin oxygenation (Appleby, 1962, 1969) and by direct measurements with microelectrodes (Ebertova, 1959; Tjepkema and Yocum, 1974). Leghemoglobin, which is believed to play a major role in buffering the intranodular concentration of dissolved oxygen and facilitating oxygen transfer to the bacteroids (e.g., Wittenberg *et al.*, 1974; Bergersen and Turner, 1975), may be present at concentrations exceeding 0.5 mM in the cytoplasm of mature bacteroid-containing cells (Smith, 1949; Bergersen and Goodchild, 1973). In addition to the 8 known leghemoglobins (Fuchsman and Appleby, 1979), as many as 20 nodule-specific plant polypeptides may be present in the cytoplasm of infected cells (Legocki and Verma, 1979, 1980).

The peribacteroid membranes probably play a large part in controlling the exchange of nutrients and metabolites between the bacteroid and plant cytoplasms, but direct evidence is lacking. The membranes are generally believed to be of plant origin (Goodchild, 1978), although the possibility that some components are contributed by the bacteroids cannot be dismissed. In plant species with enlarged bacteroids the peribacteroid membrane is closely associated with the bacteroid wall (e.g., Dart and Mercer, 1963a). Methods for purifying peribacteroid membranes from *L. angustifolius* and *G. max* nodules have been described by Robertson *et al.* (1978) and Verma *et al.* (1978).

C. Viability

The traditional view that bacteroids are nonviable (Jordan, 1962; Bergersen, 1974) was based largely on the work of Almon (1933) who used bacteroids prepared from crushed mature nodules of *P. sativum, M. Sativa,* and *T. pratense*. Almon's results have been confirmed by more recent work in which bacteroids from crushed mature nodules of *T. repens* and *L. angustifolius* gave less than 0.5% colony formation on normal media (Sutton *et al.,* 1977; Sutton and Paterson, 1980), but the interpretation has been changed as a consequence of three new results.

First, it has been established that bacteroids from crushed mature nodules of *G. max, O. sativus,* and *Phaseolus aureus* can produce 20–90% colony formation on normal media (Tsien *et al.,* 1977; Sutton and Paterson, 1980). It seems likely that this may be a general feature of bacteroids that have not undergone significant morphological changes. Second, it has been shown that immature bacteroids from *L. angustifolius* nodules may give 5–30% colony formation on normal media and up to 70% colony formation on media supplemented with 0.3 *M* mannitol or sucrose (Sutton *et al.,* 1977). Similar supplements also caused 2- to 10-fold increases in the percentage of colonies formed by mature bacteroids, but the absolute percentage remained low and colony-forming ability was largely confined to the "slowly sedimenting" bacteroid fraction (Sutton and Mahoney, 1977). Third, it has been reported that, when mature *Trifolium* spp. and *G. max* bacteroids were prepared from bacteroid-containing nodule protoplasts, they gave 5–100% colony formation on media supplemented with 0.2 *M* mannitol (Gresshoff *et al.,* 1977; Gresshoff and Rolfe, 1978; Rolfe and Gresshoff, 1980). On normal media the *G. max* bacteroids did not form colonies.

These results indicate that a high proportion of mature bacteroids retain the potential to resume cell division when isolated and cultured by appropriate methods. Low bacteroid viability appears to be associated with osmotic sensitivity, probably as a result of cell wall changes. The swollen bacteroid-like forms of *Rhizobium* bacteria induced by media supplements (Section II,C,3) also have low colony-forming ability on normal media (Jordan and Coulter, 1965; Staphorst and Strijdom, 1972) and have been reported to exhibit osmotic sensitivity (Urban, 1979).

D. The Cell Wall–Cell Membrane Complex

In addition to the structural studies described in Section II, there have been several indications that mature bacteroids may have an altered wall structure as compared with *Rhizobium* bacteria. Bacteroids from *M. sativa* nodules were said to release a higher proportion of their vitamin B12 content into the surrounding medium than bacteria from broth cultures (Levin *et al.*, 1954). Bacteroids from *L. luteus* nodules were 10-fold more permeable to chloramphenicol than cultured cells (Coventry and Dilworth, 1975). Mature bacteroids from *L. angustifolius* nodules were much more sensitive than bacteria to the detergents sodium deoxycholate and Triton X-100 (Sutton and Paterson, 1979). When bacteroids from a range of legume species were examined, there was a strong negative correlation between detergent sensitivity and colony-forming ability (Sutton and Paterson, 1980).

The structural basis for these changes is not yet clear. Van Brussel *et al.* (1977) reported that mature *P. sativum* bacteroids had reduced levels of lipopolysaccharide and heptose in their cell walls as compared with *Rhizobium* bacteria; however, a later report indicated that most of the bacteroid lipopolysaccharide had been lost during the rinsing of the cells in a low-osmolarity medium (Planque *et al.*, 1979). The lipopolysaccharide antigens from *Lotus* spp. bacteroids were more easily detached than those from cultured bacteria (Pankhurst, 1979). It was suggested by Jordan (1962) that bacteroid development might involve "a selective interference in the normal synthesis of cell wall, of cytoplasmic membrane, or of both these structures," in a manner "somewhat analogous to that observed when certain gram-negative bacteria are exposed to penicillin." This suggestion appears compatible with the results described above but would not apply to nodules from legume species in which the bacteroids retain high viability and a bacteria-like morphology.

The biological role of the presumptive cell wall modifications is unknown, but it may include control of bacteroid multiplication. Mutations that confer resistance to inhibitors of cell wall or cell membrane synthesis often result in failure of rhizobia to be released from infection threads (Jordan *et al.*, 1969; Pankhurst, 1974, 1977; Pankhurst and Craig, 1979). This may be due to altered permeability of the cell wall–cell membrane complex, preventing essential exchanges of metabolites between the host cells and bacteria (Pankhurst and Craig, 1979). Resistance to viomycin has been shown to be associated with an increase in the phospholipid content of the cell wall (MacKenzie and Jordan, 1970), resulting in decreased permeability to the antibiotic.

E. Nucleic Acids

The DNA content of bacteroids has been measured many times, and most reports have indicated that it is equal to or greater than that of *Rhizobium* bacteria.

Bisseling *et al.* (1977) found that large, swollen bacteroids from *P. sativum, M. sativa,* and *T. pratense* nodules had 4- to 8-fold more DNA than bacteria, as measured by microfluorimetry of Feulgen-stained cells. In bacteroids from five other legume species the DNA content ranged from 3-fold greater to equal to that of bacteria. Later reports using a different bacteroid preparation method showed that *P. sativum* bacteroids had 2- to 3-fold more DNA than bacteria (Van den Bos *et al.,* 1978), while *L. angustifolius* bacteroids had 1.2- to 1.6-fold more DNA than bacteria, mature bacteroids having slightly more DNA than immature ones (Sutton *et al.,* 1978).

Paau *et al.* (1979a) used bacteroids that were rinsed in a low-osmolarity buffer, fixed in 64% ethanol, digested with RNase, stained with ethidium bromide, and examined by microfluorimetry. Mature bacteroids from 18 different Eff[+] *Rhizobium*-legume combinations had nucleic acid contents equal to or greater than those of bacteria, the highest amounts being found in *M. sativa* and *T. repens* bacteroids and the lowest in *Lupinus* spp., *G. max,* and *Phaseolus vulgaris.* Nucleic acid contents lower than those of bacteria were observed in bacteroids from two out of four Eff[−] combinations. Other reports have indicated that the nucleic acid content of both immature and senescent *M. sativa* bacteroids is lower than for mature bacteroids, but that the nucleic acid content of *G. max* bacteroids is similar to that of bacteria at all stages (Paau *et al.,* 1978; Paau and Cowles, 1979). Earlier observations suggesting dispersal and perhaps digestion of DNA in senescent bacteroids were reviewed by Jordan (1962). The use of senescent nodules may perhaps account for the DNA content lower than that of bacteria reported by Sutton (1974) for one out of four strains of *Lotus* sp. bacteroids.

There have been relatively few biophysical studies with bacteroid DNA; however, Sutton (1974) found that purified DNA from *L. angustifolius* and *Lotus* sp. bacteroids had the same buoyant density in cesium chloride gradients, denaturation temperature, molecular weight, and kinetic complexity as DNA from *Rhizobium* bacteria. Both types of DNA contained 3–5% of a high-complexity plasmid DNA that appeared as a light-density satellite in cesium chloride gradients. Reijnders *et al.* (1975) found that DNA from detergent-rinsed *P. sativum* bacteroids had the same buoyant density as bacterial DNA, and the reassociation kinetics of DNA mixtures showed a sequence homology between the two types of 99 ± 4%. Cannon (1974) also found that *G. max* bacteroid DNA had the same buoyant density as bacterial DNA, but the molecular weight of bacteroid DNA was lower and the frequency of DNA circles was higher. The fourfold reduction in DNA content per cell observed for the bacteroid samples suggests that the bacteroids used may have been either senescent or damaged. Similar factors may account for the report that *P. aureus* and *Cicer arietinum* bacteroid DNAs had lower bouyant densities and denaturation temperatures than the corresponding bacterial DNAs (Agarwal and Mehta, 1974).

It was suggested by Jordan (1962) that bacteroid formation may involve "inhibition of cell division and a continued synthesis of chromatin." This suggestion has been revived by Bisseling *et al.* (1977), but it has not yet received direct experimental support. Measurements of the specific activity of DNA polymerase and coenzyme B12-dependent ribonucleotide reductase have suggested that the rate of DNA synthesis may be lower in mature *M. sativa* and *G. max* bacteroids than in *Rhizobium* bacteria (Cowles *et al.*, 1969; Paau and Cowles, 1975).

Bacteroid RNA has received little attention. Dilworth and Williams (1967) observed an eightfold reduction in RNA per cell and a threefold reduction in the RNA/DNA ratio in developing *L. luteus* bacteroids. Sutton and Robertson (1974) observed a lower RNA/DNA ratio in mature *L. angustifolius* bacteroids than in bacteria. Bisseling *et al.* (1979) observed a 25% reduction in RNA per cell and a 40% reduction in the RNA/protein ratio in developing *P. sativum* bacteroids. The bacteroid 16 S RNA was intact, but the 23 S RNA was fragmented, whereas both rRNAs from bacterial preparations were intact. It has been suggested from electron microscope observations that mature bacteroids contain few ribosomes (e.g., Ching *et al.*, 1977); however, this does not appear to be true for bacteroids fixed and stained by appropriate methods (Dart and Mercer, 1963a; Tu, 1975; W. D. Sutton and A. S. Craig, unpublished observations).

According to a recent brief report, the DNA-dependent RNA polymerase from *G. max* bacteroids has a different sensitivity to rifampicin than the enzyme from bacteria (Tierney and Schubert, 1980).

F. PROTEIN SYNTHESIS

There is a rapid synthesis of bacteroid proteins during nodule maturation, when bacteroid numbers per nodule may increase 10-fold within a few days (e.g., Sutton *et al.*, 1977; Planque *et al.*, 1978). Although few studies have been reported, it appears that the protein content per bacteroid cell may increase roughly in proportion to the DNA content (Reijnders *et al.*, 1975; Paau *et al.*, 1979b).

Studies with protein synthesis inhibitors have suggested that many and perhaps all bacteroid proteins are synthesized on bacteroid ribosomes. Coventry and Dilworth (1976) observed that the incorporation of $^{14}CO_2$-labeled photosynthate into the bacteroid proteins of *L. luteus* nodules was inhibited 40% by chloramphenicol, while the labeling of plant cytoplasmic proteins was not affected. Sutton (1980) observed that [^{35}S]methionine incorporation into bacteroid proteins of *L. angustifolius* nodules was strongly inhibited by chloramphenicol or rifampicin, while the labeling of plant cytoplasmic protein was not affected. Two inhibitors of plant cytoplasmic ribosomes, cycloheximide and anisomycin, each caused a rapid 50% inhibition of bacteroid labeling, but the effect was reversed after 12 hours. It was suggested that there was an *in vivo* mechanism for coordinating protein synthesis rates in the bacteroids and the plant cytoplasm.

Coventry and Dilworth (1975) observed a lag phase of several hours before isolated *L. luteus* bacteroids could synthesize labeled proteins, but no such lag was observed with *L. angustifolius* bacteroids protected against osmotic shock (Sutton *et al.*, 1977). There were marked differences between the sets of [^{35}S]methionine-labeled polypeptides synthesized by mature and immature bacteroids and cultured bacteria (Shaw and Sutton, 1979). Bacteroids of the same strain from different plant hosts synthesized similar sets of polypeptides.

G. NITROGENASE

Nitrogenase activity per weight of nodule tissue, measured by the acetylene reduction assay, may increase more than 1000-fold within 3–4 days during nodule maturation (e.g., Sutton and Jepsen, 1975). The increase is generally assumed to result from *de novo* synthesis of nitrogenase enzyme, and evidence supporting this view has been provided for *P. sativum* bacteroids by radioimmunoassays (Bisseling *et al.*, 1980) and for *L. angustifolius* bacteroids by gel electrophoresis of [^{35}S]methionine-labeled polypeptides (Shaw and Sutton, 1979) and experiments with protein synthesis inhibitors (Sutton, 1980).

Nitrogenase constitutes 10–12% of the total protein in mature bacteroids from *G. max* and *L. luteus* (Klucas *et al.*, 1968; Whiting and Dilworth, 1974). It is separable into C1 and C2 components with subunit molecular weights of about 60,000 and 34,000, respectively; the C1 subunits may be further resolved into α and β types (Kennedy *et al.*, 1976). In developing *P. sativum* nodules the synthesis of C1 commences before that of C2, but the ratio of ^{35}S-labeled C2 to C1 in ^{35}SO$_4^{2-}$-labeled nodules steadily increases as the nodules mature (Bisseling *et al.*, 1979, 1980).

For several years it was generally considered that the nitrogenase activity of rhizobia was confined to the bacteroid state (Bergersen, 1974); however, in 1975 it was shown that certain slow-growing *Rhizobium* strains could synthesize nitrogenase in pure cultures (Keister, 1975; Pagan *et al.*, 1975; McComb *et al.*, 1975; Kurz and LaRue, 1975; Tjepkema and Evans, 1975). Subsequent work has shown that many other slow-growing strains can produce nitrogenase (e.g., Kaneshiro *et al.*, 1978) and has achieved some progress toward establishing optimal conditions for high activities (e.g., Bergersen and Turner, 1976; Bergersen *et al.*, 1976; Pankhurst and Craig, 1978). Low oxygen concentrations appear to be essential, and it has been suggested that the level of relative adenylylation of glutamine synthetase may play a controlling role (Bergersen and Turner, 1976; see however Ranga Rao *et al.*, 1978). Studies with glutamine-requiring auxotrophic mutants of *Rhizobium*, unable to produce glutamine synthetase, have shown that they form Eff⁻ nodules lacking nitrogenase activity (Kondorosi *et al.*, 1977; Ludwig and Signer, 1977). The results certainly suggest a relationship between glutamine synthetase activity and nitrogenase synthesis, as has also been suggested for *Klebsiella pneumoniae* (e.g., Tubb, 1974), but the mechanism for control remains unclear.

It was reported by Werner (1978) that relatively low levels of the RNA polymerase inhibitor rifampicin inhibited nitrogenase activity but not growth in cultures of *Rhizobium*. These results, which have recently been confirmed (C. E. Pankhurst, D. B. Scott, C. W. Ronson, and D. White, unpublished observations), may be connected with the altered characteristics reported for the DNA-dependent RNA polymerase of bacteroids.

Attempts to demonstrate nitrogenase activity in cultures of fast-growing *Rhizobium* strains have suffered from a lack of reproducibility. It was reported that a filterable factor produced by a *G. max* cell suspension culture induced nitrogenase activity in a wide range of rhizobia (Bednarksi and Reporter, 1978), but the identity of the inducing factor(s) involved has not been established.

H. Nitrogen Metabolism

Mature bacteroid suspensions reduce nitrogen via the nitrogenase reaction and excrete the product as NH_4^+ (Bergersen and Turner, 1967; Laane *et al.*, 1980). A similar process is thought to occur in legume nodules, with NH_4^+ assimilation into amino acids and ureides largely occurring in the plant cytoplasm (reviewed by Rawsthorne *et al.*, 1980; Imsande, this volume). Ammonium excretion has also been demonstrated in nitrogenase-active *Rhizobium* cultures (O'Gara and Shanmugam, 1976; Upchurch and Elkan, 1978) and in aerobic cultures growing in histidine- or asparagine-containing media (O'Gara and Shanmugam, 1976; de Hollander *et al.*, 1979).

Brown and Dilworth (1975) observed that glutamine synthetase was less active in bacteroids from seven plant species than in nitrogen-limited bacterial cultures, and similar results have been reported by Robertson *et al.* (1975a), Bishop *et al.* (1976), and several others. Bacteroid glutamine synthetase activity decreases during nodule maturation in *L. angustifolius* (Robertson *et al.*, 1975b) and *P. sativum* (Planque *et al.*, 1978). In mature *G. max* and *P. sativum* bacteroids the enzyme is mainly in the adenylylated form (Bishop *et al.*, 1976; Planque *et al.*, 1978).

Glutamate synthase (GOGAT) and glutamate dehydrogenase have generally been reported to be present at rather low specific activities in bacteroids (e.g., Dunn and Klucas, 1973; Brown and Dilworth, 1975; Kurz *et al.*, 1975; Werner *et al.*, 1980). Alanine dehydrogenase, by way of contrast, increases 6- to 10-fold in specific activity in maturing bacteroids of *G. max* (Stripf and Werner, 1978; Werner *et al.*, 1980), and a similar increase has been observed in nitrogenase-active *Rhizobium* cultures (Werner and Stripf, 1978). Bacteroids have been shown to contain asparaginase (e.g., Lees and Blakeney, 1970; Streeter, 1977), alanine and aspartate aminotransferases (e.g., Ryan and Fottrell, 1974; Werner and Stripf, 1978), and other enzymes (tabulated by Robertson and Farnden, 1980), but the role of these enzymes in bacteroid and nodule metabolism is largely unexplored.

Useful information concerning the nutritional requirements for bacteroid development and function has been obtained through the use of *Rhizobium* auxotrophic mutants (Schwinghamer, 1969, 1970; Dénarié, 1969; Dénarié *et al.*, 1976; Pankhurst *et al.*, 1972). In many instances acquisition of auxotrophy is associated with an Inf$^+$ Eff$^-$ phenotype, although this varies with the type of mutation and the *Rhizobium* strain (Schwinghamer, 1969). Requirements for purines or pyrimidines are more frequently associated with Inf$^+$ Eff$^-$ phenotypes than requirements for amino acids (Schwinghamer, 1969; Dénarié *et al.*, 1976). Root nodules formed on *T. pratense* by a riboflavin-requiring *Rhizobium* auxotroph were blocked at the bacteroid maturation stage, but restoration of effectiveness could be achieved genetically by back-mutation to prototrophy or biochemically by adding riboflavin to the root medium of inoculated plants (Pankhurst *et al.*, 1972). A leucine-requiring *Rhizobium* auxotroph formed Eff$^-$ *M. sativa* nodules in which the rhizobia were not released from their infection threads, but normal development occurred when leucine was added to the growth medium (Truchet and Dénarié, 1973). Nodules formed on *M. sativa* by adenine-requiring *Rhizobium* auxotrophs were blocked in bacteroid maturation, while nodules formed by a uracil-requiring auxotroph were blocked at the earlier stage of bacteroid multiplication in the host cells (Dénarié *et al.*, 1975). A glutamate-requiring auxotroph, which lacked both glutamate synthase and glutamate dehydrogenase activity in culture, formed fully Eff$^+$ nodules on *M. sativa* (Kondorosi *et al.*, 1977).

I. HEME SYNTHESIS

Although the globin chains of leghemoglobin are encoded by plant DNA (Sidloi-Lumbroso *et al.*, 1978; Baulcombe and Verma, 1978) and synthesized on plant cytoplasmic ribosomes (Verma *et al.*, 1974, 1979; Verma and Bal, 1976), the heme groups are generally considered to be synthesized by the bacteroids. Bacteroid suspensions from *G. max* and *L. luteus* nodules were shown to be much more active than the plant cytoplasmic fractions in converting 5-aminolevulinic acid into heme (Cutting and Schulman, 1969; Godfrey and Dilworth, 1971), while the rate-limiting enzyme 5-aminolevulinic acid synthetase was confined to the bacteroid fraction in *G. max* nodules and increased 15-fold in activity per gram fresh weight during nodule maturation (Nadler and Avissar, 1977). One of the enzymes involved in heme biosynthesis, 5-aminolevulinic acid dehydratase, was found at higher specific activities in the plant cytoplasmic fraction of nodules than in the bacteroids (Godfrey *et al.*, 1975; Nadler and Avissar, 1977); however, although some transfer of heme precursors between the bacteroid and plant cytoplasm may occur, all the enzymes required for heme synthesis have been positively identified in bacteroids (Godfrey *et al.*, 1975).

Heme synthesis by *L. luteus* bacteroid suspensions was inhibited by anaerobic conditions (Godfrey and Dilworth, 1971), but in *Rhizobium* cultures 3 days of restricted aeration resulted in 10-fold increases in 5-aminolevulinic acid synthetase and dehydratase activities, a 20-fold increase in the intracellular heme concentration, and visible heme excretion into the growth medium (Avissar and Nadler, 1978).

J. Electron Transport Systems

Bacteroids contain an altered set of hemoproteins, as compared with aerobic *Rhizobium* cultures (reviewed by Bergersen, 1974; Dilworth and Appleby, 1979). The major features reported for *G. max* bacteroids have been the absence of cytochromes aa_3 and o and *Rhizobium* hemoglobin (not related to leghemoglobin) and the presence of cytochromes c-552, P-420, and P-450. The last of these components has been implicated in an oxidative phosphorylation pathway supporting nitrogen fixation (Kretovich *et al.*, 1974; Appleby *et al.*, 1975). Partial or complete anaerobiosis has been shown to cause reduced levels of hemoglobin and cytochrome aa_3 and the induced synthesis of cytochromes c-550, c-552, and P-450 in *Rhizobium* cultures (e.g., Daniel and Appleby, 1972; Avissar and Nadler, 1978); however, nitrogenase-active cultures of *Rhizobium* strain 32H1 have been reported to contain normal levels of cytochrome aa_3 (Evans and Crist, 1980).

The ubiquinone content of *G. max* bacteroids was reported to be about 50% higher per milligram of protein than in aerobic *Rhizobium* cultures, and 150% higher than in anaerobic nitrate cultures (Daniel, 1979).

It has been suggested by Laane *et al.* (1979, 1980) that the *in vivo* production of reducing equivalents for bacteroid nitrogenase may be driven by the electrical component of the proton motive force across the bacteroid cell membrane; however, the pathway of electron transfer to nitrogenase has not yet been fully established. It is known that bacteroids have a high flavin content (Pankhurst *et al.*, 1974), and a ferredoxin of about 6600 molecular weight has been purified from *G. max* bacteroids and shown to mediate electron transfer to nitrogenase from illuminated chloroplasts (Carter *et al.*, 1980). Two factors active in the same assay were shown to be present in anaerobic nitrate-cultured *Rhizobium* bacteria but were absent in aerobic cultures (Phillips *et al.*, 1973).

Rhizobium mutants uncoupled in oxidative phosphorylation and unable to grow aerobically on succinate have been shown to form effective nodules on *T. subterraneum* with 70% of the normal bacteroid ATPase activity (Skotnicki and Rolfe, 1979).

Under some conditions both bacteroids and *Rhizobium* cultures may perhaps derive reducing power from molecular hydrogen via the hydrogenase reaction (Dixon, 1972; Maier *et al.*, 1978; Emerich *et al.*, 1979).

K. RESPIRATION AND CARBOHYDRATE METABOLISM

Burris and Wilson (1939) found that oxygen uptake rates for bacteroid suspensions became saturated at gas phase p_{O_2} values of 0.02–0.03, as compared with 0.1–0.15 for aerobic cultures of fast-growing rhizobia. More recently, detailed studies have shown that bacteroid suspensions respond in a complex manner to alterations in p_{O_2}. The results of Bergersen and Turner (1975, 1980) with *G. max* and *V. unguiculata* bacteroids and nitrogenase-active *Rhizobium* cultures indicated that the rhizobia had at least two terminal oxidases or "oxidase affinity states," one of which was insensitive to carbon dioxide and functioned most efficiently at dissolved oxygen levels of about 0.1 μM. It was suggested that the nitrogenase activity of bacteroids was controlled via the ATP/ADP ratio; however, this was not supported by the work of Laane *et al.* (1978) who observed that *P. sativum* bacteroids maintained high ATP/ADP ratios at dissolved oxygen levels of up to 20 μM, well above the optimal range for acetylene reduction activity.

Nitrate was the only electron acceptor out of a range of compounds tested by Daniel and Appleby (1972) that permitted anaerobic growth of *Rhizobium* cultures. Bacteroids from many sources have been shown to contain nitrate reductase activity, and it has been argued that this enzyme may serve either an assimilatory function, providing an alternative source of reduced nitrogen for plant growth (e.g., Randall *et al.*, 1978), or a respiratory function as a sink for excess reducing power under anaerobic conditions (e.g., Vance *et al.*, 1979). Antoun *et al.* (1980) observed no correlation between symbiotic nitrogen-fixing activity in *M. sativa* nodules and the nitrate reductase activity of *Rhizobium* cultures; however, nodule nitrate reductase levels have been shown to increase at late stages of nodule development (Randall *et al.*, 1978; Broughton *et al.*, 1978) and after plant defoliation (Vance *et al.*, 1979), indicating a possible role during nodule senescence.

Sucrose is believed to be the main carbohydrate source for nodules (reviewed by Rawsthorne *et al.*, 1980), but the forms of carbohydrate supplied to bacteroids *in vivo* have not yet been defined. Labeling data for *P. vulgaris* plants exposed to $^{14}CO_2$ at two different light intensities were consistent with bacteroid uptake of sugars, organic acids including malate, or both (Antoniw and Sprent, 1978). Bacteroids prepared in high-osmolarity media generally respire at high rates without added substrates, but *G. max* bacteroids rinsed in low-osmolarity media respire at maximal rates when supplied with organic acids including malate and succinate (reviewed by Bergersen, 1977). The major organic acids in *G. max* nodules are malate and malonate (Stumpf and Burris, 1979). In other legumes the substrate requirements of isolated bacteroids appear to be quite variable. The acetylene reduction activity of *P. vulgaris* bacteroids is more stimulated by glucose than by succinate at gas-phase p_{O_2} values of 0.02–0.04, but at p_{O_2} values of 0.05–0.10 the relative stimulation is reversed (Trinchant and Rigaud, 1979). The acetylene reduction activity of *L. angustifolius* bacteroids at optimal p_{O_2} values is stimulated by glucose but inhibited by succinate (B. D. Shaw, unpublished observations). Bacteroids from *P. sativum* are unable to take up ^{14}C-labeled glucose, but they actively take up ^{14}C-labeled succinate (Hudman and Glenn, 1980).

Rhizobium mutants characterized as deficient in glucokinase, pyruvate carboxylase, fructose uptake, and the Entner–Doudoroff pathway all form Eff$^+$ nodules on *T. pratense* (Ronson and Primrose, 1979), while *Rhizobium* mutants deficient in C_4 organic acid uptake form Eff$^-$ nodules, although the structural development of the bacteroids appears normal (C. W. Ronson, unpublished observations). These results indicate that C_4 organic acids are essential bacteroid substrates, while sucrose, glucose, and fructose are not; however, it should be noted that *Rhizobium* cultures can utilize several carbon sources simultaneously (de Hollander and Stouthamer, 1979) and that multiple carbon sources can result in a manyfold stimulation of acetylene reduction activity (e.g., Kurz and LaRue, 1975; Wilcockson and Werner, 1979). Similar interactions may perhaps occur in nodules.

Most of the enzymes of the tricarboxylic acid (TCA) and glyoxalate cycles have been identified in bacteroids from various sources (Johnson *et al.*, 1966; Kurz and LaRue, 1977). Early results by Bergersen (1958) indicated that *G. max* bacteroids did not fully oxidize organic acids. More recently, Stovall and Cole (1978) showed that *G. max* bacteroids carried out many reactions consistent with the TCA cycle and a partial glyoxalate cycle. It was estimated that [^{14}C]acetate was 50% utilized via the citrate synthetase reaction and 50% via the malate synthetase reaction.

L. LIPID METABOLISM

The neutral lipid and phospholipid components of *Lotus pedunculatus* bacteroids were similar to those from *Rhizobium* cultures, although the bacteroids had lower PHB and free fatty acid levels and higher glyceride levels (Gerson and

Patel, 1975). Bacteroids from *G. max* nodules were reported to contain oleic acid as a major lipid component after saponification (Johnson *et al.*, 1966); however, it seems likely that this component was in fact *cis*-vaccenic acid (Gerson *et al.*, 1975). In *Vicia faba* and *L. luteus* nodules the bacteroid PHB content has been shown to decline during maturation and increase during senescence (Kretovich *et al.*, 1977). In *G. max* bacteroids the PHB content increases to 45–50% (w/w) during maturation and then stays relatively constant (Wong and Evans, 1971). Bacteroids with differing PHB contents can be separated by density gradient centrifugation (Ching *et al.*, 1977).

It has been suggested that PHB functions as a reserve polymer for bacteroid metabolism (e.g., Kretovich *et al.*, 1977; Gerson *et al.*, 1978). Bacteroids from *G. max* have been shown to contain a PHB depolymerase activity (Wong and Evans, 1971), and 3-hydroxybutyrate dehydrogenase has been observed in bacteroids from several legume species (e.g., Gerson *et al.*, 1978). Presumably, the acetoacetate formed from 3-hydroxybutyrate feeds into the glyoxalate cycle. One of the key enzymes for this cycle, isocitrate lyase, was absent in mature bacteroids of seven legume species (Johnson *et al.*, 1966) but was detected in *G. max* nodules during age-, darkness-, or detachment-induced senescence (Wong and Evans, 1971). The PHB levels in *Rhizobium* cultures have been shown to decrease during log-phase growth and increase during the stationary phase (Patel and Gerson, 1974).

M. Bacteroid Senescence

Electron microscope observations with *Trifolium* spp. (e.g., Mosse, 1964), *Medicago* spp. (e.g., Vance *et al.*, 1980), and *P. sativum* (e.g., Kijne, 1975) have shown that most bacteroids disintegrate during nodule senescence; however, published studies with *G. max* nodules have indicated that many senescent bacteroids remain intact (Tu, 1975; Werner *et al.*, 1980). It has been suggested by Thornton (1930) and Allen and Allen (1940) that many of the rhizobia from decomposing nodules reenter the soil as bacteria. Most recent workers appear to have discounted this possibility, but in view of the strong evidence for the viability of mature bacteroids (Section IV,C). The microbiology of senescing nodules should be carefully reexamined. The biological, structural, and biochemical literature relating to nodule senescence has been reviewed in detail by Sutton (1981).

REFERENCES

Agarwal, A. K., and Mehta, S. L. (1974). *Biochem. Biophys. Res. Commun.* **60**, 257–265.
Akkermans, A. D. L., Abdulkadir, S., and Trinick, M. J. (1978). *Nature (London)* **274**, 190.

Allen, O. N., and Allen, E. K. (1940). *Bot. Gaz.* **102,** 121-142.

Almon, L. (1933). *Zentralbl. Bakteriol. Parasitenk. Infektionskr. Hyg. Abt.* **87,** 289-297.

Antoniw, L. D., and Sprent, J. I. (1978). *Phytochemistry* **17,** 675-678.

Antoun, H., Bordeleau, L. M., Prevost, D., and Lachance, R. A.) (1980). *Can. J. Plant Sci.* **60,** 209-212.

Appleby, C. A. (1962). *Biochim. Biophys. Acta* **60,** 226-235.

Appleby, C. A. (1969). *Biochim. Biophys. Acta* **188,** 222-229.

Appleby, C. A., Turner, G. L., and MacNicol, P. K. (1975). *Biochim. Biophys. Acta* **387,** 461-474.

Aufeuvre, M. A. (1973). *C.R. Acad. Sci. Paris D* **277,** 921-924.

Avissar, Y. J., and Nadler, K. D. (1978). *J. Bacteriol.* **135,** 782-789.

Bassett, B., Goodman, R. N., and Novacky, A. (1977). *Can. J. Microbiol.* **23,** 573-582.

Baulcombe, D., and Verma, D. P. S. (1978). *Nucleic Acids Res.* **5,** 4141-4153.

Becking, J. H. (1979). *Plant Soil* **51,** 289-296.

Bednarski, M. A., and Reporter, M. (1978). *Appl. Environ. Microbiol.* **36,** 115-120.

Bergersen, F. J. (1955). *J. Gen. Microbiol.* **13,** 411-419.

Bergersen, F. J. (1958). *J. Gen. Microbiol.* **19,** 312-323.

Bergersen, F. J. (1974). *In* "The Biology of Nitrogen Fixation" (A. Quispel, ed.), pp. 473-498. North-Holland Publ., Amsterdam.

Bergersen, F. J. (1977). *In* "A Treatise on Dinitrogen Fixation" (R. W. F. Hardy, ed.), Sect. 3 "Biology" (W. S. Silver, ed.), pp. 519-555. Wiley (Interscience), New York.

Bergersen, F. J., and Goodchild, J. (1973). *Aust. J. Biol. Sci.* **26,** 741-756.

Bergersen, F. J., and Turner, G. L. (1967). *Biochim. Biophys. Acta* **141,** 507-515.

Bergersen, F. J., and Turner, G. L. (1975). *J. Gen. Microbiol.* **91,** 345-354.

Bergersen, F. J., and Turner, G. L. (1976). *Biochem. Biophys. Res. Commun.* **73,** 524-531.

Bergersen, F. J., and Turner, G. L. (1980). *J. Gen. Microbiol.* **118,** 235-252.

Bergersen, F. J., Turner, G. L., Gibson, A. H., and Dudman, W. F. (1976). *Biochim. Biophys. Acta* **444,** 164-174.

Bieberdorf, F. W. (1938). *J. Am. Soc. Agron.* **30,** 375-389.

Bishop, P. E., Guevara, J. G., Engelke, J. A., and Evans, H. J. (1976). *Plant Physiol.* **57,** 542-546.

Bisseling, T., Van Den Bos, R. C., Van Kammen, A., Van Der Ploeg, M., Van Duijn, P., and Houwers, A. (1977). *J. Gen. Microbiol.* **101,** 79-84.

Bisseling, T., Van Den Bos, R. C., Weststrate, M. W., Hakkaart, M. J. J., and Van Kammen, A. (1979). *Biochim. Biophys. Acta* **562,** 515-526.

Bisseling, T., Moen, A. A., Van Den Bos, R. C., and Van Kammen, A. (1980). *J. Gen. Microbiol.* **118,** 377-381.

Boland, M. J., Fordyce, A. M., and Greenwood, R. M. (1978). *Aust. J. Plant Physiol.* **5,** 553-559.

Brenchley, W. E., and Thornton, H. G. (1925). *Proc. R. Soc. London Ser. B* **98,** 373-399.

Brewin, N. J., Beringer, J. E., Buchanan-Wollaston, A. V., Johnston, A. W. B., and Hirsch, P. R. (1980). *J. Gen. Microbiol.* **116,** 261-270.

Broughton, W. J., and Dilworth, M. J. (1971). *Biochem. J.* **125,** 1075-1080.

Broughton, W. J., Hoh, C. H., Behm, C. A., and Tung, H. F. (1978). *Planta* **139,** 183-192.

Brown, C. M., and Dilworth, M. J. (1975). *J. Gen. Microbiol.* **86,** 39-48.

Brunchorst, J. (1885). *Ber. Dtsch. Bot. Ges.* **3,** 241-257.

Burris, R. H., and Wilson, P. W. (1939). *Cold Spring Harbor Symp. Quant. Biol.* **7,** 349-361.

Butler, G. W., and Bathurst, N. O. (1958). *Aust. J. Biol. Sci.* **11,** 529-537.

Caldwell, B. E., and Vest, H. G. (1968). *Crop Sci.* **8,** 680-682.

Caldwell, B. E., and Vest, H. G. (1977). *In* "A Treatise on Dinitrogen Fixation" (R. W. F. Hardy, ed.), Sect. 3 "Biology" (W. S. Silver, ed.), pp. 557-576. Wiley (Interscience), New York.

Cannon, F. C. (1974). *Heredity* **33,** 133.

Carter, K. R., Rawlings, J., Orme-Johnson, W. H., Becker, R. R., and Evans, H. J. (1980). *J. Biol. Chem.* **255,** 4213-4223.

Ching, T. M., Hedtke, S., and Newcomb, W. (1977). *Plant Physiol.* **60**, 771-774.

Coventry, D. R., and Dilworth, M. J. (1975). *J. Gen. Microbiol.* **90**, 69-75.

Coventry, D. R., and Dilworth, M. J. (1976). *Biochim. Biophys. Acta* **447**, 1-10.

Coventry, D.R., Trinick, M. J., and Appleby, C. A. (1976). *Biochim. Biophys. Acta* **420**, 105-111.

Cowles, J. R., Evans, H. J., and Russell, S. A. (1969). *J. Bacteriol.* **97**, 1460-1465.

Craig, A. S. (1974). *Histochemistry* **42**, 141-144.

Craig, A. S., and Williamson, K. I. (1972). *Arch. Mikrobiol.* **87**, 165-171.

Craig, A. S., Greenwood, R. M., and Williamson, K. I. (1973). *Arch. Mikrobiol.* **89**, 23-32.

Cutting, J. A., and Schulman, H. M. (1969). *Biochim. Biophys. Acta* **192**, 486-493.

Daniel, R. M. (1979). *J. Gen. Microbiol.* **110**, 333-337.

Daniel, R. M., and Appleby, C. A. (1972). *Biochim. Biophys. Acta* **275**, 347-354.

Dart, P. J. (1974). *In* "The Biology of Nitrogen Fixation" (A. Quispel, ed.), pp. 381-429. North-Holland Publ., Amsterdam.

Dart, P. J. (1977). *In* "A Treatise on Dinitrogen Fixation" (R. W. F. Hardy, ed.), Section 3 "Biology" (W. S. Silver, ed.), pp. 367-472. Wiley (Interscience), New York.

Dart, P. J., and Mercer, F. V. (1963a). *Arch. Mikrobiol.* **46**, 382-401.

Dart, P. J., and Mercer, F. V. (1963b). *Arch. Mikrobiol.* **47**, 1-18.

Dart, P. J., and Mercer, F. V. (1966). *J. Bacteriol.* **91**, 1314-1319.

Dazzo, F. B. (1980). *In* "Advances in Legume Science" (R. J. Summerfield and A. H. Bunting, eds.), pp. 49-59. R. Botanic Gardens, Kew.

Degenhardt, T. L., LaRue, T. A., and Paul, E. A. (1976). *Can. J. Bot.* **54**, 1633-1636.

De Hollander, J. A., and Stouthamer, A. H. (1979). *FEMS Lett.* **6**, 57-59.

De Hollander, J. A., Bettenhaussen, C. W., and Stouthamer, A. H. (1979). *Antonie van. Leeuwenboek* **45**, 401-415.

Dénarié, J. (1969). *C.R. Acad. Sci. Paris D* **269**, 2464-2466.

Dénarié, J., Truchet, G., and Bergeron, B. (1976). *In* "Symbiotic Nitrogen Fixation in Plants" (P. S. Nutman, ed.), pp. 47-61. Cambridge Univ. Press, London and New York.

Devine, T. E., and Weber, D. F. (1977). *Euphytica* **26**, 527-535.

Dilworth, M. J. (1969). *Biochim. Biophys. Acta* **184**, 432-441.

Dilworth, M. J., and Appleby, C. A. (1979). *In* "A Treatise on Dinitrogen Fixation (R. W. F. Hardy, ed.), Section 2 "Biochemistry" (R. C. Burns, ed.), pp. 691-764. Wiley (Interscience), New York.

Dilworth, M. J., and Williams, D. C. (1967). *J. Gen. Microbiol.* **48**, 31-36.

Dixon, R. O. D. (1964). *Arch. Mikrobiol.* **48**, 166-178.

Dixon, R. O. D. (1967). *Arch. Mikrobiol.* **56**, 156-166.

Dixon, R. O. D. (1972). *Arch. Mikrobiol.* **85**, 193-201.

Dunn, S. D., and Klucas, R. V. (1973). *Can. J. Microbiol.* **19**, 1493-1499.

Ebertova, H. (1959). *Nature (London)* **184**, 1046-1047.

Emerich, D. W., Ruiz-Arguesco, T., Ching, T. M., and Evans, H. J. (1979). *J. Bacteriol.* **137**, 153-160.

Evans, W. R., and Crist, D. K. (1980). *Plant Physiol. Suppl.* **65**, 109.

Fred, E. B., Baldwin, I. L., and McCoy, E. (1932). "Root Nodule Bacteria and Leguminous Plants." University of Wisconsin Press, Madison, Wisconsin.

Freidburg, I., and Avigad, G. (1968). *J. Bacteriol.* **96**, 544-553.

Fuchsman, W. H., and Appleby, C. A. (1979). *Biochim. Biophys. Acta* **579**, 314-324.

Gerson, T., and Patel, J. J. (1975). *Appl. Microbiol.* **30**, 193-198.

Gerson, T., Patel, J. J., and Nixon, L. N. (1975). *Lipids* **10**, 134-139.

Gerson, T., Patel, J. J., and Wong, M. N. (1978). *Physiol. Plant.* **42**, 420-424.

Gibson, A. H. (1968). *Aust. J. Agric. Res.* **19**, 907-918.

Gibson, A. H. (1979). *In* "Advances in Legume Science" (R. J. Summerfield and A. H. Bunting, eds.), pp. 69-76. R. Botanic Gardens, Kew.

Gibson, A. H., and Brockwell, J. B. (1968). *Aust. J. Agric. Res.* **19**, 891–905.

Godfrey, C. A., and Dilworth, M. J. (1971). *J. Gen. Microbiol.* **69**, 385–390.

Godfrey, C. A., Coventry, D. R., and Dilworth, M. J. (1975). *In* "Nitrogen Fixation by Free-living Microorganisms" (W. D. P. Stewart, ed.), pp. 311–332. Cambridge Univ. Press, London and New York.

Goodchild, D. J. (1978). *Int. Rev. Cytol. Suppl.* **6**, 235–288.

Goodchild, D. J., and Bergersen, F. J. (1966). *J. Bacteriol.* **92**, 204–213.

Gourret, J. P. (1975). Thesis, Université de Reuue, France.

Gourret, J. P. (1978). *Biol. Cell.* **32**, 299–308.

Gourret, J. P., and Fernandez-Arias, H. (1974). *Can. J. Microbiol.* **20**, 1169–1181.

Greenwood, R. M., and Bathurst, N. O. (1968). *N.Z. J. Sci.* **11**, 280–283.

Greenwood, R. M., and Bathurst, N. Ô. (1978). *N.Z. J. Sci.* **21**, 107–120.

Gresshoff, P. M., and Rolfe, B. G. (1978). *Planta* **142**, 329–333.

Gresshoff, P. M., Skotnicki, M., Eadie, J., and Rolfe, B. G. (1977). *Plant Sci. Lett.* **10**, 299–304.

Guyon, P., Chilton, M. D., Petit, A., and Tempe, J. (1980). *Proc. Natl. Acad. Sci. U.S.A.* **77**, 2693–2697.

Harold, F. M. (1966). *Bacteriol. Rev.* **30**, 772–794.

Higashi, S. (1967). *J. Gen. Appl. Microbiol.* **13**, 391–403.

Holl, F. B., and LaRue, T. A. (1976). *In* "Proceedings of the First International Symposium on Nitrogen Fixation" (W. E. Newton and C. J. Nyman, eds.), pp. 391–399. Washington State Univ. Press, Pullman, Washington.

Hornez, J. P., Courtois, B., Defives, C., and Derieux, J. C. (1974). *C.R. Acad. Sci. Paris D* **278**, 157–160.

Hudman, J. F., and Glenn, A. R. (1980). *Arch. Microbiol.*, in press.

Johnson, G. V., Evans, H. J., and Ching, T. M. (1966). *Plant Physiol.* **41**, 1330–1336.

Johnston, A. W. B., Beynon, J. L., Buchanan-Wollaston, A. V., Setchell, S. M., Hirsch, P. R., and Beringer, J. E. (1978). *Nature (London)* **276**, 635–636.

Jones, G. D., and Hardarson, G. (1979). *Ann. Appl. Biol.* **92**, 221–228.

Jordan, D. C. (1962). *Bacteriol. Rev.* **26**, 119–141.

Jordan, D. C., and Coulter, W. H. (1965). *Can. J. Microbiol.* **11**, 709–720.

Jordan, D. C., and Garrard, E. H. (1951). *Can. J. Bot.* **29**, 360–372.

Jordan, D. C., and Grinyer, I. (1965). *Can. J. Microbiol.* **11**, 721–725.

Jordan, D. C., Yamaura, Y., and MacKague, M. E. (1969). *Can. J. Microbiol.* **15**, 1005–1012.

Kaneshiro, T., Crowall, C. D., and Hanrahan, R. F. (1978). *Int. J. Syst. Bacteriol.* **28**, 27–31.

Keister, D. L. (1975). *J. Bacteriol.* **123**, 1265–1268.

Kennedy, C., Eady, R. R., Kondorosi, E., and Rekosh, D. K. (1976). *Biochem. J.* **155**, 383–389.

Kennedy, L. D. (1981). (In press).

Kidby, D. K., and Goodchild, D. J. (1966). *J. Gen. Microbiol.* **45**, 147–152.

Kijne, J. W. (1975). *Physiol. Plant Pathol.* **7**, 17–21.

Kucas, R. V., Koch, B., Russell, S. A., and Evans, H. J. (1968). *Plant Physiol.* **43**, 1906–1912.

Kondorosi, A., Svab, Z., Kiss, G. B., and Dixon, R. A. (1977). *Mol. Gen. Genet.* **151**, 221–226.

Kretovich, W. L., Melik-Sarkissian, S. S., and Raikchistein, M. V. (1974). *FEBS Lett.* **44**, 305–308.

Kretovich, W. L., Romanov, V. I., Yushkova, L. A., Shramko, V. I., and Fedulova, N. G. (1977). *Plant Soil* **48**, 291–302.

Kurz, W. G. W., and LaRue, T. A. (1975). *Nature (London)* **256**, 407–409.

Kurz, W. G. W., and LaRue, T. A. (1977). *Can. J. Microbiol.* **23**, 1197–1200.

Kurz, W. G. W., Rokosh, D. A., and LaRue, T. A. (1975). *Can. J. Microbiol.* **21**, 1009–1012.

Laane, C., Haaker, H., and Veeger, C. (1978). *Eur. J. Biochem.* **87**, 147–153.

Laane, C., Krone, W., Konings, W. N., Haaker, H., and Veeger, C. (1979). *FEBS Lett.* **103**, 328–332.

Laane, C., Krone, W., Konings, W., Haaker, H., and Veeger, C. (1980). *Eur. J. Biochem.* **103,** 39–46.

Lechtova-Trnka, M. (1931). *Botaniste* **23,** 301–530.

Lees, E. M., and Blakeney, A. B. (1970). *Biochim. Biophys. Acta* **215,** 145–151.

Legocki, R. P., and Verma, D. P. S. (1979). *Science* **205,** 190–193.

Legocki, R. P., and Verma, D. P. S. (1980). *Cell* **20,** 153–163.

Levin, A. P., Funk, H. B., and Tendler, M. D. (1954). *Science* **120,** 784.

Lie, T. A. (1971). *Plant Soil* **34,** 751–752.

Ludwig, R. A., and Signer, E. R. (1977). *Nature (London)* **267,** 245–248.

McComb, J. A., Elliot, J., and Dilworth, M. J. (1975). *Nature (London)* **256,** 409–410.

MacKenzie, C. R., and Jordan, D. C. (1970). *Biochem. Biophys. Res. Commun.* **40,** 1008–1012.

MacKenzie, C. R., Vail, W. J., and Jordan, D. C. (1973). *J. Bacteriol.* **113,** 387–393.

Maier, R. J., Campbell, N. E. R., Hanus, F. J., Simpson, F. B., Russell, S. A., and Evans, H. J. (1978). *Proc. Natl. Acad. Sci., U.S.A.* **75,** 3258–3262.

Marks, I., and Sprent, J. I. (1974). *J. Cell Sci.* **16,** 623–637.

Mosse, B. (1964). *J. Gen. Microbiol.* **36,** 49–66.

Nadler, K. D., and Avissar, Y. J. (1977). *Plant Physiol.* **60,** 433–436.

Newcomb, W. (1976). *Can. J. Bot.* **54,** 2163–2186.

Newcomb, W., Sippell, D., and Peterson, R. L. (1979). *Can. J. Bot.* **57,** 2603–2616.

Nuti, M. P., Ledeboer, A. M., Lepidi, A. A., and Schilperoort, R. A. (1977). *J. Gen. Microbiol.* **100,** 241–248.

Nuti, M. P., Lepidi, A. A., Prakash, R. K., Schilperoort, R. A., and Cannon, F. C. (1979). *Nature (London)* **282,** 533–535.

Nutman, P. S. (1949). *Heredity* **3,** 263–271.

Nutman, P. S. (1969). *Proc. R. Soc. London Ser. B* **172,** 417–437.

O'Gara, F., and Shanmugam, K. T. (1976). *Biochim. Biophys. Acta* **437,** 313–321.

Paau, A. S. (1978). *J. Theoret. Biol.* **74,** 139–142.

Paau, A. S., and Cowles, J. R. (1975). *Plant Physiol.* **56,** 526–528.

Paau, A. S., and Cowles, J. R. (1979). *J. Gen. Microbiol.* **111,** 101–107.

Paau, A. S., Cowles, J. R., and Raveed, D. (1978). *Plant Physiol.* **62,** 526–530.

Paau, A. S., Oro, J., and Cowles, J. R. (1979a). *Plant Physiol.* **63,** 402–405.

Paau, A. S., Oro, J., and Cowles, J. R. (1979b). *Plant Sci. Lett.* **15,** 63–68.

Pagan, J. D., Child, J. J., Scowcroft, W. R., and Gibson, A. H. (1975). *Nature (London)* **256,** 406–407.

Pankhurst, C. E. (1974). *J. Gen. Microbiol.* **82,** 405–413.

Pankhurst, C. E. (1977). *Can. J. Microbiol.* **23,** 1026–1033.

Pankhurst, C. E. (1979). *Microbios* **24,** 19–28.

Pankhurst, C. E., and Craig, A. S. (1978). *J. Gen. Microbiol.* **106,** 207–219.

Pankhurst, C. E., and Craig, A. S. (1979). *J. Gen. Microbiol.* **110,** 177–184.

Pankhurst, C. E., Schwinghamer, E. A., and Bergersen, F. J. (1972). *J. Gen. Microbiol.* **70,** 161–177.

Pankhurst, C. E., Schwinghamer, E. A., Thorne, S. W., and Bergersen, F. J. (1974). *Plant Physiol.* **53,** 198–205.

Patel, J. J. (1978). *Plant Soil.* **49,** 251–257.

Patel, J. J., and Gerson, T. (1974). *Arch. Mikrobiol.* **101,** 211–220.

Peterson, J. B., and Evans, H. J. (1978). *Plant Physiol.* **61,** 909–914.

Phillips, D. A., Daniel, R. M., Appleby, C. A., and Evans, H. J. (1973). *Plant Physiol.* **51,** 136–138.

Planque, K., De Vries, G. E., and Kijne, J. W. (1978). *J. Gen. Microbiol.* **106,** 173–178.

Planque, K., Van Nierop, J. J., and Burgers, A. (1979). *J. Gen. Microbiol.* **110,** 151–159.

Prasade, D. N., and De, D. N. (1971). *Microbios* **4,** 13–20.

Randall, D. D., Russell, W. J., and Johnson, D. R. (1978). *Physiol. Plant.* **44**, 325–328.

Ranga Rao, V., Darrow, R. A., and Keister, D. L. (1978). *Biochem. Biophys. Res. Commun.* **81**, 224–231.

Rawsthorne, S., Minchin, F. R., Summerfield, R. J., Cookson, C., and Coombs, J. (1980). *Phytochemistry* **19**, 341–355.

Reijnders, L., Visser, L., Aalbers, A. M. J., Van Kammen, A., and Houwers, A. (1975). *Biochim. Biophys. Acta* **414**, 206–216.

Robertson, J. G., and Farnden, K. J. F. (1980). *In* "The Biochemistry of Plants: A Comprehensive Treatise" (P. K. Stumpf and E. E. Cohn, eds.), Vol. 5 "Amino Acids: Their Synthesis and Metabolism" (B. J. Miflin, ed.), pp. 65–113. Academic Press, New York.

Robertson, J. G., Farnden, K. J. F., Warburton, M. P., and Banks, J. M. (1975a). *Aust. J. Plant Physiol.* **2**, 265–272.

Robertson, J. G., Warburton, M. P., and Farnden, K. J. F. (1975b). *FEBS Lett.* **55**, 33–37.

Robertson, J. G., Warburton, M. P., Lyttleton, P., Fordyce, A. M., and Bullivant, S. (1978). *J. Cell Sci.* **30**, 151–174.

Rolfe, B. G., and Gresshoff, P. M. (1980). *Aust. J. Biol. Sci.* **33**, 491–504.

Rolfe, B. G., Gresshoff, P. M., Shine, J., and Vincent, J. M. (1980). *Appl. Environ. Microbiol.* **39**, 449–452.

Ronson, C. W., and Primrose, S. B. (1979). *J. Gen. Microbiol.* **112**, 77–88.

Ryan, E., and Fottrell, P. F. (1974). *Phytochemistry* **13**, 2647–2652.

Schwinghamer, E. A. (1967). *Antonie Van Leeuwenboek* **33**, 121–136.

Schwinghamer, E. A. (1968). *Can. J. Microbiol.* **14**, 355–367.

Schwinghamer, E. A. (1969). *Can. J. Microbiol.* **15**, 611–622.

Schwinghamer, E. A. (1970). *Aust. J. Biol. Sci.* **23**, 1187–1196.

Schwinghamer, E. A., and Dennis, E. S. (1979). *Aust. J. Biol. Sci.* **32**, 651–662.

Shaw, B. D., and Sutton, W. D. (1979). *Biochim. Biophys. Acta* **563**, 216–226.

Sidloi-Lumbroso, R., Kleiman, L., and Schulman, H. M. (1978). *Nature (London)* **273**, 558–560.

Skinner, F. A., Roughley, R. J., and Chandler, M. R. (1977). *J. Appl. Bacteriol.* **43**, 287–297.

Skotnicki, M. L., and Rolfe, B. G. (1979). *Aust. J. Biol. Sci.* **32**, 501–517.

Smith, J. D. (1949). *Biochem. J.* **44**, 585–591.

Spitzbarth, M., Puhler, A., and Heumann, W. (1979). *Arch. Microbiol.* **121**, 1–7.

Staphorst, J. L., and Strijdom, B. W. (1971). *Phytophylactica* **3**, 131–136.

Staphorst, J. L., and Strijdom, B. W. (1972). *Phytophylactica* **4**, 29–32.

Stovall, I., and Cole, M. (1978). *Plant Physiol.* **61**, 787–790.

Streeter, J. G. (1977). *Plant Physiol.* **60**, 235–239.

Stripf, R., and Werner, D. (1978). *Z. Naturforsch. C* **33**, 373–381.

Stumpf, D. K., and Burris, R. H. (1979). *Anal. Biochem.* **95**, 311–315.

Sutton, W. D. (1974). *Biochim. Biophys. Acta* **366**, 1–10.

Sutton, W. D. (1980). *Aust. J. Plant Physiol.* **7**, 261–270.

Sutton, W. D. (1981). *In* "Nitrogen Fixation, Vol. 3, Legumes" (W. J. Broughton, ed.). Oxford Univ. Press, London and New York.

Sutton, W. D., and Jepsen, N. M. (1975). *Plant Physiol.* **56**, 665–670.

Sutton, W. D., and Mahoney, P. (1977). *Plant Physiol.* **60**, 800–802.

Sutton, W. D., and Paterson, A. D. (1979). *Plant Sci. Lett.* **16**, 377–385.

Sutton, W. D., and Paterson, A. D. (1980). *Planta* **148**, 287–292.

Sutton, W. D., and Robertson, J. G. (1974). *In* "Mechanisms of Regulation of Plant Growth" (R. L. Bieleski, A. R. Ferguson, and M. M. Cresswell, eds.), pp. 23–30. Bull. 12, R. Soc. New Zealand.

Sutton, W. D., Jepsen, N. M., and Shaw, B. D. (1977). *Plant Physiol.* **59**, 741–744.

Sutton, W. D., Van Den Bos, R. C., and Bisseling, T. (1978). *Plant Sci. Lett.* **12**, 145–149.

Thornton, H. G. (1930). *Proc. R. Soc. London. Ser. B* **106,** 110–122.

Tierney, M., and Schubert, K. R. (1980). *Fed. Proc. Fed. Am. Soc. Exp. Biol.* **39,** 2110.

Tjepkema, J. D., and Evans, H. J. (1975). *Biochem. Biophys. Res. Commun.* **65,** 625–628.

Tjepkema, J. D., and Yocum, C. S. (1974). *Planta* **119,** 351–360.

Trinchant, J. C., and Rigaud, J. (1979). *Physiol. Veg.* **17,** 547–556.

Trinick, M. J. (1973). *Nature (London)* **244,** 459–460.

Trinick, M. J. (1979). *Can. J. Microbiol.* **25,** 565–578.

Trinick, M. J., and Galbraith, J. (1976). *Arch. Microbiol.* **108,** 159–166.

Truchet, G., and Denarie, J. (1973). *C.R. Acad. Sci. Paris D* **277,** 925–928.

Tsien, H. C., and Schmidt, E. L. (1977). *Can. J. Microbiol.* **23,** 1274–1284.

Tsien, H. C., Cain, P. S., and Schmidt, E. L. (1977). *Appl. Environ. Microbiol.* **34,** 854–856.

Tu, J. C. (1975). *Phytopathology* **65,** 447–454.

Tu, J. C. (1977). *Can. J. Bot.* **55,** 35–43.

Tubb, R. S. (1974). *Nature (London)* **251,** 481–485.

Urban, J. E. (1979). *Appl. Environ. Microbiol.* **38,** 1173–1178.

Van Brussell, A. A. N., Planque, K., and Quispel, A. (1977). *J. Gen. Microbiol.* **101,** 51–56.

Van Brussell, A. A. N., Costerton, J. W., and Child, J. J. (1979). *Can. J. Microbiol.* **25,** 352–361.

Vance, C. P., Heichel, G. H., Barnes, D. K., Bryan, J. W., and Johnson, L. E. (1979). *Plant Physiol.* **64,** 1–8.

Vance, C. P., Johnson, L. E. B., Halvorsen, A. M., Heichel, G. H., and Barnes, D. K. (1980). *Can. J. Bot.* **58,** 295–309.

Van Den Bos, R. C., Bisseling, T., and Van Kammen, A. (1978). *J. Gen. Microbiol.* **109,** 131–139.

Verma, D. P. S., and Bal, A. K. (1976). *Proc. Natl. Acad. Sci. U.S.A.* **73,** 3843–3847.

Verma, D. P. S., Nash, D. T., and Schulman, H. M. (1974). *Nature (London)* **251,** 74–77.

Verma, D. P. S., Kazazian, V., Zogbi, V., and Bal, A. K. (1978). *J. Cell Biol.* **78,** 919–936.

Verma, D. P. S., Ball, S., Guerin, C., and Wanamaker, L. (1979). *Biochemistry* **18,** 476–483.

Viands, D. R., Vance, C. P., Heichel, G. H., and Barnes, D. K. (1979). *Crop Sci.* **19,** 905–908.

Vincent, J. M. (1980). *In* "Nitrogen Fixation," pp. 103–129, Vol. II. (W. E. Newton and W. H. Orme-Johnson, eds.). Univ. Park Press, Baltimore, Maryland.

Watson, B., Currier, T. C., Gordon, M. P., Chilton, M. D., and Nester, E. W. (1975). *J. Bacteriol.* **123,** 255–264.

Werner, D. (1978). *Z. Naturforsch. C* **33,** 859–862.

Werner, D., and Morschel, E. (1978). *Planta* **141,** 169–177.

Werner, D., and Stripf, R. (1978). *Z. Naturforsch. C* **33,** 245–252.

Werner, D., Morschel, E., Stripf, R., and Winchenbach, B. (1980). *Planta* **147,** 320–329.

Whiting, M. J., and Dilworth, M. J. (1974). *Biochim. Biophys. Acta* **371,** 337–351.

Widra, A. (1959). *J. Bacteriol.* **78,** 664–670.

Wilcockson, J., and Werner, D. (1979). *Arch. Microbiol.* **122,** 153–159.

Williams, L. F., and Lynch, D. L. (1954). *Agron. J.* **46,** 28–29.

Wilson, M. H. M., Humphrey, B. A., and Vincent, J. M. (1975). *Arch. Microbiol.* **103,** 151–154.

Wittenberg, J. B., Bergersen, F. J., Appleby, C. A., and Turner, G. L. (1974). *J. Biol. Chem.* **249,** 4057–4066.

Wong, P. P., and Evans, H. J. (1971). *Plant Physiol.* **47,** 750–755.

Zurkowski, W., and Lorkiewicz, Z. (1979). *Arch. Microbiol.* **123,** 195–201.

INTERNATIONAL REVIEW OF CYTOLOGY, SUPPLEMENT 13

Exchange of Metabolites and Energy between Legume and *Rhizobium*

JOHN IMSANDE

Department of Genetics, Iowa State University, Ames, Iowa

I.	Introduction	179
II.	Root–Shoot Interactions	180
	A. Translocation	180
	B. Hormonal Regulation of Root–Shoot Interactions	180
III.	Basic Reactions of Dinitrogen Fixation	181
	A. Components	181
	B. Chemistry of Dinitrogen Fixation	182
IV.	Ammonia Incorporation and Translocation	182
V.	Cost of Dinitrogen Fixation	183
VI.	Mass and Composition of Nodules	185
VII.	Summary and Prospects	187
	References	188

I. Introduction

Nitrogen, in a metabolizable form, is often the rate-limiting nutrient in vegetative plant growth. Nodulated legumes, however, can convert atmospheric nitrogen to ammonia, hence provide much of their own dietary nitrogen. Because natural nodulation generally enhances plant growth and yield, agriculturists have long attempted to unravel and exploit the mechanisms controlling nodulation and nitrogen fixation. The alternative to enhancing and diversifying nitrogen fixation is of course application of industrially prepared ammonia or nitrate fertilizer. Industrial production of ammonia, however, requires enormous amounts of energy, basically fossil fuels. Consequently, in the past few years political leaders, too, have come to see the virtue of investigating how nodulated plants can be used to increase nitrogen fixation. There are, however, many important questions that must be answered before optimum nitrogen fixation by nodulated plants can be realized. Four of these questions are central to this chapter: (1) What is the basic mechanism for the conversion of dinitrogen to ammonia? (2) How does the plant process and transport the newly fixed nitrogen? (3) What does it cost the plant, in terms of photosynthate and energy, to reduce and process atmospheric nitrogen? (4) How much tax, in terms of lost photosynthate, can the plant pay for fixing atmospheric nitrogen before plant growth and seed yield are reduced?

Definitive answers to these questions cannot be provided at this time, but possible approaches in obtaining answers to some of these questions are discussed.

II. Root–Shoot Interactions

A. Translocation

A recall of some basic aspects of plant physiology and morphology will hasten significantly our understanding of nitrogen fixation and assimilation in legumes. Vegetative plant growth occurs primarily at the shoot tips and root tips. Obviously, good supplies of photosynthate, minerals, nitrogenous compounds, etc., are required at the sites of rapid growth. Consequently photosynthate, which is formed predominantly in the leaves, must be translocated down to the root tips by the phloem. Likewise, minerals taken up from the soil by the roots must be translocated to the shoot tips by the xylem. Hence most plants can readily transport, in an orderly fashion, nutrients from one segment of the plant to another. In this manner, nitrogenous compounds formed in the nodules are translocated by the xylem to the regions of rapid growth in the shoot, while the energy required to drive nitrogen fixation is provided by photosynthate brought through the phloem to the nodules (Pate, 1980). Of course, the translocation of nutrients in the plant must be carefully coordinated so that the optimum concentration of each nutrient appears at the appropriate site of synthesis. Recent reviews on plant physiology will provide a detailed description of translocation in plants (Zimmermann and Milburn, 1975).

B. Hormonal Regulation of Root–Shoot Interactions

As stated above, translocation of nutrients in plants must be, and is, carefully controlled and coordinated. Although the exact mechanism(s) for the regulation of translocation of specific compounds is generally unknown, plant hormones are thought to be involved. There are five known classes of plant hormones: auxins, cytokinins, gibberellins, abscisic acid (ABA)-like compounds, and ethylene (Letham et al., 1978). One simplistic theory on the regulation of plant growth stipulates that auxins formed at the shoot tip are translocated to the root, while cytokinins formed at the root tips are translocated to the shoot (Sachs, 1972). According to Sachs' (1972) theory, shoot and root growth are coordinated by movement of the appropriate concentrations of these hormones. Gibberellins, which also stimulate plant cell growth, are frequently found in both the phloem and xylem sap (Goodwin et al., 1978). Whether this class of hormones is synthesized primarily in the root, shoot, or both root and shoot remains to be established. ABA, which seems to be formed predominantly in the leaf, plays an important role in reducing water loss during water stress and may inhibit root growth during water stress (Walton, 1980). Although ethylene is often referred to as the "fruit-ripening hormone," it also influences root growth and stimulates stem thickening. The important point for our purpose is that plant hormones do exist and that, by mechanisms that have not been completely elucidated, they help to coordinate and direct both the movement of nutrients within the plant and general plant growth. A precise role, if any, of plant hormones in nitrogen fixation is unknown. There is evidence, however, that auxins, cytokinins, and gibberellins may be involved in the infection process (Libbenga *et al.,* 1973; Syōno *et al.,* 1976), while ABA may inhibit nodulation (Phillips, 1971). Much evidence suggests that plant hormones are present in root nodules (Henson and Wheeler, 1976; Newcomb *et al.,* 1977).

III. Basic Reactions of Dinitrogen Fixation

A. COMPONENTS

The relationship that exists between the legume host and its *Rhizobium* symbiont is dependent upon several highly specific interactions. Once the appropriate *Rhizobium* has recognized its legume host (Graham, this volume), the infection process may begin. Following invasion, the infection thread is formed and nodule development soon follows (Newcomb, this volume). Of course, during nodule development the invading bacteria are transformed into bacteroids (Sutton *et al.,* this volume). In order for a newly formed nodule to be effective (i.e., actively engaged in dinitrogen fixation), it must produce large amounts of leghemoglobin. In fact, approximately 40% of the protein in an effective nodule can be leghemoglobin (Nash and Schulman, 1976). Because effective nodules can account for approximately 5% of the total weight of a soybean plant in full bloom (Latimore *et al.,* 1977), it is clear that a legume must expend a significant proportion of its potential growth resources for the formation of effective nodules. Once a sufficient quantity of leghemoglobin has been synthesized, the globin portion by the host plant (Verma, 1980) and the heme portion by the bacteroids (Godfrey *et al.,* 1975; Nadler and Avissar, 1977), nitrogenase and nitrogenase reductase are synthesized by the bacteroids (Bisseling *et al.,* 1978; Verma *et al.,* 1979). Although the exact function(s) of leghemoglobin is unknown, much evidence suggests that it provides a low level of oxygen to the bacteroid cytochrome chain, while keeping the intracellular level of oxygen sufficiently low to prevent the nitrogenase complex from being inactivated by oxygen (Wittenberg, 1980). Also, the rate of dinitrogen fixation by pure bacterial cultures is known to be influenced inversely by the partial pressure of oxygen (Dilworth, 1980). It is possible therefore that the interactions between oxygen and leghemoglobin may be both directly and indirectly involved in controlling dinitrogen fixation in the symbiotic relationship.

B. CHEMISTRY OF DINITROGEN FIXATION

Dinitrogen is converted into ammonia by the nitrogenase complex contained within the bacteroid. The reaction is $N_2 + 6e^- + 6H^+ \rightarrow 2NH_3$. Ammonia thus formed is excreted immediately by the bacteroid into the nodule cytoplasm. The exact cost, in terms of energy required to drive dinitrogen reduction, can be estimated reasonably accurately. *In vitro* studies using the purified nitrogenase–nitrogenase reductase complex suggest that 12 moles of ATP and 6 electrons are consumed in the production of 2 moles of ammonia (Tso and Burris, 1973; Hageman and Burris, 1978). Passage of 2 moles of electrons through the cytochrome chain normally generates 3 moles of ATP. Hence the 6 moles of electrons are the equivalent of 9 moles of ATP. If dinitrogen fixation were totally energy-efficient, reduction of 1 mole of dinitrogen would consume 21 moles of ATP and produce 2 moles of ammonia. Unfortunately, the nitrogenase system is rather energy-inefficient. The best estimates suggest that perhaps 40% of the energy directed into the nitrogenase complex (i.e., 8.4 ATPs) is lost to protons in the generation of hydrogen gas (Evans *et al.*, 1980; Gutschick, 1980); however, approximately 30% of the energy lost in the formation of hydrogen (i.e., 2.5 ATPs) is subsequently recaptured (Evans *et al.*, 1980). Thus it can be estimated that approximately 27 moles of ATP are directly consumed for each 2 moles of ammonia formed from dinitrogen. In addition, energy must be spent to build and maintain effective nodules. This includes energy and photosynthate provided for bacteroid reproduction and growth. It should be noted also that approximately 0.50 mole-equivalents of acid (H^+) is produced for each mole of nitrogen assimilated (Ashmed and Evans, 1960; Raven and DeMichelis, 1979). Because ammonia assimilation occurs primarily in the nodule, the excess acid can be readily excreted into the soil by the roots. It is interesting to speculate, however, that the spontaneous evolution of hydrogen by the nodules may be promoted by the hydrogen ions produced during ammonia assimilation.

IV. Ammonia Incorporation and Translocation

As stated above, conversion of dinitrogen to ammonia occurs in the bacteroids. Immediately after its formation, the ammonia is excreted into the nodule cytoplasm (Robertson *et al.*, 1975). Although the pathway for ammonia incorporation has not been established for all legumes, the following sequence of reactions is known to take place in some legumes (Scott *et al.*, 1976; Meeks *et al.*, 1978) and probably occurs in most other legumes.

$$2NH_3 + 2 \text{ glutamate} \rightarrow 2 \text{ glutamine} \tag{1}$$

$$\text{Glutamine} + 2\text{-oxoglutarate} \rightarrow 2 \text{ glutamate} \tag{2}$$

$$\text{Glutamate} + \text{oxaloacetate} \rightarrow \text{aspartate} + 2\text{-oxoglutarate} \tag{3}$$

$$\text{Aspartate} + \text{glutamine} \rightarrow \text{asparagine} + \text{glutamate} \tag{4}$$

Net: $2NH_3 + \text{oxaloacetate} \rightarrow \text{asparagine}$

In most legumes asparagine is the most abundant amino acid found in the xylem sap (Streeter, 1972). Furthermore, in many legumes asparagine is also the most abundant nitrogenous compound translocated from the root. In soybean and cowpea, however, ureides (i.e., allantoin and allantoic acid) are the major carriers of ammonia from the nodule (Matsumoto *et al.*, 1977; Streeter, 1979; Pate, 1980). Ureide biosynthesis presumably utilizes glutamine and glycine and proceeds via the standard pathway for *de novo* synthesis of purines (Triplett *et al.*, 1980). During pod fill (i.e., seed formation) the ureides are presumably converted into amino acids to support the formation of seed proteins. Of course energy, in the form of photosynthate, is required for the incorporation and metabolism of the nitrogenous compounds. Also, carbon skeletons (e.g., oxaloacetate and, or related compounds) must be translocated from the leaf to the nodule so that the newly formed inorganic ammonia can be converted to organic nitrogen.

V. Cost of Dinitrogen Fixation

Photosynthate, produced predominantly in the leaf, must be translocated from the leaf to the major sites of biosynthetic activity. Thus photosynthate is provided to the root of both nodulated and nonnodulated legumes. Likewise nitrogen, which is generally either taken up as nitrate from the soil by the roots or obtained from dinitrogen by the nodules, must be translocated from the roots to the growth regions of the stem. Nitrate is used readily by both nodulated and nonnodulated legumes. In fact, most nodulated legumes will restrict their dinitrogen fixation activity if the supply of nitrate is plentiful (Harper and Hageman, 1972; Imsande, 1981). This latter statement implies that legumes have evolved preferentially to use nitrate and secondarily to fix dinitrogen. Restated, it seems that it is economically advantageous for the legume to use soil nitrate rather than to fix dinitrogen. How, then, does the cost of nitrate utilization compare to the cost of dinitrogen fixation? As stated in Section III,B, approximately 27 moles of ATP is consumed for each 2 moles of ammonia formed from dinitrogen. Reduction of 1 mole of nitrate requires 8 moles of electrons (Pate *et al.*, 1979) and generates approximately 0.75 mole-equivalent of hydroxide anion (Raven and DeMichelis, 1979). The 8 moles of electrons consumed per mole of nitrate reduced may be equated to 12 moles of ATP. Thus the cost of nitrate reduction (12 moles ATP/mole of ammonia formed) is comparable to the cost of dinitrogen fixation (approximately 13.5 moles ATP/mole of ammonia formed). When nitrate is the major source of nitrogen provided to the legume, asparagine is still generally the most abundant amino acid in the xylem sap (Streeter, 1979). The concentration of ureides in the xylem sap, however, is greatly reduced (Matsumoto *et al.*, 1977; McClure and Israel, 1979; Pate *et al.*, 1980).

Because asparagine is usually the major vector for the translocation of organic nitrogen from the root, it seems that the cost of ammonia assimilation should be similar regardless of whether nitrate or dinitrogen is the major source of nitrogen. Hence, when nitrate reduction occurs in the roots, the cost of both nitrate reduction and assimilation of the ammonia resulting from this reduction may be very similar to the cost of dinitrogen reduction and utilization (Minchin and Pate, 1973). On the other hand, some legumes transport a large portion of their nitrate directly to the stem and leaves where it is subsequently reduced to ammonia (Pate, 1973; Atkins *et al.*, 1979). Nitrate reduction in the leaves may, at first glance, appear to be less costly because of the availability of inexpensive photosynthetic electrons provided by the ferredoxin system present in the leaves. It has been shown, however, that the reduction of 1 mole of nitrate produces approximately 0.75 mole of base (Raven and DeMichelis, 1979). When nitrate reduction occurs in the roots, the excess base is simply excreted into the soil by the roots (Raven and Smith, 1976). When nitrate reduction occurs in the leaf, energy must be spent in order to maintain the proper pH within the leaf (Raven and Smith, 1976). The cost of maintaining a neutral pH is largely unknown, but it very well may exceed the cost of reducing nitrate in the root.

Why then do legumes prefer to use nitrate rather than fix dinitrogen? Does this suggest that the cost of nodule growth and maintenance may contribute heavily to the total cost of dinitrogen fixation? Recent experiments by Pate and his collaborators (1979) suggest that 69% of the net photosynthate produced by a nitrate-fed white lupine (*Lupinus albus* L.) is converted into plant dry weight. On the other hand, only 57% of the net photosynthate of a nodulated white lupine is converted to dry weight (Pate *et al.*, 1979). If, as suggested above, the cost of dinitrogen reduction is approximately equal to the cost of nitrate reduction, then it might be concluded that 12% of the net photosynthate produced by white lupine is spent for nodule formation and maintenance.

Earlier experiments described the cost of dinitrogen fixation in clover (Gibson, 1966) and cowpea (Herridge and Pate, 1977). For clover it was concluded that only during the early stages of nodule formation and nitrogen fixation was consumption of carbohydrate by nodulated plants significantly greater than that observed in nonnodulated control plants fed ammonium nitrate. For cowpea (Herridge and Pate, 1977) it was concluded that 5 and 12% of the net photosynthate produced by the shoot is consumed by nodule growth and nodule respiration, respectively. Nodules on the garden pea consumed even more photosynthate (Minchin and Pate, 1973). Furthermore, it was concluded that the respiratory efficiency of a nodulated root fixing dinitrogen was very similar to that of a nonnodulated root assimilating nitrate (Minchin and Pate, 1973). Thus it seems much more economical to produce the nitrate and nitrite reductase systems than to construct and maintain nodules. As noted above, however, the actual energy spent in driving dinitrogen reduction is very similar to that spent in driving nitrate reduction. In order to account for the actual cost of dinitrogen fixation, therefore, one must compare the mass and composition of nodules to the mass of the nitrate–nitrite reductase systems.

VI. Mass and Composition of Nodules

Nodules on a seedling are usually first visible 2–3 weeks after plant emergence. On the other hand, nodule senescence usually precedes generalized plant senescence. It is clear therefore that the ratio of nodule mass to total plant mass must change during the course of plant development. For a well-nodulated soybean plant at the beginning pod stage, nodules constitute approximately 4.8% of the total dry weight of the plant (Table I). It should be noted that, while nodules constitute only approximately 5% of the total dry weight of the plant (Latimore *et al.*, 1977), nodule growth and respiration reportedly consume approximately 12% (Pate *et al.*, 1979) to 17% (Minchin and Pate, 1973) of the net

TABLE I
COMPOSITION OF A SOYBEAN PLANT[a]

Plant part	Nitrogen (mg/plant)	Dry weight (gm/plant)	Percentage of total		Nitrogen (%)
			Nitrogen	Dry weight	
Leaves	289	7.95	43.9	29.7	3.6
Petioles	35	2.75	5.3	10.3	1.3
Stem	141	8.85	21.4	33.0	1.6
Roots	120	5.95	18.2	22.2	2.0
Nodules	74	1.30	11.2	4.8	5.7
Total	659	26.80	100.0	100.0	

[a] All values listed were calculated from data presented by Latimore *et al.* (1977).

photosynthate produced by the plant. Why is nodule growth and respiration so costly? Soybean nodules are very rich in protein, especially leghemoglobin (Nash and Schulman, 1976; Legocki and Verma, 1980). Furthermore, it is estimated that 1.2 gm of photosynthate is consumed during the synthesis of 1.0 gm of plant carbohydrate, while 2.5 gm of photosynthate is consumed for the synthesis of 1.0 gm of plant protein (Penning de Vries, 1975). Thus the cost of synthesizing 1 gm of plant protein is approximately twice the cost of synthesizing 1 gm of plant carbohydrate.

At beginning pod stage, nodules contain approximately 11.2% of the total plant nitrogen, while constituting approximately 4.8% of the plant dry weight (Table I). Hence nodules are approximately twice as rich in nitrogen, and presumably protein, as the overall soybean plant. According to Nash and Schulman (1976) 1 gm fresh weight of mature nodules contains approximately 23 mg of soluble protein. It is reported that 1 gm dry weight of soybean nodules is equal to 7 gm fresh weight (Streeter, 1979). Therefore, it can be calculated that 1 gm dry weight of nodules contains approximately 161 mg of soluble protein. An average protein is approximately 16% nitrogen; therefore, 1 gm dry weight of mature soybean nodules contains at least 26 mg of protein nitrogen. Restated, at least 2.6% of the dry weight of a mature soybean nodule is protein nitrogen. Most parts of the soybean plant contain less than 2% total nitrogen (Table I), much of which is free amino acids and ureide (Streeter, 1979). Consequently, soybean nodules are relatively rich in protein, and protein synthesis is relatively costly in terms of photosynthate consumption.

In a soybean nodule actively engaged in nitrogen fixation, the concentration of leghemoglobin may reach 1.5 mM (Bergersen and Goodchild, 1973) and constitute 40% of the total soluble protein in the nodule (Nash and Schulman, 1976). Thus synthesis of leghemoglobin, a nodule-specific protein, is very costly to the plant. Likewise, bacteroid growth and metabolism are energy-demanding. Consequently, nodule construction and maintenance consume a disproportionately large amount of energy and photosynthate. Fortunately, during nodule senescence much of this protein nitrogen is given up to the soil.

Nitrate reduction, on the other hand, occurs entirely within normal plant cells. In both the leaf and the root, nitrate reductase is a cytosolic enzyme, while nitrite reductase is associated with plastids in both tissues (Dalling et al., 1972a,b; Miflin, 1974). Together, nitrate and nitrite reductases constitute only a small percentage of the total plant protein. Hence nitrate reduction requires fewer alterations in the plant than dinitrogen fixation. Consequently, even though the energy required to drive nitrate reduction is similar to that required to drive dinitrogen fixation, the overall cost of dinitrogen fixation is much higher than that for nitrate reduction because of the enormous cost for nodule construction and maintenance.

VII. Summary and Prospects

Recently much emphasis has been placed on increasing the ability of the nodulated legume to fix dinitrogen (Hardy and Havelka, 1975). A major approach has been to engineer new strains of *Rhizobia* that produce increased amounts of nitrogenase activity (Ausubel *et al.*, 1977). The rationale seems to be that the bacterium, or bacteroid, determines the rate of dinitrogen fixation during the symbiotic relationship (Lim and Shanmugan, 1979; Lim *et al.*, 1979). While it is clear that some rhizobial strains fix dinitrogen at a higher rate than others, it is also clear that the rate of dinitrogen fixation in the nodule is not directly proportional to the rate of dinitrogen fixation by the derepressed culture (Lindemann and Ham, 1979). Instead, it seems clear that the legume host controls the rate of dinitrogen fixation. Indeed, dinitrogen fixation is slowly but gradually repressed when a small amount of nitrate is added to a well-nodulated plant that has been actively fixing dinitrogen (Imsande, 1981). Even more striking, the daily rate of dinitrogen fixation by an individual plant, under carefully controlled conditions, may fluctuate severalfold (Imsande, 1981). Hence, if one is to enhance dinitrogen fixation in legumes, one must understand how the legume host controls the nitrogenase activity of the bacteroid.

Dinitrogen fixation by pure bacterial cultures derepressed for nitrogenase activity is greatly influenced by the partial pressure of oxygen, by the presence of ammonia (Dilworth, 1980), and by the presence of nitrate (Hom *et al.*, 1980). Once the mechanisms for inhibition by these materials have been established for derepressed bacterial cultures, perhaps the plant physiologist can attempt to establish whether or not the same mechanisms are operative in the nodulated legume and, if so, how the plant controls them.

It is also abundantly clear that nodule construction and maintenance are major costs in dinitrogen fixation by the nodulated legume (Pate *et al.*, 1979). It should be recalled, however, that the roots of many plants possess two distinct pathways for electron transport. Each pair of electrons passed through the standard electron transport pathway generates 3 molecules of ATP whereas each pair of electrons passed through the alternate pathway produces only 1 molecule of ATP. Lambers (1980) has proposed that the function of the alternate pathway is to dispose of excess photosynthate that has been translocated into the roots. If the alternate pathway functions solely to rid the root of excess photosynthate, then perhaps the energy required for nodulation and dinitrogen fixation could be provided with little real expense to the plant (Lambers *et al.*, 1980). Regardless of the source of photosynthate, the cost to the plant for dinitrogen fixation is minimized when the plant contains relatively few nodules, each of which functions at or near a maximum rate in dinitrogen fixation. Only by producing a few large nodules per plant and allowing each nodule to function at or near its maximum rate can the plant afford to provide more of its dietary nitrogen without reducing plant size and plant yield. Hence, in the author's view, it is these areas that must be exploited if the legume is to assume a larger role in producing organic nitrogen for a hopeful world. The alternative to enhanced dinitrogen fixation is increased commercial manufacture of ammonia and/or, nitrate. Unfortunately, both the chemical production and the application of these compounds to the soil are very expensive processes. Thus an intensified effort to enhance dinitrogen fixation seems wise.

ACKNOWLEDGMENT

The author is indebted to Dr. Edward J. Ralston for helpful suggestions provided during the preparation of this chapter.

REFERENCES

Ashmed, S., and Evans, H. J. (1960). *Soil Sci.* **90**, 205–210.
Atkins, C. A., Pate, J. S., and Layzell, D. B. (1979). *Plant Physiol.* **64**, 1078–1082.
Ausubel, F., Riedel, G., Cannon, F., Peskin, A., and Margolskee, R. (1977). *In* "Genetic Engineering for Nitrogen Fixation" (A. Hollaender, R. H. Burris, P. R. Day, R. W. F. Hardy, D. R. Helinski, M. R. Lamborg, L. Owens, and R. C. Valentine, eds.), Vol. 9, pp. 111–128. Plenum, New York.
Bergersen, F. J., and Goodchild, D. J. (1973). *Aust. J. Biol. Sci.* **26**, 741–756.
Bisseling, T., Van Den Bos, R. C., and Van Kammen, A. (1978). *Biochim. Biophys. Acta* **539**, 1–11.
Dalling, M. J., Tolbert, N. E., and Hageman, R. H. (1972a). *Biochim. Biophys. Acta* **283**, 505–512.
Dalling, M. J., Tolbert, N. E., and Hageman, R. H. (1972b). *Biochim. Biophys. Acta* **283**, 513–519.
Dilworth, M. J. (1980). *In* "Nitrogen Fixation" (W. E. Newton and W. H. Orme-Johnson, eds.), Vol. II, pp. 3–31. Univ. Park Press, Baltimore, Maryland.
Evans, H. J., Emerich, D. W., Ruiz-Argueso, T., Maier, R. J., and Albrecht, S. L. (1980). *In* "Nitrogen Fixation" (W. E. Newton and W. H. Orme-Johnson, eds.), Vol. II, pp. 69–86. Univ. Park Press, Baltimore, Maryland.
Gibson, A. H. (1966). *Aust. J. Biol. Sci.* **19**, 499–515.
Godfrey, C. A., Coventry, D. R., and Dilworth, M. J. (1975). *In* "Nitrogen Fixation by Free-Living Microorganisms" (W. D. P. Stewart, ed.), pp. 311–332. Cambridge Univ. Press, London and New York.
Goodwin, P. B., Gollnow, B. I., and Letham, D. S. (1978). *In* "Phytohormones and Related Compounds: A Comprehensive Treatise" (D. S. Letham, P. B. Goodwin, and T. J. V. Higgins, eds.), Vol. II, pp. 215–249. Elsevier, Amsterdam.
Gutschick, V. P. (1980). *In* "Nitrogen Fixation" (W. E. Newton and W. H. Orme-Johnson, eds.), Vol. I, pp. 17–27. Univ. Park Press, Baltimore, Maryland.
Hageman, R. V., and Burris, R. H. (1978). *Proc. Natl. Acad. Sci. U.S.A.* **75**, 2699–2702.
Hardy, R. W. F., and Havelka, V. D. (1975). *Science* **188**, 633–643.
Harper, J. E., and Hageman, R. H. (1972). *Plant Physiol.* **49**, 146–154.
Henson, I. E., and Wheeler, C. T. (1976). *New Phytol.* **76**, 433–439.
Herridge, D. F., and Pate, J. S. (1977). *Plant Physiol.* **60**, 759–764.

Hom, S. S. M., Hennecke, H., and Shanmugam, K. T. (1980). *J. Gen. Microbiol.* **117,** 169–179.
Imsande, J. (1981). *Plant Physiol.* In press.
Lambers, H. (1980). *Plant, Cell Environ.* **3,** 293–302.
Lambers, H., Layzell, D. B., and Pate, J. S. (1980). *Physiol. Plant.* **45,** 351–356.
Latimore, M., Jr., Giddens, J., and Ashley, D. A. (1977). *Crop Sci.* **17,** 399–404.
Legocki, R. P., and Verma, D. P. S. (1980). *Cell* **20,** 153–163.
Letham, D. S., Goodwin, P. B., and Higgins, T. J. V. (1978). "Phytohormones and Related Compounds: A Comprehensive Treatise," Vol. I and II. Elsevier, Amsterdam.
Libbenga, K. R., Van Iren, F., Bogers, R. J., and Schraag-Lamers, M. F. (1973). *Planta* **114,** 29–39.
Lim, S. T., and Shanmugam, K. T. (1979). *Biochim. Biophys. Acta* **584,** 479–492.
Lim, S. T., Hennecke, H., and Scott, D. B. (1979). *J. Bacteriol.* **139,** 256–263.
Lindemann, W. C., and Ham, G. E. (1979). *Soil Sci. Soc. Am. J.* **43,** 1134–1137.
McClure, P. R., and Israel, D. W. (1979). *Plant Physiol.* **64,** 411–416.
Matsumoto, T., Yatazawa, M., and Yamamoto, Y. (1977). *Plant Cell Physiol.* **18,** 613–624.
Meeks, J. C., Wolk, C. P., Schilling, N., and Shaffer, P. W. (1978). *Plant Physiol.* **61,** 980–983.
Miflin, B. (1974). *Plant Physiol.* **54,** 550–555.
Minchin, F. R., and Pate, J. S. (1973). *J. Exp. Bot.* **24,** 259–271.
Nadler, K. D., and Avissar, Y. J. (1977). *Plant Physiol.* **60,** 433–436.
Nash, D. T., and Schulman, H. M. (1976). *Can. J. Bot.* **54,** 2790–2797.
Newcomb, W., Syono, K., and Torrey, J. G. (1977). *Can. J. Bot.* **55,** 1891–1907.
Pate, J. S. (1973). *Soil Biol. Biochem.* **5,** 109–119.
Pate, J. S. (1980). *In* "Annual Review of Plant Physiology" (W. R. Briggs, P. B. Green, and R. L. Jones, eds.), Vol. 31, pp. 313–340. *Annual Rev.*
Pate, J. S., Layzell, D. B., and Atkins, C. A. (1979). *Plant Physiol.* **64,** 1083–1088.
Pate, J. S., Atkins, C. A., White, S. T., Rainbird, R. M., and Woo, K. C. (1980). *Plant Physiol.* **65,** 961–965.
Phillips, D. A. (1971). *Planta* **100,** 181–190.
Penning de Vries, F. W. T. (1975). *In* "Photosynthesis and Productivity in Different Environments" (J. P. Cooper, ed.), pp. 459–480. Cambridge Univ. Press, London and New York.
Raven, J. A., and DeMichelis, M. I. (1979). *Plant Cell Environ.* **2,** 245–257.
Raven, J. A., and Smith, F. A. (1976). *New Phytol.* **76,** 415–431.
Robertson, J. G., Farnden, K. J. F., Warbarton, M. P., and Banke, J. M. (1975). *Aust. J. Plant Physiol.* **2,** 265–272.
Sachs, T. (1972). *J. Theor. Biol.* **37,** 353–361.
Scott, D. B., Farnden, K. J. F., and Robertson, J. G. (1976). *Nature (London)* **263,** 703–705.
Streeter, J. G. (1972). *Agron. J.* **64,** 315–319.
Streeter, J. G. (1979). *Plant Physiol.* **63,** 478–480.
Syōno, K., Newcomb, W., and Torrey, J. G. (1976). *Can. J. Bot.* **54,** 2155–2162.
Triplett, E. W., Blevins, D. G., and Randall, D. D. (1980). *Plant Physiol.* **65,** 1203–1206.
Tso, M. Y. W., and Burris, R. H. (1973). *Biochim. Biophys. Acta* **309,** 263–270.
Verma, D. P. S. (1980). *In* "Genome Organization and Expression in Plants" (C. J. Leaver, ed.), pp. 439–452. Plenum, New York.
Verma, D. P. S., Ball, S., Guerin, C., and Wanamaker, L. (1979). *Biochemistry* **18,** 476–483.
Walton, D. C. (1980). *In* "Annual Review of Plant Physiology" (W. R. Briggs, P. B. Green, and R. L. Jones, ed.), Vol. 31, 453–489. *Annual Rev.*
Wittenberg, J. B. (1980). *In* "Nitrogen Fixation" (W. E. Newton and W. H. Orme-Johnson, eds.), Vol. II, pp. 53–67. Univ. Park Press, Baltimore, Maryland.
Zimmermann, M. H., and Milburn, J. A. (1975). *In* "Encyclopedia of Plant Physiology" (A. Pirson and M. H. Zimmermann, eds.), Vol. I, pp. 3–473. Springer-Verlag, Berlin and New York.

The Genetics of *Rhizobium*

Adam Kondorosi

*Institute of Genetics, Biological Research Center, Hungarian Academy of Sciences,
Szeged, Hungary*

Andrew W. B. Johnston

John Innes Institute, Colney Lane, Norwich, England

I.	Introduction	191
II.	Mutant Isolation	192
III.	Gene Transfer Systems	196
	A. Conjugation	196
	B. Transformation	205
	C. Transduction	207
IV.	Chromosomal Mapping	208
	A. Construction of Chromosomal Linkage Maps	208
	B. Homology of Chromosomes of Different *Rhizobium* Species	210
	C. Localization of Nonselectable Mutations	211
	D. Replication Mapping	212
V.	Arrangement of Genes in *Rhizobium*	212
	A. Chromosomal Genes	212
	B. Plasmid Genes	213
VI.	Genetics of Rhizobiophages	214
	A. Phage–Bacterium Interactions	214
	B. Genetic and Physical Maps of the Phage Genome	216
	C. Transfection	217
VII.	Strain Construction in *Rhizobium*	217
VIII.	Conclusions	218
	References	219

I. Introduction

In recent years there has been an increasing awareness of the importance of the process of biological nitrogen fixation, due largely to the fact that the production of nitrogenous fertilizer by chemical means is an energetically demanding procedure. Of the nitrogen-fixing bacteria, *Rhizobium* can be considered the most important genus. Because it forms an intimate symbiotic relationship with leguminous plants and is thus provided with energy in the form of photosynthate, *Rhizobium* in the root nodules fixes nitrogen at a higher rate than free-living nitrogen-fixing bacteria or fixing bacteria that form loose associations with plants. The fact that among the legumes nodulated by *Rhizobium* are included some of the world's most important crop plants further emphasizes the importance of this group of bacteria.

This increased interest in biological nitrogen fixation has occurred at a time during which there have been advances in the methods available for genetic analysis of bacteria, so it is not surprising that *Rhizobium* genetics has developed fairly rapidly.

One of the most important aspects of this research is the demonstration of the crucial role of indigenous *Rhizobium* plasmids as determinants of symbiotic functions (see Dénarié *et al.*, this volume). There has also been progress in the development of methods of chromosomal gene transfer and mapping and *Rhizobium* bacteriophage genetics, and in this chapter we will concentrate on these latter aspects of *Rhizobium* genetics.

II. Mutant Isolation

For the purposes of gene mapping it is clearly important to have available strains of *Rhizobium* which carry readily selectable markers, i.e., antibiotic resistance and auxotrophic mutations.

Spontaneous mutations leading to resistance to various antibiotics have been isolated in a number of *Rhizobium* species. Those that have been used in construction of the genetic maps of *Rhizobium meliloti* and *R. leguminosarum* can be seen in Fig. 1. *Rhizobium* can grow on simple defined media, and auxotrophs of several types have been isolated in several of the fast-growing species following standard techniques of chemical mutagenesis using mutagens such as ethylmethanesulfonate (EMS), nitrous acid, and N-methyl-N'-nitro-N-nitrosoguanidine (NTG). The range of auxotrophs obtained can be gauged from Fig. 1 which shows those that have been used for mapping the chromosomes of strains of *R. meliloti* and *R. leguminosarum* (Meade and Signer, 1977; Kondorosi *et al.*, 1977a; Beringer *et al.*, 1978b). It is noticeable that on these maps there are no representatives of mutations leading to certain nutritional requirements; e.g., there are no Ala⁻, Asp⁻, Asn⁻, Hom⁻, Ile⁻, Lys⁻, Pro⁻, Thr⁻, or Val⁻ strains. Following mutagenesis of *R. leguminosarum*, the most frequent auxotrophs are Pur⁻, Trp⁻, Cys⁻, and Met⁻, and this is the case for both NTG-

---

FIG. 1.   *Rhizobium* linkage maps. Linkage maps for *R. meliloti* 2011 (Meade and Signer, 1977), for *R. meliloti* 41 (Kondorosi *et al.*, 1977a; Sváb *et al.*, 1978; Kiss *et al.*, 1979, 1980; Forrai *et al.*, unpublished results), and for *R. leguminosarum* 300 (Beringer *et al.*, 1978b) were constructed by means of plasmid-mediated recombination. Only one allele is given at each locus. Distances between markers are in additive map units (*d*). Markers which probably correspond to each other are joined by dashed lines; alleles suppressed by the same R-prime carrying Rm2011 chromosomal DNA are connected by solid lines (Kondorosi *et al.*, 1980).

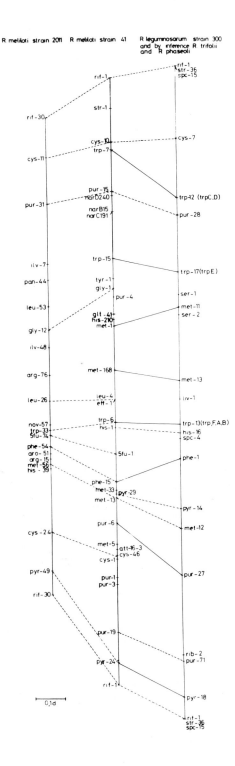

R meliloti strain 2011   R meliloti strain 41   R leguminosarum strain 300
                                                  and by inference R trifolii
                                                  and R phaseoli

0,1 d

and nitrous acid-induced mutations (J. E. Beringer, personal communication). The same observation was made for *R. meliloti* (A. Kondorosi, unpublished result).

There appears to be strain variation in the mutagenicity of particular agents in different strains (Cunningham, 1980). For example, Meade and Signer (1977) used EMS to induce auxotrophs in *R. meliloti* strain 2011, but in *R. leguminosarum* EMS was found to be a very inefficient mutagen (J. E. Beringer, personal communication). The reasons for these various cases of mutation specificity in *Rhizobium* are not known. In slow-growing species of *Rhizobium,* such as *R. japonicum,* very few auxotrophic mutants have been isolated, perhaps because the cells have a tendency to clump (L. D. Kuykendall, personal communication).

More recently a method has been developed which allows transposons to be used as mutagens in *Rhizobium.* One advantage of this is that, when the transposon inserts into a gene, it not only mutates it but also marks it with the phenotype (such as antibiotic resistance) determined by the tranposon. Thus, by mapping the resistance marker, the site of the mutation can be located. This is of particular value for mutants which do not have a phenotype that is recognizable on petri dishes, an obvious example of such a class being symbiotically defective mutants. Other advantages of transposons are that their sites of insertion can also be located by physical methods, such as hybridization with radioactive probes, heteroduplex analysis, and restriction endonuclease mapping. This is of relevance in view of observations that many "symbiotic" genes are located on plasmids and are not amenable to mapping by standard genetic techniques.

The method for the insertion of a transposon into *Rhizobium* relies upon the fact that, although plasmids of the P1 incompatibility group can be stably transferred from *Escherichia coli* to strains of *Rhizobium* at frequencies of $\sim 10^{-2}$ (see below), when the bacteriophage Mu is inserted into such plasmids their establishment is prevented (Boucher *et al.,* 1977). The failure of such hybrid plasmids to become established in *Agrobacterium tumefaciens* was reported by van Vliet *et al.* (1978). They took advantage of this to introduce the transposon Tn7 into this species using the following procedure. A tripartite plasmid comprising the P1 group plasmid RP4, bacteriophage Mu, and Tn7 was constructed in *E. coli.* Selection was then made for the transfer of streptomycin resistance (specified by Tn7) to *A. tumefaciens.* Because the hybrid plasmid was not able to survive in this species, an expected type of progeny from such a cross would be one in which Tn7 had been transposed into the genome of *A. tumefaciens,* and indeed this was found to be the case (van Vliet *et al.,* 1978). Similarly Beringer *et al.* (1978a) constructed another plasmid, pJB4JI, comprising pPH1JI (another P1 group plasmid specifying gentamicin resistance), bacteriophage Mu, and Tn5 (specifying kanamycin resistance). Following the selection of *kan* transfer from an *E. coli* strain carrying this plasmid to strains of *Rhizobium leguminosarum, R.*

trifolii, and *R. phaseoli,* approximately 0.5% of the resistant progeny were auxotrophic, and for those cases studied, the auxotrophic mutation and kanamycin resistance were very closely linked on the chromosome, indicating that insertion of the transposon was responsible for the mutation.

In addition to these mutations used primarily in mapping studies, the isolation of strains defective in their symbiotic phenotype, using two approaches, has been reported. One involves the direct screening of survivors of mutagenesis on plants for symbiotic defects; in the other, selection for particular classes of mutants is made on petri dishes, followed by the screening of such mutants on plants.

The former method was used by Maier and Brill (1976) who isolated NTG-induced mutants of *R. japonicum* which were either unable to nodulate soybeans (Nod⁻) or which induced nodules that failed to fix nitrogen (Fix⁻). Included in the latter group was one mutant that was defective in the enzyme nitrogenase. Also, following NTG mutagenesis Beringer *et al.* (1977) isolated a number of Fix⁻ mutants of *R. leguminosarum,* some of which were temperature-sensitive, being Fix⁺ on peas grown at 15°C but Fix⁻ on peas grown at 26°C. Using Tn5 as a mutagen, Buchanan-Wollaston *et al.* (1980) isolated a number of Fix⁻ mutants in *R. leguminosarum* following the screening of strains in which the transposon had been inserted into a transmissible plasmid known to carry genes concerned with a symbiotic function (Johnston *et al.,* 1978a; Brewin *et al.,* 1980).

Biochemical analysis of the lesions in mutations isolated in this way is no simple matter, since our knowledge of the biochemistry of infection and nodule development is at a very rudimentary stage and consequently there are real difficulties in knowing where to look in any attempts to identify the defective gene product in a particular mutant strain. This problem may be exacerbated by the fact that at least some symbiotic genes may be expressed only during nodule formation, and for these cases it would be impossible to detect any differences between mutant and wild-type bacteria unless they were isolated from the particular stage in nodule development at which the gene was normally expressed. Despite such problems it is to be hoped that a more detailed understanding of the biochemical changes associated with nodule development will be forthcoming and that this knowledge will be used to define more precisely the nature of the defects in *Rhizobium* mutants. Conversely, the characterization of symbiotic mutants will indicate the biochemical changes required for successful nodule development.

The second strategy used for the isolation of symbiotic mutants, in which prior selection is made for a mutant with a phenotype recognizable on petri dishes, has the advantages of relative ease of mutant isolation and biochemical and genetic characterization, but the disadvantage that only a limited range of mutants can be isolated.

Most reports have described the symbiotic characteristics of auxotrophic mutants, mutants resistant to antibiotics, antimetabolites, or bacteriophages, and mutants defective in nitrate reduction, and representatives of such types of mutants have been shown to be symbiotically defective in one or more *Rhizobium* species (reviewed by Dénarié *et al.*, 1976; Beringer *et al.*, 1980). In some cases, reasonable models can account for the pleiotropic effect of such mutants.

For example, the fact that glutamine auxotrophs of *R. meliloti* (Kondorosi *et al.*, 1977b) and of a cowpea *Rhizobium* strain (Ludwig and Signer, 1977) induced nonfixing nodules supports the view that glutamine synthetase may control nitrogenase activity, just as it appears to do in the well-studied system in *Klebsiella pneumoniae* (Streicher *et al.*, 1974; Tubb, 1974).

The importance of the cell surface in the ability of *Rhizobium* to infect and nodulate, possibly due to interactions with plant lectins (Bohlool and Schmidt, 1974; Dazzo and Hubell, 1975), has been substantiated by the use of mutants defective in various components of the cell surface. Mutants of *R. meliloti* resistant to viomycin are Fix⁻ and failed to differentiate into bacteroids (Hendry and Jordan, 1969). Analysis of the cell walls of these strains showed that they had increased amounts of phospholipid and that their transport of cations was also affected (MacKenzie and Jordan, 1970, 1972; Yu and Jordan, 1971). Mutants of *R. leguminosarum* have been isolated directly on the basis of having reduced amounts of exopolysaccharide. The fact that such mutants are defective in the nodulation of peas and that the severity of the defect is inversely proportional to the amount of exopolysaccharide made by a given mutant implicates this component as an important one in the infection process (Sanders *et al.*, 1978; Napoli and Albersheim, 1980).

III. Gene Transfer Systems

In *Rhizobium* the three main types of bacterial gene transfer systems, conjugation, transformation, and transduction, have all been demonstrated. Historically, transformation was described much earlier, and its literature is more abundant than that of the other two methods. Since the genetics of *Rhizobium* was extensively reviewed recently by Schwinghamer (1977), the major emphasis of this section is on the recent progress of this field, achieved mainly in studies on conjugational gene transfer and mapping.

A. Conjugation

Since Dénarié *et al.* (this volume) give a detailed description of *Rhizobium* plasmids, this section deals mainly with the chromosome-mobilizing ability of plasmids in *Rhizobium*. Conjugation is the most suitable method for chromosome mapping in bacteria, since in this process large DNA pieces are transferred. For the establishment of a conjugation system with chromosome mobilization ability (Cma) in a microorganism, one can either look for naturally existing systems where pair formation and gene transfer may occur, or one can assay known conjugative plasmids (either indigenous or exogenous) for chromosome-mobilizing ability.

1. *Naturally Existing Conjugation Systems*

Heumann (1968) described a natural conjugation system in a star-forming, pigmented derivative of *Rhizobium lupini*. Auxotrophic, antibiotic resistance, and pigmentation markers have been located on a circular chromosome, and the nature of the fertility system has been investigated in some detail (Heumann *et al.*, 1971, 1973, 1974; Heumann and Springer, 1977; Schöffl *et al.* 1974). However, because this strain is unable to nodulate and its relatedness to other *Rhizobium* strains has still to be established, these studies have not been useful in the analysis of symbiotic functions.

Naturally existing conjugation systems have been described in other *Rhizobium* species. Higashi (1967) reported that the nodulation ability of a *R. trifolii* strain was lost spontaneously at a fairly high frequency and that this frequency was further increased by treatment with acridine orange, a known plasmid-eliminating agent. Moreover, nodulation ability and host specificity of nodulation was transferred by conjugation from *R. trifolii* to *R. phaseoli*. These results suggested that in this strain of *R. trifolii* some of the genes required for nodulation were located on a plasmid. The chromosome-mobilizing ability of this plasmid has not been reported.

Conjugative gene transfer in *R. trifolii* was reported by Lorkiewicz *et al.* (1971), but no further study on this conjugation system was published. In *R. japonicum* an R plasmid coding for multiple resistance (penicillin, neomycin, chloramphenicol) was detected (Cole and Elkan, 1973). It could be transferred by conjugation with *A. tumefaciens*. Transfer of a resident *R. meliloti* plasmid of 59.6 megadaltons, conferring sensitivity to phage AL1, was reported by Bedmar and Olivares (1980). The Cma property of these plasmids has not been reported.

Recently, the presence of indigenous conjugative plasmids in three *R. leguminosarum* strains was clearly demonstrated (Hirsch, 1979). These plasmids coded for bacteriocin production, and one of them carried genes that determined the ability to induce nitrogen-fixing nodules on peas (Johnston *et al.*, 1978a; Brewin *et al.*, 1980). Each of these plasmids could mobilize chromosomal markers at frequencies of 10^{-7}–10^{-8} per recipient (Hirsch, 1979).

The insertion of transposons into resident *Rhizobium* plasmids will greatly facilitate the detection of transmissible plasmids; this has already been done for some *R. leguminosarum* plasmids (Beringer *et al.*, 1978a; Brewin *et al.*, 1980).

2. *Exogenous Plasmids with Chromosome-Mobilizing Ability in Rhizobium*

Another approach in establishing a chromosome mobilization system in *Rhizobium* is to identify conjugative plasmids which are transmissible to *Rhizobium* and can be maintained in the new host and whose transfer genes are expressed. Once such a plasmid is found, its chromosome mobilizing ability can be tested.

Chromosomal gene transfer promoted by various plasmids has been reported in several bacterial genera (see a review by Holloway, 1979). Although conjugative transfer of exogenous plasmids, such as F-prime factors of *E. coli* K12 to *R. lupini* (Datta and Hedges, 1972; Mergeay *et al.*, 1973) and transfer of an F-like R factor, Rl-19drd, by transformation (Dunican and Tierney, 1973) and by conjugation (Kowalczuk and Lorkiewicz, 1977), has been reported, the breakthrough in gene transfer for mapping of *Rhizobium* was achieved by the use of P1 group plasmids.

R plasmids originally found in *Pseudomonas aeruginosa* and belonging to the P1 incompatibility class were shown to be transmissible between a wide range of gram-negative bacteria, including *R. trifolii* and *R. meliloti* (Datta *et al.*, 1971; Datta and Hedges, 1972). The properties of the P1-type plasmids in *R. leguminosarum* have been described by Beringer (1974). They were transferred by conjugation in liquid medium at a frequency of 10^{-3}–10^{-5}, they were stably maintained and, with the exception of carbenicillin, their antibiotic resistance genes were expressed. However, R-plasmid-mediated recombination could not be demonstrated.

The introduction of plate mating techniques in conjugation experiments resulted in much higher transfer frequencies of these R plasmids (5×10^{-1}–10^{-2} per donor) in different *Rhizobium* strains (Dixon *et al.*, 1976; Jacob *et al.*, 1976; Beringer and Hopwood, 1976; Kondorosi *et al.*, 1977a; Meade and Signer, 1977). In conjugation experiments on a solid agar surface, chromosome mobilization mediated by plasmid RP4 was detected at a very low frequency (10^{-9} per recipient) in different *R. leguminosarum* strains, e.g., R1300 (Beringer and Hopwood, 1976), but at a much higher frequency (10^{-5}–10^{-6}) in *R. meliloti* strain 2011 (Rm2011) (Meade and Signer, 1977). It is not known why RP4 is inefficient in mobilizing the chromosome of *R. meliloti* strain 41 (Rm41) (A. Kondorosi and E. Vincze, unpublished results), although the chromosomes of these two *R. meliloti* strains are highly homologous (Kondorosi *et al.*, 1980). A similar observation was described for *P. aeruginosa:* P1-type plasmids promoted chromosome transfer in strain PAT but were unable to do so in strain PAO (Stanisich and Holloway, 1971).

In a search for plasmids with better Cma properties in *R. leguminosarum,* Beringer and Hopwood (1976) found that a derivative of another P1 plasmid had the required properties. This plasmid (R68.45) was isolated by Haas and Holloway (1976) as a variant of R68 with enhanced chromosome-mobilizing ability in *P. aeruginosa* strain PAO. Physical examination of R68.45 showed that it had acquired an extra piece of DNA of about 1.5 megadaltons (Jacob *et al.*, 1977). This DNA fragment is similar in structure to a transposable element: It consists of a 1.2-megadalton repeat of adjacent R68 sequences and a 0.3-megadalton foreign

DNA fragment linking the direct repetition sequences (J. Leemans, A. Depicker, R. Villarroel, J. Schell, and M. Van Montagu, unpublished results). In view of the involvement of insertion sequences in chromosome mobilization by F plasmids in *E. coli* (Davidson *et al.*, 1975; Ohtsubo and Ohtsubo, 1977), it has been suggested that this DNA fragment acts as an insertion sequence element in the mechanism of chromosome mobilization by R68.45 (Holloway, 1979). The Cma region of R68.45 is rather unstable. The spontaneous loss of Cma is sometimes accompanied by loss of the kanamycin resistance marker (Holloway, 1979; A. Kondorosi, unpublished observation). This is not unexpected, since deletion formation is a common property of insertion sequence elements (Reif and Saedler, 1975).

R68.45 has the ability to mobilize the chromosomes of a series of bacteria, including several *Pseudomonas* species (Martinez and Clarke, 1975; Holloway, 1979), *E. coli* (Beringer and Hopwood, 1976; Kondorosi *et al.*, 1977a), *Rhodopseudomonas sphaeroides* (Sistrom, 1977), and *Rhizobium leguminosarum* (Beringer and Hopwood, 1976), *R. trifolii*, *R. phaseoli* (Johnston and Beringer, 1977), and *R. meliloti* (Kondorosi *et al.*, 1977a; Casadesus and Olivares, 1979a). Although plasmid transfer and chromosome mobilization in slow-growing rhizobia have not yet been reported, recent results indicate that R68.45 can be transferred to *R. japonicum* (Kuykendall, 1979; C. Kennedy, personal communication), and it is to be hoped that this plasmid can mediate chromosomal gene transfer also in these bacteria.

Recently, R68.45-like plasmids were derived from other P1 group plasmids, and all those tested possessed an additional inserted DNA segment, denoted IS*P* (see review by Holloway *et al.*, 1979). It was proposed to refer to such plasmids as enhanced chromosome mobilization (ECM) plasmids. The IS*P* region on ECM plasmids acts as a recognition site for some sequences on the chromosome of several *Pseudomonas* species. It is possible that some ECM plasmids also promote chromosome transfer in *Rhizobium*.

The mechanism by which R68.45 and other ECM plasmids mobilize the chromosome is not known. Although these plasmids should interact somehow with the chromosome during chromosome mobilization and R-prime formation can occur (see Section III,A,3), stable integration of these plasmids into the chromosome has not been observed (Holloway, 1979; Holloway *et al.*, 1979).

R68.45-mediated recombination had three important properties which allowed fairly rapid mapping of the *Rhizobium* chromosome:

1. R68.45 can interact with a large number of chromosomal sites. This was deduced from the observation that recombinants for various markers were produced at about the same frequency (10^{-6} per donor cell) in *R. leguminosarum* (Beringer *et al.*, 1978b) and at frequencies of 10^{-3}–10^{-5} in *R. meliloti* (Kondorosi *et al.*, 1977a; Casadesús and Olivares, 1979a).

2. Large segments of chromosomal DNA can be transferred. These segments were estimated to be as much as 30% (Kondorosi *et al.*, 1977a) and 15% of the chromosomes (Beringer *et al.*, 1978b) of *R. meliloti* and *R. leguminosarum*, respectively.

3. The recombinants were haploid, since they did not segregate and recessive markers were also recovered.

RP4-mediated recombination in *R. meliloti* 2011 showed similar characteristics, with a recombination frequency of $10^{-5}-10^{-6}$ per donor (Meade and Signer, 1977).

3. *Construction of Other Plasmids with High Chromosome-Mobilizing Ability in Rhizobium*

a. *Construction of R-prime Plasmids.*   F plasmids carrying sections of the *E. coli* chromosome (F-primes) are known to promote enhanced chromosome transfer from the homologous site (Pittard and Adelberg, 1963; Low, 1972). It is to be expected that, by analogy, R-primes carrying sections of the *Rhizobium* chromosome would also have improved chromosome mobilizing abilities.

Jacob *et al.* (1976) constructed RP4-primes by inserting *R. leguminosarum* DNA fragments, digested with restriction endonuclease *Eco*RI, into the *Eco*RI cleavage site of RP4. Two such R-primes were isolated and tested for chromosome mobilization in R1300, but no increase in recombinant formation for any marker was detected. In a similar experiment carried out by Julliot and Boistard (1979), 16 R-primes of Rm2011 were constructed by inserting *Rhizobium* DNA, digested with *Hin*dIII, into the *Hin*dIII site of RP4 and 4 of these R-primes promoted enhanced, polarized gene transfer. Since both *R. leguminosarum* and *R. meliloti* strains harbor large indigenous plasmids, the inability of several R-primes to promote gene transfer could be explained by assuming that these primes carried segments of the *Rhizobium* plasmids.

Barth (1978) and Julliot and Boistard (1979) demonstrated that RP4-primes constructed *in vitro* using *E. coli* DNA promoted high-frequency polarized transfer in *E. coli*, although the original RP4 vector had only very weak chromosome mobilizing ability in this host. RP4-primes were constructed also from DNA of Rm41, where RP4 was a very poor sex factor (recombination frequency was $10^{-7}-10^{-9}$ per donor) and several of the R primes promoted enhanced chromosome transfer (E. Vincze, C. Koncz, and A. Kondorosi, unpublished results). These results suggest that, by creating homology between the conjugative plasmid and the host chromosome, a gene transfer system with enhanced and polarized chromosome transfer ability can be established.

F-prime plasmids can be isolated by a number of *in vivo* methods (Low, 1972). In principle these techniques can be applied in constructing prime derivatives of other conjugative plasmids. R-prime derivatives of R68.45 containing chromosomal regions of *P. aeruginosa* have been isolated *in vivo* (Hedges *et al.*, 1977; Hedges and Jacob, 1977; Holloway, 1978) using recombination-deficient *P. aeruginosa* or *E. coli trp* mutants as recipients to prevent the formation of chromosomal recombinants; therefore, transconjugants inheriting both the selected chromosomal marker and the R plasmid were R-primes.

A similar approach was used to isolate R68.45-primes in *Rhizobium* by Johnston *et al.* (1978c). R-primes were generated in matings between *R. meliloti* 2011 and *R. leguminosarum* 300, where the homology of the two chromosomes is so low that chromosomal recombinants are not formed. These R-primes carried different sections of the Rm2011 chromosome. The physical characterization of R-primes showed differences in the size of the inserted DNA (13, 20, and 43 megadaltons, respectively), and the *Rhizobium* DNA was inserted at two different sites in R68.45 (Johnston *et al.*, 1978b). Hu *et al.* (1975) found that in different F-primes the chromosomal DNA was integrated at different plasmid sites, suggesting that several plasmid regions could interact with the chromosome. Although the presence of the Cma region on R68.45 is unambigously responsible for its Cma phenotype, other DNA sequences of the plasmid may also be involved in the chromosome–plasmid interaction during R-prime formation.

The isolation and characterization of an R-prime from *R. meliloti* 41 was described by Kiss *et al.* (1980). This R-prime, pGY1, obtained from intrastrain mating by a simple enrichment procedure, contained a 17.8-megadalton chromosomal fragment carrying the *cys-46*$^+$ marker and the attachment site for bacteriophage 16-3. Integration of phage 16-3 into pGY1 could occur readily, resulting in a fairly stable cointegrate molecule (pGY2). When cells were cured of pGY1, loss of the antibiotic resistance markers was always accompanied by loss of the *cys-46*$^+$ marker, indicating that integration of the *cys-46*$^+$ region into the chromosome did not occur at high frequency. As demonstrated by chromosome mobilization experiments, pGY1 promoted enhanced gene transfer from the *cys-46* region, and a gradient of marker transmission was observed in an anticlockwise direction on the chromosome. On the other hand, recombinants for markers located distantly or clockwise from *cys-46* were formed only at barely detectable frequencies, indicating that pGY1 had lost the Cma function of R68.45. The physical characterization of pGY1 showed that pGY1 lacked two *Sma*I fragments of R68.45, one of which probably contained the Cma region.

The R-primes carrying sections of the Rm2011 chromosome did not promote enhanced gene transfer in R1300 (Johnston *et al.*, 1978c). This is consistent with the demonstrated lack of recombination between the two chromosomes (Johnston and Beringer, 1977; Johnston *et al.*, 1978c; Kondorosi *et al.*, 1980). However, chromosome mobilization experiments involving strains Rm2011 and Rm41 indicated that the two chromosomes were highly homologous (Kondorosi *et al.*, 1980), so the R-primes containing sections of the Rm2011 chromosome were expected to promote enhanced gene transfer from the corresponding site in

Rm41. Eight such R-primes were tested, and all showed increased gene transfer in the vicinity of the respective allele suppressed by the R-prime and, as for pGY1, in all cases a gradient of marker transmission counterclockwise from the homologous region was observed. It is not clear why the R-primes constructed *in vivo* differ from F-primes which can promote polarized transfer in both directions in *E. coli.* There are also several other differences between F-mediated and R68.45-mediated chromosome mobilization (Holloway, 1979; Kiss *et al.,* 1980). In contrast to R68.45-primes constructed *in vivo,* RP4-primes constructed by *in vitro* techniques promoted chromosome transfer either in a clockwise or a counterclockwise direction in Rm2011 (Julliot and Boistard, 1979) and in Rm41 (E. Vincze, C. Koncz, and A. Kondorosi, unpublished results).

b. *Chromosome-Mobilizing Ability of RP4::Mu Replicons.* Another approach in developing oriented gene transfer systems involved the use of RP4::Mu hybrid plasmids. Dénarié *et al.* (1977) and Faelen *et al.* (1977) developed techniques for the insertion of phage Mu into RP4. It was shown in *E. coli* carrying RP4::Mu that, when Mu was also present in the chromosome, the two Mu DNAs provided homology between the two replicons, allowing RP4::Mu-mediated polarized chromosome transfer at frequencies as high as 10^{-4} per donor (Dénarié *et al.,* 1977). RP4::Mu was transferred to *Rhizobium,* and the production of infectious phage particles was detected but only at a very low titer (Boucher *et al.,* 1977). Therefore, this approach may not be successful in this host.

c. *Chromosome-Mobilizing Ability of dnaG-Suppressing Derivative of R68.45.* Recently Ludwig and Johansen (1980) isolated a derivative of R68.45 which could suppress an *E. coli dnaG* mutation. Although this plasmid was no better at chromosome mobilization in *E. coli* than R68.45, in *R. meliloti* it promoted the formation of recombinants at frequencies about 10^2-fold greater than R68.45. The basis of this enhanced chromosome-mobilizing ability is not known.

4. *Interspecific and Intergeneric Conjugative Gene Transfer*

a. *Transfer of Naturally Occurring Plasmids.* As discussed above, R plasmids of the P1 incompatibility group are readily transmissible to both fast- and slow-growing rhizobia, and most of the antibiotic resistance genes transferred are expressed (e.g., Datta *et al.,* 1971; Beringer, 1974; Kuykendall, 1979). In a few cases F or F-like plasmids were also transferred (Datta and Hedges, 1972; Mergay *et al.,* 1973; Dunican *et al.,* 1976; Kowalczuk and Lorkiewicz, 1977), but the transmissibility of these plasmids seems to be restricted to only a few strains. Dunican and Tierney (1974) reported that, when the F-like R factor Rl-19drd was transferred from *R. trifolii* strain T1 to a non-nitrogen-fixing strain of *K. aerogenes,* nitrogen-fixing transconjugants appeared, suggesting the transfer and expression of *Rhizobium nif* genes in *Klebsiella*. Further experiments indicated that an indigenous *Rhizobium* plasmid carrying *nif* genes was mobilized by the R factor (Dunican *et al.*, 1976). Experiments by Skotnicki and Rolfe (1978) showed that this *R. trifolii* strain was able to transfer *nif* genes to *E. coli* K12 together with genetic material resembling the *gal-chlA* region of *E. coli*, and that it also formed tumors on tomato stems. Thus this strain appears to be a bacterium with properties of both *A. tumefaciens* and *R. trifolii*. Recently a P1 group plasmid RP1 was reported to have mobilized *nif* genes from *R. trifolii* T1 (Stanley and Dunican, 1979) to *nif* mutants of *K. pneumoniae* and to *A. tumefaciens*. Interestingly in both cases *nif* genes were expressed.

The tumor-inducing (Ti) plasmid of *A. tumefaciens* has been transferred to another strain of *R. trifolii* (Hooykaas *et al.*, 1977). Although the Ti plasmid is self-transmissible, plasmid RP4 was probably involved in its mobilization during the intergeneric transfer. In the Ti plasmid-carrying *R. trifolii* strain the expression of two Ti plasmid-coded functions was observed, namely, the ability to catabolize octopine and the ability to induce tumors on plants. However, the nodulation specificity of this strain was altered: It was unable to nodulate *Trifolium parviflorum*, but on *T. pratense* its nodulation ability was retained (Hooykaas *et al.*, 1977).

b. *Transfer of R-primes.* Plasmid primes are very useful not only for mapping purposes but also for other aspects of bacterial genetics. The introduction of R-primes into the species in which they were isolated results in the formation of partial diploids, and this can allow analysis of gene function, dominance, complementation, etc.

In *Rhizobium*, however, no complementation studies have been reported, and there are only a few communications on the intraspecific or interspecific transfer of R-primes. This is probably because only a limited number of *Rhizobium* R-primes have been constructed, mainly from *R. meliloti* (Johnston *et al.*, 1978b,c; Julliot and Boistard, 1979; Kiss *et al.*, 1980; Kondorosi *et al.*, 1980).

F-primes can be stably maintained in *recA* derivatives of *E. coli* or in mutants with large chromosomal deletions (Low, 1972). Recombination-deficient or deletion mutants of *Rhizobium* are not yet available; however, the lack of recombination between *R. meliloti* and *R. leguminosarum* chromosomes (Johnston *et al.*, 1978c) suggests that R-primes of one of these species can be maintained in a strain of the other. Johnston *et al.* (1978c) found that the expression of some *R. meliloti* genes in *R. leguminosarum* was not always complete. For instance, the *trp-19* allele was suppressed by pAJ24JI, an R-prime carrying the appropriate *R. meliloti trp*+ allele, but the transconjugants grew slowly in the absence of tryptophan.

Rhizobium R-primes have been used for mapping alleles by suppression tests (Johnston *et al.*, 1978b) or by polarized gene transfer (Julliot and Boistard, 1979;

Kiss *et al.*, 1980; Kondorosi *et al.*, 1980), for studying the organization of *trp* genes (Johnston *et al.*, 1978b), and for comparing the chromosomes of different *Rhizobium* species (Kondorosi *et al.*, 1980).

R-primes carrying *trp* genes of *R. meliloti* were used to study the organization of *trp* genes in *R. leguminosarum* (Johnston *et al.*, 1978b). It was found that all Trp⁻ mutations investigated so far mapped in one of three different regions on the *R. leguminosarum* chromosome. Three different R-primes suppressing all the *trp* alleles in a particular region were isolated. The *trp* genes on each R-prime were identified by transferring the R-primes to strains of *P. aeruginosa* carrying mutations in different *trp* genes. On this basis, it was established that one R-prime, pAJ24JI, carried *trpA, -B,* and *-F*, pAJ73JI carried *trpC* and *-D,* and pAJ88JI contained *trpE*, although the *trpA, -B,* and *-D* genes did not show full expression. When these R-primes were transferred to *E. coli trp* mutants, they failed to suppress any of the *trp* mutations tested. Derivatives of pAJ24JI were selected which could suppress the *trpA* and *-F* alleles. The absence of production of phage 16-3 of *R. meliloti* in *E. coli*, introduced on pGY2 in the prophage state, has also been noted (Kiss *et al.*, 1980). There have, in contrast, been reports that the *nif* genes of *Rhizobium* can be expressed in *K. pneumoniae* (Dunican and Tierney, 1974; Skotnicki and Rolfe, 1978; Stanley and Dunican, 1979) and in *Azotobacter* (Sen *et al.*, 1969; Bishop *et al.*, 1977; Page, 1978).

It is probable that the expression of *Rhizobium* genes in other genera depends both on the gene in question and on the recipient bacterium.

P1 group plasmid primes carrying genes of different gram-negative bacteria have been used in investigating the expression of foreign genes in *Rhizobium*. Dixon *et al.* (1976) studied the expression of *his* and *nif* genes from *K. pneumoniae* in *R. meliloti* with the use of plasmid pRD1, a derivative of RP4. When pRD1 was transferred to the *his-1* mutant of Rm41, the *his* allele was suppressed, indicating the expression of at least one *his* gene of *K. pneumoniae*. The His⁺ transconjugants were tested for free-living nitrogen fixation. In cells grown under conditions allowing derepression of *nif* genes in *K. pneumoniae*, no free-living nitrogen fixation occurred, but the production of one of the *Klebsiella* nitrogenase components was detected immunologically. Several other His⁻ mutants with different map locations were also suppressed by pRD1 (A. Kondorosi, unpublished result). pRD1 was rather unstable in intergeneric matings, and the majority of the segregants contained the original RP4. When selection was made for the *his* marker, the *nif* region was usually retained (Dixon *et al.*, 1976). Therefore, in studies on *Klebsiella nif* gene expression in unrelated bacteria it is preferable to use His⁻ hosts if pRD1 is being used.

Johnston *et al.* (1978b) reported that a P1 group R-prime carrying the *trpA* and *-B* genes of *P. aeruginosa*, isolated by Hedges *et al.* (1977), suppressed a number of *R. leguminosarum* Trp⁻ mutants, presumably defective in the *trpA* or *-B* gene. Nagahari *et al.* (1979) transferred an RP4-*trp* plasmid containing the whole tryptophan operon of *E. coli* to four *R. leguminosarum* strains carrying mutations in different *trp* genes, and the suppression of each *trp* allele was demonstrated. Moreover, enzymic activities of two enzymes involved in tryptophan biosynthesis were much higher in *R. leguminosarum* carrying RP4-*trp* than in wild-type cells and were not repressed by the addition of exogenous tryptophan. When RP4-*trp* was introduced into *P. aeruginosa*, the *trp* genes were also constitutively expressed. In view of the lack of expression of *Rhizobium trp* genes in *E. coli* (Johnston *et al.*, 1978b) it was suggested that, because of a supposedly more stringent specificity of *E. coli* RNA polymerase for its own promoter regions, *E. coli* was less able to express foreign genes than *R. leguminosarum* or *P. aeruginosa*.

B. TRANSFORMATION

1. *Intra- and Interspecific Transformation*

The first report on the transformation of *Rhizobium* dates back to 1941 when Krasilnikov (1941) described the transfer of nodulation ability to a strain of *Rhizobium* accomplished by growing this strain during several passages in the sterile filtrate of a culture of another strain. With the same method interspecific transfer of nodulation ability was also found (Krasilnikov, 1945). Several laboratories made attempts to repeat interspecific transformation of nodulation specificity with culture filtrates or purified DNA of donor bacteria, but these results were not convincing because of technical difficulties preventing the unambiguous demonstration of successful transformation. For instance, a large number of factors which could have greatly influenced the results were not excluded, and the strains used did not carry appropriate genetic markers. Other reports on intra- or interspecies transformation of symbiotic functions are also difficult to assess. Since this problem and in general the transformation of *Rhizobium* were extensively and critically reviewed recently by Gabor-Hotchkiss (1972) and by Schwinghamer (1977), this topic is treated very briefly in this section.

Transformation of antibiotic resistance, antibiotic dependence, or auxotrophic markers by purified DNA is a more reliable and reproducible procedure allowing the optimal conditions of transformation in *Rhizobium* to be worked out. Extensive work on the transformation of *R. meliloti, R. lupini*, and *R. japonicum* was done by R. Balassa (1956, 1957, 1960; Balassa and Gabor, 1961) and has been reviewed by G. Balassa (1963).

Following Balassa's work, several other laboratories reported gene transfer by transformation in *R. trifolii* (Ellis *et al.*, 1962; Zelazna, 1964), *R. meliloti* (Szende *et al.*, 1961; Zelazna-Kowalska and Lorkiewicz, 1971) and *R. japonicum* (Rain and Modi, 1969).

Despite the relatively large number of publications on transformation, only a few reports presented linkage data, and these were only for a few marker pairs (Balassa and Gabor, 1965; Gabor, 1965; Doctor and Modi, 1976; Drozanska and Lorkiewicz, 1978). Therefore, the contribution of these transformation studies to the field of *Rhizobium* mapping has been limited.

2. Intergeneric Gene Transfer by Transformation

A very early report on gene transfer in *Rhizobium* dealt with intergeneric transformation, namely, transfer of nodulation ability to *Pseudomonas* (Krasilnikov, 1945). It is again rather difficult to assess this finding. Gene transfer between *A. tumefaciens* and *Ř. leguminosarum* was reported by Klein and Klein (1953) and Kern (1965a), where the tumor-inducing ability of *A. tumefaciens* was transferred to *R. leguminosarum*. Kern (1965b) showed that the *Rhizobium* transformants exhibited properties intermediate between those of the two parents. Antibiotic resistance markers were also transferred by transformation (Kern, 1969). The recent demonstration of the transfer of oncogenic properties from *A. tumefaciens* to *Rhizobium* by RP4-mediated conjugation (Hooykaas *et al.*, 1977) seems to support these studies. Expression of *Agrobacterium* genes in fast-growing rhizobia is conceivable in view of the considerable DNA homology between these bacteria (Gibbins and Gregory, 1972) and of the proposed common ancestor (Reijnders, 1976).

Intergeneric transformation between *Rhizobium* and *Azotobacter* was first reported by Sen *et al.* (1969). With different species of both *Rhizobium* and *Azotobacter,* crystal violet- or streptomycin-resistant transformants of *Azotobacter* were obtained (Sen *et al.*, 1969; Sadasivam and Pandian, 1978). Recently the transformation of *Azotobacter* Nif⁻ mutants to Nif⁺ by *R. trifolii* DNA was reported (Bishop *et al.*, 1977; and Page, 1978). In addition, 13% of the Nif⁺ transformants were agglutinated by trifoliin, indicating that the gene(s) coding for a surface antigen involved in the initial steps of legume infection was also transferred. Similar observations were made when DNA from *R. japonicum* was used (Maier *et al.*, 1978a): In this case 3 out of 50 Nif⁺ transformants tested contained the O-antigen-related polysaccharide specifically required for the nodulation of *R. japonicum*. As discussed in Section V,B, both genetic and physical evidence indicated that at least some of the *nif* and *nod* genes were located on the *Rhizobium* plasmids. However, in the Nif⁺ *Azotobacter* transformants no additional extrachromosomal elements were detected (Bishop *et al.*, 1977).

The introduction of plasmids from unrelated bacteria into *Rhizobium* by transformation was also reported. An F-like R factor, Rl-19drd, from *P. aeruginosa* (Dunican and Tierney, 1973) and the P1 group plasmid RP4 from *E. coli* (O'Gara and Dunican, 1973) were transferred to *R. trifolii* by transformation.

In view of the current interest in applying genetic engineering techniques in various genetic studies, the use of transformation for the introduction of potential cloning vehicles into different *Rhizobium* strains is expected.

C. Transduction

Transduction, a very useful technique in fine-linkage mapping of bacteria, has been reported only for a few *Rhizobium* strains.

Generalized transduction in *R. meliloti* was first described by Kowalski (1967). Several temperate phage isolates were able to mediate transduction at frequencies of 10^{-5}–10^{-7} (Kowalski, 1967, 1970a). Phage L5, an isolate showing the highest transduction frequency in *R. meliloti* strain L5-30 was used to transduce streptomycin resistance, auxotrophy, and effectiveness (Kowalski, 1970b, 1971a,b). Cotransduction for a few marker pairs in *R. meliloti* L5-30 was also observed (Kowalski, 1976; Malek and Kowalski, 1977; Zelazna-Kowalska and Kowalski, 1978). In Leu⁻ mutants effectiveness was found to be correlated with leucine dependence, since all Leu⁺ transductants regained effectiveness (Kowalski and Dénarié, 1972).

Several *Rhizobium* phages have a rather narrow host range. Therefore, studies on different *R. meliloti* strains necessitated the isolation of transducing phages for each one. The apparently virulent phage DF2 (Corral *et al.*, 1978) mediated transduction of both chromosomal and plasmid markers at a frequency of 10^{-6} per plaque-forming unit (PFU) in *R. meliloti* GR4 (Casadesús and Olivares, 1979a). Transductants were stable and were sensitive to phage DF2. With this phage a fine-structure analysis of the linkage map of the RmGR4 chromosome constructed by means of R68.45-mediated recombination was carried out for marker pairs which showed high linkage in conjugative gene transfer experiments. A thermosensitive derivative of this phage is now available; at higher temperatures no lytic growth of phages occurred, and in this way the lysis of transductants by reinfection was prevented (Casadesús and Olivares, 1979b). Phage 11 of Rm41 (a close relative of phage 16-3; M. Mink, L. Orosz, and T. Sik, unpublished results) transduced several markers of Rm41 at frequencies of from 10^{-5} to 10^{-7} (Sik *et al.*, 1980).

Generalized transduction in *R. leguminosarum* 300 was reported by Buchanan-Wollaston (1979). The transducing phages RL38 and RL39 were virulent; therefore, the killing potential of these phage lysates had to be decreased by ultraviolet irradiation. The frequency of transduction was high enough to use these phages for fine-structure mapping. The complete coinheritance of Tn5-induced auxotrophy with kanamycin resistance was demonstrated by the use of this transduction system (Beringer *et al.*, 1978a). Phage RL38 formed plaques on *R. trifolii* strain 6621 and could be propagated on this host. Transduction between strains of *R. trifolii* and from *R. leguminosarum* to *R. trifolii*, but not in the reverse direction, occurred (Buchanan-Wollaston, 1979). In view of the homology between the chromosomes of *R. leguminosarum* and *R. trifolii* shown by plasmid-mediated conjugation (Johnston and Beringer, 1977) interspecific transduction was not surprising. Phage RL38 may allow fine-linkage mapping of host specificity genes, although by definition these will differ in the two species or may have only limited homology.

Transduction in *R. japonicum* mediated by a temperate phage was briefly reported (Doctor and Modi, 1976). Cotransduction of *ura* and *arg* alleles was observed at a very high frequency (3×10^{-3}), but transductants for the other two markers tested were not obtained, which might indicate that this was a case of specialized transduction. Transduction was also demonstrated in the nonnodulating *R. lupini* strain studied by Heumann *et al.* (1974).

Specialized transduction in Rm41 was demonstrated by Sváb *et al.*, 1978). The *cys-46+* allele was transduced by phage 16-3 at a frequency of $10^{-6}-10^{-7}/$ PFU. With the exception of other *cys* markers located in the *cys-46* region on the linkage map of *R. meliloti* (Kondorosi *et al.*, 1977a) no other markers were transduced by phage 16-3. As determined by R68.45-mediated recombination, the prophage integration site on the *R. meliloti* chromosome is between the *cys-46* and *met-5* markers (Sváb *et al.*, 1978). An analysis of this transduction system revealed that about 65% of the *cys-46+* transductants yielded high-frequency transducing (HFT) lysates ($10^{-2}/$PFU). Some of the HFT lysates contained plaque-forming transducing particles, and others contained defective *cys+*-transducing particles in which the two distal ends of the vegetative phage chromosome were lacking.

IV. Chromosomal Mapping

A. Construction of Chromosomal Linkage Maps

As discussed above, there have been some reports describing cotransformation or cotransduction of a few pairs of alleles in different *Rhizobium* species. However, it has been due to the use of P1 plasmid-mediated chromosomal transfer that linkage maps have been constructed for several *Rhizobium* strains (Beringer and Hopwood, 1976; Meade and Signer, 1977; Kondorosi *et al.*, 1977a; Beringer *et al.*, 1978b; Casadesús and Olivares, 1979a). In these analyses auxotrophic and antibiotic resistance markers were mapped by means of R-plasmid-mediated (R68.45 or RP4) recombination based on the determination of coinheritance (linkage) frequencies between different marker pairs. Since very large fragments of the chromosome were transferred and integrated into the recipient chromosome linkage, analysis of relatively few markers was required to define a circular linkage map; 20 markers for Rm2011 (Meade and Signer, 1977), 19 for Rm41 (Kondorosi *et al.*, 1977a), 16 for R1300 (Beringer *et al.*,

1978b), and 19 for RmGR4 (Casadesús and Olivares, 1979a). Recently the linkage map of the Rm41 chromosome was supplemented with 15 new markers, and the sequence of markers in the *met-1-tyr-1* region was corrected (Fig. 1) (Sváb *et al.*, 1978; Kiss *et al.*, 1979, 1980; Kondorosi *et al.*, 1980; Forrai *et al.*, unpublished results); four new map locations were determined on the R1300 map (Kondorosi *et al.*, 1980) (Fig. 1). These mapped markers now provide a frame for the localization of new mutations, for instance, those affecting the symbiotic properties of *Rhizobium*.

The linkage maps were constructed on the assumption of inverse proportionality between coinheritance frequency and physical distance between markers. Marker distances on the maps for Rm2011 (Meade and Signer, 1977), for R1300 (Beringer *et al.*, 1980b), and for RmGR4 (Casadesús and Olivares, 1979a) were given in coinheritance frequency values. Although the coinheritance frequencies unambigously determine the marker order, these values are not additive. Kondorosi *et al.* (1977a) transformed the coinheritance frequency values (c) for Rm41 into approximately additive map distances (d in arbitrary units) by applying the following empirical mapping function:

$$c = (1 - d)^3 \qquad\qquad (1)$$

This resembles the equation derived by Wu (1966) for transduction and is based on the assumption that the donor fragments have nearly constant length and that the almost randomly formed donor fragments are incorporated into the chromosome of the recipient by pairs of crossover events. It is very probable that these assumptions are valid only on a statistical basis. Variations for certain chromosomal regions or in certain experiments may exist. Nevertheless, data from a series of three-factor crosses indicated that the d values were approximately additive in most cases. However, for short distances these values were too inaccurate to give the marker order. This may be a consequence of the fact that very large fragments of the donor chromosome are incorporated into the recipient chromosome (Kondorosi *et al.*, 1977a; Beringer *et al.*, 1978b; Casadesús and Olivares, 1979a).

It is obvious that for fine-structure mapping transduction or transformation systems are required. The map of RmGR4 was established by the joint application of conjugation and transduction techniques (Casadesús and Olivares, 1979a). According to previous studies on the R1300 map (Beringer *et al.*, 1978b) *spc*, *str*, and *rif* alleles were coinherited at a very high frequency in R68.45-mediated recombination. Three-factor transduction of these alleles revealed the order *spc-str-rif* (Buchanan-Wollaston, 1979). On the Rm41 map linkage relationships for three new marker pairs were determined (Sik *et al.*, 1980). In these three strains coinheritance data obtained by transduction and data obtained by conjugation were compared for a few marker pairs, and a direct correlation was observed. However, Casadesús and Olivares (1979a) reported some discrepancies in the linkage data obtained by the two methods and suggested that in these cases the conjugational data were distorted.

B. Homology of Chromosomes of Different *Rhizobium* Species

The map function described for the mapping of Rm41 was also applicable for the linkage data of Rm2011 and R1300 maps, since in about 80% of the cases where additivity of the *d* values could be tested, these values were approximately additive (Kondorosi *et al.*, 1980). On this basis a comparison of the Rm41, Rm2011, and R1300 linkage maps was made. As seen in Fig. 1, the three maps show striking similarities, as indicated by the lines joining the corresponding markers, although in most cases the enzyme deficiences caused by these mutations are not known. To substantiate this comparison two approaches were followed.

The homology between the chromosomes of the two *R. meliloti* strains was proved in R68.45-mediated chromosome mobilization experiments. It was found that the recombination frequency for all markers tested and the linkage for marker pairs along the entire length of the chromosome in crosses between Rm2011 and Rm41 were the same as in intrastrain crosses. In contrast in crosses between strains of *R. leguminosarum* and *R. meliloti* haploid recombinants arose at very low frequencies (Johnston *et al.*, 1978c; Kondorosi *et al.*, 1980), suggesting little chromosomal homology. However, the similarity between the arrangement of alleles for particular nutritional requirements on the Rm41 and R1300 maps was substantiated by the use of eight R68.45-primes carrying different prototrophic alleles of Rm2011. Each of the eight R-primes suppressed the auxotrophic alleles of Rm41 and R1300 which mapped at similar positions on the chromosomes of the two species. It was concluded therefore that, in spite of the lack of recombination between the two chromosomes, the organization of the genes appeared to be similar (Kondorosi *et al.*, 1980).

Johnston and Beringer (1977) reported that R68.45 promoted gene transfer among three fast-growing *Rhizobium* species, namely, *R. leguminosarum, R. trifolii,* and *R. phaseoli.* In a series of crosses between these species they demonstrated that the frequencies of recombination and linkage for several marker pairs in matings of *R. leguminosarum* 300 with *R. phaseoli* and *R. trifolii* were the same as between different mutants of R1300, indicating that the chromosomes of these species were highly homologous. They tested the transfer of symbiotic host specificity of these species in these crosses, but the recombinants retained the host specificity of the recipient, suggesting that either these genes were not expressed in the new host or they did not reside on the transferred fragments. Later it was shown (Johnston *et al.*, 1978a) that these genes were located on a plasmid of *R. leguminosarum.* Apparently these species have very similar chromosomes but different plasmids, at least with respect to host specificity. The above results indicate that with respect to chromosomal homology there are at least two groups of fast-growing bacteria: *R. leguminosarum, R. phaseoli,* and *R. trifolii* represent one group, and *R. meliloti* the other. This is consistent with the taxonomic studies of Graham (1964). In view of the demonstrated similarities between linkage maps of related bacterium species such as *E. coli* and *Salmonella typhimurium* (Sanderson, 1972; Bachmann *et al.,* 1976) or *Streptomyces coelicolor* and *S. rimosus* (Friend and Hopwood, 1971) the conservation of linkage maps of fast-growing rhizobia is not surprising.

C. Localization of Nonselectable Mutations

The linkage analyses revealed an interesting property of R-plasmid-mediated recombination in *Rhizobium*. In three-factor crosses it was observed that, when two donor markers were coinherited, in most cases other markers located between them were also transferred together. These observations may indicate that the number of multiple crossover events is surprisingly low in *Rhizobium,* at least when plasmid-mediated recombination occurs. However, in *E. coli* the transferred fragments mediated by F or R68.45 seem to take part in multiple crossover events, resulting in the segregation of markers within relatively short distances (Hayes, 1968; Z. Bánfalvi and A. Kondorosi, unpublished results). These observations might suggest that there is some type of ''end effect'' favoring the inheritance of entire transferred fragments (Beringer *et al.,* 1978b). The rare occurrence of multiple crossovers and the additivity of recombination frequency values even for intracistronic distances were observed in studies on recombination and heterozygote formation of phage 16-3 of *R. meliloti.* It was suggested that the effectiveness of mismatch repair in *Rhizobium* was low (Orosz, 1980; Dudás and Orosz, 1980; Orosz *et al.,* 1980a,b).

This property of the *Rhizobium* recombination system facilitated the location of markers with no easily selectable phenotype, such as symbiotic mutations. When a chromosomal fragment flanked by two donor markers is transferred, the overwhelming majority of the recombinants selected for both markers inherit other donor alleles between the two markers.

Therefore, it is enough to test a few double recombinants to see whether the marker to be mapped is located between the two donor markers. This approach was applied in mapping nonselectable mutations in Rm41, and using eight marker pairs the entire length of the chromosome was covered (Kondorosi *et al.,* 1980). Several symbiotic mutations were located on the Rm41 chromosome in this way, but others, including all *nod* mutations, did not map on the chromosome (Forrai *et al.,* unpublished results).

D. Replication Mapping

Attempts to map the chromosome by determining the frequency of mutations after pulsed mutagenic treatment of synchronized cell populations has been reported for *R. meliloti* (Szende, 1971; Al-Bayatti and Al-Ani, 1977) and for *R. trifolii* (Al-Ani and Farouk, 1976; Zurkowski and Lorkiewicz, 1978). Chromosomal replication was shown to be bidirectional (Zurkowski and Lorkiewicz, 1977), and a circular map of *R. trifolii* T37 was constructed consisting of 15 markers (Zurkowski and Lorkiewicz, 1978). It should be emphasized that mapping by such a method is much less precise than the techniques described above, but this procedure can be used to localize the replication origin on a linkage map already established.

V. Arrangement of Genes in *Rhizobium*

Although our knowledge of the organization of genes in fast-growing rhizobia is still very limited, some general features have already been established.

A. Chromosomal Genes

At present the linkage maps of *R. meliloti* and of *R. leguminosarum* contain relatively few alleles. However, it can be stated that genes for several biosynthetic pathways are scattered in *Rhizobium*. The absence of clustering of functionally related genes was demonstrated also in *P. aeruginosa* (Fargie and Holloway, 1965), unlike the arrangement in *E. coli* (Bachmann et al., 1976). For instance, the tryptophan genes of *E. coli* are located in one cluster on the *E. coli* chromosome, whereas in both *R. leguminosarum* and *R. meliloti* three distinct regions (Johnston et al., 1978b; Kondorosi et al., 1980), containing *trpA, -B, -F, trpC, -D,* and *trpE,* respectively (Johnston et al., 1978b), were determined. Similarly, genes coding for histidine biosynthesis are located in one contiguous region of the *E. coli* chromosome; a comparison of mapping data of *his* alleles in *R. meliloti* suggests several scattered *his* loci in *Rhizobium* (Meade and Signer, 1977; Kondorosi et al., 1977a; Casadesús and Olivares, 1979a).

Because of recent developments in gene transfer and mapping systems for *Rhizobium*, progress in the field of biochemical genetics in *Rhizobium* is expected. Enzyme deficiencies have been demonstrated so far only in a very few mapped mutants. A mutant of Rm41 lacking glutamate synthase activity, and consequently unable to assimilate NH_4^+ (Kondorosi et al., 1977b), was mapped (allele *glt-41,* Kondorosi et al., 1977a). Biochemical and genetic studies on *R. meliloti* mutants affected in nitrate reduction identified four genes, namely, *narA, -B, -C,* and *-D,* determining the utilization of NO_3^- as sole nitrogen source

(Kiss *et al.*, 1979). The effects of these mutations on both assimilatory and respiratory nitrate reduction have been determined. Mutations in *narA* seem to affect the electron transport pathway serving NO_3^- assimilation; *narB* probably represents the structural gene of the nitrate reductase apoprotein; and the *narC* and *narD* mutants are deficient in an active molybdenum cofactor common for nitrate reductase and xanthine dehydrogenase. The *nar* alleles mapped at four different sites; *narB* and -*C* were closely linked. These mutants were not affected in symbiotic nitrogen fixation abilities, in contrast to other mutants possessing very low nitrate reductase activity and defective in respiratory nitrate reduction (Kondorosi *et al.*, 1973; Barabás and Sik, 1979). It was therefore concluded that nitrogenase and nitrate reductase did not share the same molybdenum cofactor.

Currently there is interest in studying various aspects of *Rhizobium* metabolism and symbiotic functions with the use of biochemical mutants (e.g., Dénarié *et al.*, 1976; Gibson and Pagan, 1977; Ludwig and Signer, 1977; Niel *et al.*, 1977; Shanmugam *et al.*, 1978; Maier *et al.*, 1978b; Sanders *et al.*, 1978; Duncan and Fraenkel, 1979; Skotnicki and Rolfe, 1979; Arias *et al.*, 1979; Ronson and Primrose, 1979; Vincent, 1980). The mapping of these mutants will certainly expand our knowledge on the organization of *Rhizobium* genes.

B. PLASMID GENES

The genetics of *Rhizobium* plasmids is discussed by Dénarié *et al.* (this volume), therefore genes carried by indigenous *Rhizobium* plasmids will be summarized only briefly.

The most important finding was that genes required for the establishment of symbiosis or for the nitrogen fixation process may be carried by plasmids, including genes responsible for nodulation and its specificity (Higashi, 1967; Parijskaya, 1973; Zurkowski *et al.*, 1973; Johnston *et al.*, 1978a; Zurkowski and Lorkiewicz, 1979; Brewin *et al.*, 1980; Prakash *et al.*, 1980; Buchanan-Wollaston *et al.*, 1980; Bánfalvi *et al.*, unpublished results) and for nitrogen fixation (*nif* genes) (Dunican and Tierney, 1974; Nuti *et al.*, 1979; Koncz *et al.*, unpublished), as well as other genes affecting symbiotic effectiveness (Brewin *et al.*, 1980) or the ability to induce polygalacturonase in legume root cells (Palomares *et al.*, 1978). There is very little information about other plasmid-coded functions. Recently, Hirsch (1979) has demonstrated that many *R. leguminosarum* strains carry bacteriocinogenic plasmids. The sensitivity of *Rhizobium* to certain phages is suggested also to be coded by plasmids (Corral *et al.*, 1978). It appears that, in one strain of *R. phaseoli*, the ability to produce melanin is determined by a plasmid (Beynon *et al.*, 1980).

The presence of symbiotic genes on plasmids will undoubtedly focus the attention on these plasmids, and other genetic determinants coded by them will be intensively sought.

VI. Genetics of Rhizobiophages

The frequent occurrence of rhizobiophages in the nodules of leguminous plants and in the surrounding soil greatly facilitated the isolation of phages for different *Rhizobium* species and strains (Allen and Allen, 1950; Kleczkowska, 1957). The general properties of rhizobiophages were recently reviewed by Vincent (1977). This section will deal with the genetic aspects of phage–*Rhizobium* interactions and with genetic studies on rhizobiophages.

A. PHAGE–BACTERIUM INTERACTIONS

The host range of different phage isolates is rather variable. While certain phages are strain- or species-specific, others have a very wide host range and are able to lyse rhizobia of different species (Laird, 1932; Conn *et al.*, 1945; Bruch and Allen, 1957; Kleczkowska, 1957; Szende and Ördögh, 1960; Schwinghamer and Reinhardt, 1963; Staniewski, 1970; Barnet, 1972; Patel, 1976; Buchanan-Wollaston, 1979). The large collection of phage isolates can be used in typing *Rhizobium* isolates and in correlating genetic alterations of a *Rhizobium* strain with its change in phage sensitivity pattern (Staniewski, 1970; Staniewski and Kowalska, 1974; Patel, 1976). This approach can be especially useful in studies on the symbiotic properties of *Rhizobium* or on indigenous *Rhizobium* plasmids where plasmid genes are responsible for sensitivity to certain phages (Corral *et al.*, 1978). Recent reports (Bohlool and Schmidt, 1974; Dazzo and Hubbel, 1975; Wolpert and Albersheim, 1976) suggest that certain cell surface components of *Rhizobium* such as lipopolysaccharides are responsible for the initiation of plant infections. On the other hand, lipopolysaccharides may function as phage receptors in many bacteria (Weidel, 1958). One may expect that *Rhizobium* mutants, resistant to certain phages, may have altered nodulation properties.

Lysogenization by temperate phages has been reported for almost all *Rhizobium* groups (Marshall, 1956; Ördögh and Szende, 1961; Takahashi and Quadling, 1961; Schwinghamer and Reinhardt, 1963; Moskalenko and Rautenstein, 1969; Barnet and Vincent, 1970; Heumann *et al.*, 1974). The integration of prophage 16-3 into the *R. meliloti* chromosome was demonstrated, and it was located between the *cys-46* and *met-5* markers on the map of *R. meliloti* 41 by specialized transduction and by conjugation (Sváb *et al.*, 1978).

Lysogenic conversion of a somatic antigen in *R. trifolii* has been observed (Barnet and Vincent, 1970), but no alteration in symbiotic properties following lysogenization has been demonstrated. Host-controlled modification–restriction of phage DNA was suggested to exist in *R. leguminosarum* (Schwinghamer, 1966) and in *R. trifolii* (Schwinghamer, 1971).

Many *Rhizobium* isolates produced bacteriocins which resembled defective

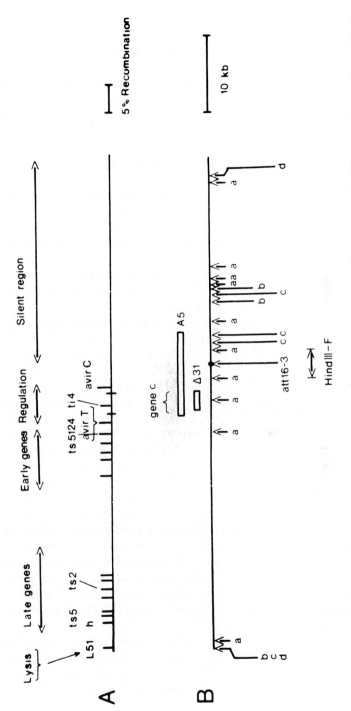

FIG. 2. Genetic (A) and physical (B) map of *R. meliloti* phage 16-3. On the genetic map only reference markers for different regions are shown. *A5* and *Δ31* are deletions (Orosz *et al.*, 1973, 1980a,b; Dallman *et al.*, 1979, 1980; Dorgai *et al.*, unpublished results). Recognition sites: a, *Hind*III; b, *Hpa*I; c, *Kpn*I; d, *Bgl*II.

prophages (Roslicky, 1967; Lotz and Mayer, 1972; Schwinghamer *et al.*, 1973; Pfister and Lotz, 1974; Lotz *et al.*, 1974).

B. Genetic and Physical Maps of the Phage Genome

As surveyed below, genetic studies have been carried out only on temperate phage 16-3 of *R. meliloti* 41 and on virulent phage c of *R. trifolii* W19.

The genetic map of phage 16-3 (Fig. 2) was established by complementation and mapping of a number of temperature-sensitive, lysis-deficient, and virulent mutants (Orosz and Sik, 1970; Sik and Orosz, 1971; Szende, 1971; Orosz *et al.*, 1973; Kondorosi *et al.*, 1974; Dallmann *et al.*, 1980). Recombination between phage DNAs occurred readily, with a maximum recombination frequency of 21–23%. The following sequence of genes was determined: regulatory (immunity), early functions, late functions, lysis (Orosz *et al.*, 1973). This sequence was supported by linkage analysis of markers in transfection (Kondorosi *et al.*, 1974). The structure of the regulatory region, especially cistron *c* of 16-3, the functional analog of cistron *cI* of phage λ was studied in detail (Orosz and Sik, 1970; Dudás and Orosz, 1980). Mutations affecting the virulence (immunity insensitivity) of phage 16-3 mapped at two loci in the sequence *avirC-c-avirT*. One may say that this and the overall organization of 16-3 genes show striking similarities to that of phage λ of *E. coli*. Mutations in the vicinity of *avirT* had the ability to alter recombination frequency (Dallmann *et al.*, 1980). The recent isolation of a derivative of *R. meliloti* 41 carrying an amber suppressor allele allows the analysis of amber mutations not only in phage 16-3 but also in the host bacterium (Zs. Palágyi and L. Orosz, unpublished results).

Phage 16-3 DNA is a 61.57-kilobase (kb)-long linear molecule with complementary single-stranded sequences at each end allowing the phage DNA to form circles or concatenates, a process analogous to that observed for phage λ. The comparison of the genetic map of phage 16-3 with the physical map constructed using four restriction enzymes indicated a long (20 kb), genetically silent chromosomal arm probably containing nonessential genes and preceding the immunity region on the vegetative phage map (Dallmann *et al.*, 1979, 1980). The physical analysis of pGY2 (R68.45::Rm *cys-46*$^+$ att_{16-3}::16-3; Kiss *et al.*, 1980) indicates that the phage integration site resides on this silent region in the *Hind*III-F fragment (Fig. 2) (L. Dorgai, M. Berényi and L. Orosz, unpublished results). Prophage 16-3 can be introduced into unrelated gram-negative bacteria on plasmid pGY2 (Kiss *et al.*, 1980). In this way elimination of restriction sites on the phage DNA could be attempted, allowing the possibility of developing a cloning vector from phage 16-3.

A preliminary genetic map (Atkins, 1973b) for virulent phage c (Atkins, 1973a; Atkins and Hayes, 1972) of *R. trifolii* W19, based on complementation analysis, measurement of phage products, and mapping of markers representing different complementation groups, has also been constructed.

C. TRANSFECTION

Transfection of competent *R. meliloti* cells has been demonstrated at a very low frequency (10^{-8} per phage-equivalent DNA) (Sik and Orosz, 1971; Staniewski *et al.*, 1971; Kondorosi *et al.*, 1974). Competence was found in the early log phase of growth (Kondorosi *et al.*, 1974). The use of spheroplasts resulted in somewhat higher transfection efficiency (Kondorosi *et al.*, 1974; Staniewski and Rugala, 1975).

Transfection in the presence of helper phages resulted in rescuing markers of the transfecting DNA at a much higher frequency ($1-3 \times 10^{-6}$; Kondorosi *et al.*, 1974). The formation of transfectants was governed by early gene functions of the helper phage and was independent of the physiological state of the host bacterium. With the use of helper phages, transformation of the *cys-46* allele of *R. meliloti* 41 was achieved with DNA isolated from the HFT lysate of the *cys-46*+-transducing derivative of 16-3. In view of the present attempts to derive a phage-cloning vehicle for *Rhizobium* the transfection frequency should be increased.

VII. Strain Construction in *Rhizobium*

One of the potential benefits of studies on *Rhizobium* genetics apparently is the construction of strains that are better suited as inoculants than those presently used. We stress the word "potential," since to our knowledge there have been no reports of *Rhizobium* strains constructed by genetic manipulation being used commercially as inoculants.

Nevertheless we will consider how it may be possible to exploit the genetic systems that have been, and may continue to be, developed in efforts aimed at *Rhizobium* strain improvement. The end products of such a program would be strains of *Rhizobium* which possess various symbiotically desirable characters that have been introduced into the constructed strain from field isolates, each one of which may possess one particular attribute.

In some cases it is fairly obvious that certain characters are desirable; the strain should fix nitrogen at high rates and should do so efficiently in terms of the amount of photosynthate utilized. (In connection with this last point one character that appears to be useful is the possession of an uptake hydrogenase, since in Hup+ strains much of the energy that is lost in the hydrogen liberated by nitrogenase can be recouped.) The desirability of other characters may, however, be conditional. Thus persistence is an important attribute for a strain of (for example) *R. trifolii* which is to be used on clover grown in a permanent pasture, but not for a strain of *Rhizobium* to be used as an annual grain legume; indeed, in the latter case it could be argued that lack of persistence is desirable since it would allow the more ready introduction of a different strain in subsequent years. If there are present in the soil large numbers of indigenous strains of a particular species, then it is important that any inoculant strain of the same species be competitive for nodule formation, but if a legume is to be grown on soil lacking indigenous *Rhizobium,* then competitiveness is of little or no relevance. The conditional nature of such traits as the ability to nodulate and fix nitrogen in extreme environments (such as high or low pH, high salinity) or to form an effective symbiosis with a particular variety of host plant is self-evident.

It will be advantageous, perhaps obligatory, to be able to correlate particular symbiotic characters with characters that, at least initially, can be scored in laboratory conditions. Thus, if the ability to nodulate in acid conditions in the field is correlated with acid tolerance *in vitro,* then selection for the transfer of this character would obviously be facilitated.

It would also be desirable to know something of the basis of important symbiotic characters. For example, there is no knowledge of why one strain is more competitive in nodulating then another. Is it faster growth rate, production of bacteriocin, speed of binding to the root, or what? Not only would this be of interest in its own right, but it might also provide phenotypes that are more selectable following genetic transfer than competitiveness would be per se.

It seems to be inevitable that any attempts at strain construction will involve the transfer of *Rhizobium* plasmid-borne genes, given the importance of indigenous plasmids as determinants of symbiotic function. We illustrate this with a single example. It was mentioned above that the possession by a strain of an uptake hydrogenase (Hup$^+$) could result in an energetically more efficient symbiosis. It has been found that Hup$^+$ can be transferred from a Hup$^+$ strain of *R. leguminosarum* to a Hup$^-$ strain of the same species and that this is due to the transfer of plasmid-borne genes (N. J. Brewin and D. A. Phillips, personal communication). The significance of this, in terms of an improved symbiosis between Hup$^+$ transconjugants and peas, has still to be established.

VIII. Conclusions

In the last few years significant developments have been made in the field of *Rhizobium* genetics. Systems of chromosomal gene transfer and mapping have been established, and the crucial role of plasmids in the symbiosis is clear. However, we are still at an early stage in using this genetic information as a tool for understanding the symbiosis. The success or failure of *Rhizobium* genetics will be judged on the extent to which it helps in elucidating the details of nodule development and nitrogen fixation and on the way in which such information is used to improve the symbiotic interaction.

ACKNOWLEDGMENTS

We wish to thank Nick Brewin, David Hopwood, Csaba Kari, László Orosz, Tibor Sik, and Zóra Sváb for a critical reading of the manuscript.

REFERENCES

Al-Bayatti, K., and Al-Ani, F. (1977). *Bull. Coll. Sci. Univ. Baghdad* **18**, 3–26.
Al-Ani, F., and Farouk, Y. (1976). *Bull. Coll. Sci. Univ. Baghdad* **17**, 329–352.
Allen, E. K., and Allen, O. N. (1950). *Bacteriol. Rev.* **14**, 273–330.
Arias, A., Cervenansky, C., Gardiol, A., and Martinez-Drets, G. (1979). *J. Bacteriol.* **137**, 409–414.
Atkins, G. J. (1973a). *J. Virol.* **12**, 149–156.
Atkins, G. J. (1973b). *J. Virol.* **12**, 157–164.
Atkins, G. J., and Hayes, A. H. (1972). *J. Gen. Microbiol.* **43**, 273–278.
Bachmann, B. J., Low, K. B., and Taylor, A. L. (1976). *Bacteriol. Rev.* **40**, 116–167.
Balassa, G. (1963). *Bacteriol. Rev.* **27**, 228–241.
Balassa, R. (1956). *Naturwissenschaften* **43**, 133.
Balassa, R. (1957). *Acta Microbiol. Acad. Sci. Hung.* **4**, 85–95.
Balassa, R. (1960). *Nature (London)* **188**, 246–247.
Balassa, R., and Gábor, M. (1961). *Mikrobiologiya* **30**, 457–463.
Balassa, R., and Gábor, M. (1965). *Acta Microbiol. Acad. Sci. Hung.* **11**, 329–339.
Barabás, I., and Sik, T. (1979). *Can. J. Microbiol.* **25**, 298–301.
Barnet, Y. M. (1972). *J. Gen. Virol.* **15**, 1–15.
Barnet, Y. M., and Vincent, J. M. (1970). *J. Gen. Microbiol.* **61**, 319–325.
Barth, P. T. (1978). *Plasmid* **2**, 130–136.
Bedmar, E. J., and Olivares, J. (1980). *Mol. Gen. Genet.* **177**, 329–331.
Beringer, J. E. (1974). *J. Gen. Microbiol.* **84**, 188–198.
Beringer, J. E., and Hopwood, D. A. (1976). *Nature (London)* **264**, 291–293.
Beringer, J. E., Johnston, A. W. B., and Wells, B. (1977). *J. Gen. Microbiol.* **98**, 339–343.
Beringer, J. E., Beynon, J. L., Buchanan-Wollaston, A. V., and Johnston, A. W. B. (1978a). *Nature (London)* **276**, 633–634.
Beringer, J. E., Hoggan, S. A., and Johnston, A. W. B. (1978b). *J. Gen. Microbiol.* **104**, 201–207.
Beringer, J. E., Brewin, N. J., and Johnston, A. W. B. (1980). *Heredity* **45**, 161–186.
Beynon, J. L., Beringer, J. E., and Johnston, A. W. B. (1980). *J. Gen. Microbiol.* **120**, 421–429.
Bishop, P. E., Dazzo, F. B., Applebaum, E. R., Maier, R. J., and Brill, W. J. (1977). *Science* **198**, 938–940.
Bohlool, B. B., and Schmidt, E. L. (1974). *Science* **185**, 269–271.
Boucher, C., Bergeron, B., Barate de Bertalmio, M., and Dénarié, J. (1977). *J. Gen. Microbiol.* **98**, 253–263.
Brewin, N. J., Beringer, J. E., Buchanan-Wollaston, A. V., Johnston, A. W. B., and Hirsch, P. R. (1980). *J. Gen. Microbiol.* **116**, 261–270.

Bruch, C. W., and Allen, O. N. (1957). *Can. J. Microbiol.* **3**, 181–189.
Buchanan-Wollaston, V. (1979). *J. Gen. Microbiol.* **112**, 135–142.
Buchanan-Wollaston, A. V., Beringer, J. E., Brewin, N. J., Hirsch, P. R., and Johnston, A. W. B. (1980). *Mol. Gen. Genet.* **178**, 185–190.
Casadesús, J., and Olivares, J. (1979a). *Mol. Gen. Genet.* **174**, 203–209.
Casadesús, J., and Olivares, J. (1979b). *J. Bacteriol.* **139**, 316–317.
Cole, M. A., and Elkan, G. H. (1973). *Antimicrob. Chemother. Agents* **4**, 248–253.
Conn, H. J., Bottcher, E. J., and Randall, C. (1945). *J. Bacteriol.* **49**, 359–373.
Corral, E., Montoya, E., and Olivares, J. (1978). *Microbios* **5**, 77–80.
Cunningham, D. A. (1980). Ph.D. Thesis. University of Edinburgh.
Dallmann, G., Orosz, L., and Sain, B. (1979). *Mol. Gen. Genet.* **176**, 439–448.
Dallman, G., Olasz, F., and Orosz, L. (1980). *Mol. Gen. Genet.* **178**, 443–446.
Datta, N., and Hedges, R. W. (1972). *J. Gen. Microbiol.* **70**, 453–460.
Datta, N., Hedges, R. W., Shaw, E. J., Sykes, R. B., and Richmond, M. H. (1971). *J. Bacteriol.* **108**, 1244–1249.
Davidson, N., Deonier, R. C., Hu, S., and Ohtsubo, E. (1975). *In* "Microbiology- 1974" (D. Schlessinger, ed.), pp. 56–65. ASM, Washington, D.C.
Dazzo, F. B., and Hubbell, D. H. (1975). *Appl. Microbiol.* **30**, 172–177.
Dénarié, J., Truchet, G., and Bergeron, B. (1976). *In* "Symbiotic Nitrogen Fixation in Plants" (P. S. Nutman, ed.), pp. 47–61. Cambridge Univ. Press, London and New York.
Dénarié, J., Rosenberg, C., Bergeron, B., Boucher, C., Michel, M., Barate, and de Bertalmio, M. (1977). *In* "DNA Insertion Elements, Plasmids and Episome" (A. I. Bukhari, J. A. Shapiro, and S. L. Adhya, eds.), pp. 507–520. Cold Spring Harbor Laboratory, Cold Spring Harbor, New York.
Dixon, R. A., Cannon, F. C., and Kondorosi, A. (1976). *Nature (London)* **260**, 268–271.
Doctor, F., and Modi, V. V. (1976). *In* "Symbiotic Nitrogen Fixation in Plants" (P. S. Nutman, ed.), pp. 69–76. Cambridge Univ. Press, London and New York.
Drozanska, D., and Lorkiewicz, Z. (1978). *Acta Microbiol. Pol.* **10**, 81–88.
Dudas, B., and Orosz, L. (1980). *Genetics* **96**, 321–329.
Duncan, M., and Fraenkel, D. G. (1979). *J. Bacteriol.* **137**, 415–419.
Dunican, L. K., and Tierney, A. B. (1973). *Mol. Gen. Genet.* **126**, 187–190.
Dunican, L. K., and Tierney, A. B. (1974). *Biochem. Biophys. Res. Commun.* **57**, 62–72.
Dunican, L. K., O'Gara, F., and Tierney, A. B. (1976). *In* "Symbiotic Nitrogen Fixation in Plants" (P. S. Nutman, ed.), pp. 77–90. Cambridge Univ. Press, London and New York.
Ellis, N. J., Kalz, G. G., and Doncaster, J. J. (1962). *Can. J. Microbiol.* **8**, 835–840.
Faelen, M., Toussaint, A., Van. Montagu, M., Van der Elsacker, S., Engler, G., and Schell, J. (1977). *In* "DNA Insertion Elements, Plasmids and Episomes" (A. I. Bukhari, J. A. Shapiro, and S. L. Adhya, eds.), pp. 521–530. Cold Spring Harbor Laboratory, Cold Spring Harbor, New York.
Fargie, B., and Hooloway, B. W. (1965). *Genet. Res.* **6**, 284–299.
Friend, E. J., and Hopwood, D. A. (1971). *J. Gen. Microbiol.* **68**, 187–197.
Gabor, M. (1965). *Genetics* **52**, 905–913.
Gabor-Hotchkiss, M. (1972). *In* "Uptake of Informative Molecules by Living Cells" (L. Ledoux, ed.), pp. 252–234. North-Holland Publ., Amsterdam.
Gibbins, A. M., and Gregory, K. F. (1972). *J. Bacteriol.* **111**, 129–141.
Gibson, A. H., and Pagan, J. D. (1977). *Planta* **134**, 17–22.
Graham, P. H. (1964). *J. Gen. Microbiol.* **35**, 511–517.
Haas, D., and Holloway, B. W. (1976). *Mol. Gen. Genet.* **144**, 243–251.
Hayes, W. (1968). "The Genetics of Bacteria and Their Viruses," 2nd ed. Blackwell, Oxford.
Hedges, R. W., and Jacob, A. E. (1977). *FEMS Microbiol. Lett.* **2**, 15–19.
Hedges, R. W., Jacob, A. E., and Crawford, I. P. (1977). *Nature (London)* **267**, 283–284.

Hendry, G. S., and Jordan, D. C. (1969). *Can. J. Microbiol.* **15,** 671-675.

Heumann, W. (1968). *Mol. Gen. Genet.* **102,** 132-144.

Heumann, W., and Springer, R. (1977). *Mol. Gen. Genet.* **150,** 73-79.

Heumann, W., Pühler, A., and Wagner, E. (1971). *Mol. Gen. Genet.* **113,** 308-315.

Heumann, W., Pühler, A., and Wagner, E. (1973). *Mol. Gen. Genet.* **126,** 267-274.

Heumann, W., Kamberger, W., Pühler, A., Rörsch, A., Springer, R., and Burkardt, H. J. (1974). *In* "Proc. 1st Int. Symp. Nitrogen. Fixation" (W. E. Newton and C. J. Nyman, eds.), pp. 383-390. Washington State Univ. Press, Pullman, Washington.

Higashi, S. (1967). *J. Gen. Appl. Microbiol.* **13,** 391-403.

Hirsch, P. R. (1979). *J. Gen. Microbiol.* **113,** 219-228.

Holloway, B. W. (1978). *J. Bacteriol.* **133,** 1078-1082.

Holloway, B. W. (1979). *Plasmid* **2,** 1-19.

Holloway, B. W., Krishnapillai, V., and Morgan, A. F. (1979). *Microbiol. Rev.* **43,** 73-102.

Hooykaas, P. J. J., Klapwijk, P. M., Nuti, M. P., Schilperoort, R. A., and Rörsch, A. (1977). *J. Gen. Microbiol.* **98,** 477-484.

Hu, S., Ohtsubo, E., and Davidson, N. (1975). *J. Bacteriol.* **122,** 749-763.

Jacob, A. E., Cresswell, J. M., Hedges, R. W., Coetzee, J. N., and Beringer, J. E. (1976). *Mol. Gen. Genet.* **147,** 315-323.

Jacob, A. E., Cresswell, J. M., and Hedges, R. W. (1977). *FEMS Microbiol. Lett.* **1,** 71-74.

Johnston, A. W. B., and Beringer, E. J. (1977). *Nature (London)* **267,** 611-613.

Johnston, A. W. B., Beynon, J. L., Buchanan-Wollaston, A. V., Setchell, S. M., Hirsch, P. R., and Beringer, J. E. (1978a). *Nature (London)* **276,** 635-636.

Johnston, A. W. B., Bibb, M. J., and Beringer, J. E. (1978b). *Mol. Gen. Genet.* **165,** 323-330.

Johnston, A. W. B., Setchell, S. M., and Beringer, J. E. (1978c). *J. Gen. Microbiol.* **104,** 209-218.

Julliot, J. S., and Boistard, P. (1979). *Mol. Gen. Genet.* **173,** 289-298.

Kern, H. (1965a). *Arch. Mikrobiol.* **51,** 140-155.

Kern, H. (1965b). *Arch. Mikrobiol.* **52,** 206-223.

Kern, H. (1969). *Arch. Mikrobiol.* **66,** 63-68.

Kiss, G. B., Vincze, E., Kalman, Z., Forrai, T., and Kondorosi, A. (1979). *J. Gen. Microbiol.* **113,** 105-118.

Kiss, G. B., Dobo, K., Dusha, I., Breznovits, A., Orosz, L., Vincze, E., and Kondorosi, A. (1980). *J. Bacteriol.* **141,** 121-128.

Kleczkowska, J. (1957). *Can. J. Microbiol.* **3,** 171-180.

Klein, T., and Klein, R. M. (1953). *J. Bacteriol.* **66,** 220-228.

Kondorosi, A., Barabas, I., Svab, Z., Orosz, L., Sik, T., and Hotchkiss, R. D. (1973). *Nature (London) New Biol.* **246,** 153-154.

Kondorosi, A., Orosz, L., Svab, Z., and Sik, T. (1974). *Mol. Gen. Genet.* **132,** 153-163.

Kondorosi, A., Kiss, G. B., Forrai, T., Vincze, E., and Banfalvi, Z. (1977a). *Nature (London)* **268,** 525-527.

Kondorosi, A., Svab, Z., Kiss, G. B., and Dixon, R. A. (1977b). *Mol. Gen. Genet.* **151,** 221-226.

Kondorosi, A., Vincze, E., Johnston, A. W. B., and Beringer, J. E. (1980). *Mol. Gen. Genet.* **178,** 403-408.

Kowalczuk, E., and Lorkiewicz, Z. (1977). *Acta Microbiol. Pol.* **26,** 9-18.

Kowalski, M. (1967). *Acta Microbiol. Pol. Ser. A.* **16,** 7-12.

Kowalski, M. (1970a). *Acta Microbiol. Pol. Ser. A.* **2,** 109-114.

Kowalski, M. (1970b). *Acta Microbiol. Pol. Ser. A.* **2,** 115-122.

Kowalski, M. (1971a). *Genet. Pol.* **12,** 201-204.

Kowalski, M. (1971b). *Plant Soil* Special Vol., pp. 63-66.

Kowalski, M. (1976). *In* "Symbiotic Nitrogen Fixation in Plants" (P. S. Nutman, ed.), pp. 63-67. Cambridge Univ. Press, London and New York.

Kowalski, M., and Dénarié, J. (1972). *C.R. Acad. Sci. Paris Ser. D.* **275,** 141-144.

Krasilnikov, N. A. (1941). *C.R. Acad. Sci. USSR* **31**, 90–92.

Krasilnikov, N. A. (1945). *Mikrobiologya* **14**, 230–236.

Kuykendall, L. D. (1979). *Appl. Environ. Microbiol.* **37**, 862–866.

Laird, D. G. (1932). *Arch. Mikrobiol.* **3**, 159–193.

Lie, T. A., and Winarno, R. (1979). *Plant Soil* **51**, 135–142.

Lorkiewicz, Z., Zurkowski, W., Kowalczuk, E., and Gorska-Melke, A. (1971). *Acta Microbiol. Pol. Ser. A.* **3**, 101–107.

Lotz, W., and Mayer, F. (1972). *J. Virol.* **9**, 160–173.

Lotz, W., Pfister, H., Gissmann, L., and Lurz, R. (1974). *Zentralbl. Bakteriol. Hyg. I* **228**, 175–178.

Low, K. B. (1972). *Bacteriol. Rev.* **36**, 587–607.

Ludwig, R. A., and Johansen, E. (1980). *Plasmid* **3**, 359–361.

Ludwig, R. A., and Signer, E. Ŗ. (1977). *Nature (London)* **267**, 245–248.

Mackenzie, C. R., and Jordan, D. C. (1970). *Biochem. Biophys. Res. Commun.* **40**, 1008–1012.

Mackenzie, C. R., and Jordan, D. C. (1972). *Can. J. Microbiol.* **18**, 1168–1170.

Maier, R. J., and Brill, W. J. (1976). *J. Bacteriol.* **127**, 763–769.

Maier, R. J., Bishop, P. E., and Brill, W. J. (1978a). *J. Bacteriol.* **134**, 1199–1201.

Maier, R. J., Postgate, J. R., and Evans, H. J. (1978b). *Nature (London)* **276**, 494–495.

Malek, W., and Kowalski, M. (1977). *Acta Microbiol. Pol.* **9**, 345–350.

Marshall, K. C. (1956). *Nature (London)* **177**, 92.

Martinez, J., and Clarke, P. H. (1975). *Proc. Soc. Gen. Microbiol.* **3**, 51.

Meade, H. M., and Signer, E. R. (1977). *Proc. Natl. Acad. Sci. U.S.A.* **74**, 2076–2078.

Mergeay, M., Tshitenge, G., Jacquemin, J. M., Gerits, Y., and Ledoux, L. (1973). *Arch. Int. Physiol. Biochim.* **81**, 805.

Moskalenko, L. N., and Rautenstein, Y. I. (1969). *Mikrobiologiya* **38**, 340–345.

Nagahari, K., Koshikawa, T., and Sakaguchi, K. (1979). *Mol. Gen. Genet.* **171**, 115–119.

Napoli, C., and Albersheim, P. (1980). *J. Bacteriol.* **141**, 1454–1456.

Niel, C., Guillaume, J. B., and Bechet, M. (1977). *Can. J. Microbiol.* **23**, 1178–1181.

Nuti, M. P., Lepidi, A. A., Prakash, R. K., Schilperoort, R. A., and Cannon, F. C. (1979). *Nature (London)* **282**, 533–535.

O'Gara, F., and Dunican, L. K. (1973). *J. Bacteriol.* **116**, 1177–1180.

Ohtsubo, H., and Ohtsubo, E. (1977). *In* "DNA Insertion Elements, Plasmids and Episomes" (A. I. Bukhari, J. A. Shapiro, and J. L. Adhya, eds.), pp. 49–63. Cold Spring Harbor Laboratory, Cold Spring Harbor, New York.

Orosz, L. (1980). *Genetics* **94**, 265–276.

Orosz, L., and Sik, T. (1970). *Acta Microbiol. Acad. Sci. Hung.* **17**, 185–194.

Orosz, L., Svab, Z., Kondorosi, A., and Sik, T. (1973). *Mol. Gen. Genet.* **125**, 341–350.

Orosz, L., Pay, A., and Dallmann, G. (1980a). *Mol. Gen. Genet.* **179**, 163–167.

Orosz, L., Rostas, K., and Hotchkiss, R. D. (1980b). *Genetics* **94**, 249–263.

Ördögh, F., and Szende, K. (1961). *Acta Microbiol. Acad. Sci. Hung.* **8**, 65–71.

Page, W. J. (1978). *Can. J. Microbiol.* **24**, 209–214.

Palomares, A., Montoya, E., and Olivares, J. (1978). *Microbios* **21**, 33–39.

Parijskaya, A. N. (1973). *Microbiologiya* **42**, 119–121.

Patel, J. J. (1976). *Can. J. Microbiol.* **22**, 204–212.

Pfister, H., and Lotz, W. (1974). *Zantralbl. Bakteriol. Hyg. I.* **228**, 179–182.

Pittard, J., and Adelberg, E. A. (1963). *J. Bacteriol.* **85**, 1402–1408.

Prakash, R. K., Hooykaas, P. J. J., Ledeboer, A. M., Kijne, J., Schilperoort, R. A., Nuti, M. P., Lepidi, A. A., Casse, F., Boucher, C., Julliot, J. S., and Dénarié, J. (1980). *In* "Nitrogen Fixation" (W. E. Newton and W. H. Orme-Johnson, eds.), Vol. II, pp. 139–163. Univ. Park Press, Baltimore, Maryland.

Pühler, A., and Burkardt, H. J. (1978). *Mol. Gen. Genet.* **162**, 163-171.

Raina, J. K., and Modi, V. V. (1969). *J. Gen. Microbiol.* **57**, 125-130.

Reif, H. J., and Saedler, H. (1975). *Mol. Gen. Genet.* **137**, 17-28.

Reijnders, L. (1976). *J. Theor. Biol.* **61**, 245-248.

Ronson, C. W., and Primrose, S. B. (1979). *J. Gen. Microbiol.* **112**, 77-88.

Roslycky, E. B. (1967). *Can. J. Microbiol.* **13**, 431-432.

Sadasivam, S., and Pandian, S. (1978). *Curr. Sci.* **42**, 863.

Sanders, R. E., Carlson, R. W., and Albersheim, P. (1978). *Nature (London)* **271**, 240-242.

Sanderson, K. E. (1972). *Bacteriol. Rev.* **36**, 558-586.

Schöffl, F., Wunder, M., Tucher, R., Heumann, W., and Pühler, A. (1974). *Zentralbl. Bakteriol. Hyg. I.* **228**, 155-161.

Schwinghamer, E. A. (1966). *Can. J. Microbiol.* **12**, 395-407.

Schwinghamer, E. A. (1971). *Soil Biol. Biochem.* **3**, 355-363.

Schwinghamer, E. A. (1977). *In* "A Treatise on Dinitrogen Fixation" (R. W. F. Hardy and W. S. Silver, eds.), Section III, pp. 577-622. Wiley (Interscience), New York.

Schwinghamer, E. A., and Reinhardt, D. J. (1963). *Aust. J. Biol. Sci.* **16**, 597-605.

Schwinghamer, E. A., Pankhurst, C. E., and Whitfeld, P. R. (1973). *Can. J. Microbiol.* **19**, 359-368.

Sen, M., Pal, K. T., and Sen, S. P. (1969). *Antonie van. Leeuwenbroek* **35**, 533-540.

Shanmugam, K. T., O'Gara, F., Andersen, K., and Valentine, R. C. (1978). *Annu. Rev. Plant Physiol.* **29**, 263-276.

Sik, T., and Orosz, L. (1971). *Plant Soil* Special Vol., pp. 57-62.

Sik, T., Horvath, J., and Chatterjee, S. (1980). *Mol. Gen. Genet.* **178**, 511-516.

Sistrőm, W. (1977). *J. Bacteriol.* **131**, 526-532.

Skotnicki, M. L., and Rolfe, B. G. (1978). *J. Bacteriol.* **133**, 518-526.

Skotnicki, M. L., and Rolfe, B. G. (1979). *Aust. J. Biol. Sci.* **32**, 501-517.

Springer, R., and Heumann, W. (1978). *Mol. Gen. Genet.* **165**, 57-63.

Staniewski, R. (1970). *Can. J. Microbiol.* **16**, 1003-1009.

Staniewski, R., and Kowalska, W. (1974). *Acta Microbiol. Pol. Ser. A* **6**, 183-186.

Staniewski, R., and Rugala, A. (1975). *Acta Microbiol. Pol. Ser. A* **8**, 151-160.

Staniewski, R., Lorkiewicz, Z., and Chomicka, Z. (1971). *Acta Microbiol. Pol. Ser. A* **3**, 97-100.

Stanisich, V. A., and Holloway, B. W. (1971). *Genet. Res. Cambridge* **17**, 169-172.

Stanley, J., and Dunican, L. K. (1979). *Mol. Gen. Genet.* **174**, 211-220.

Streicher, S. L., Shanmugam, K. T., Ausubel, F., Morandi, C., and Goldberg, R. B. (1974). *J. Bacteriol.* **129**, 815-821.

Svab, Z., Kondorosi, A., and Orosz, L. (1978). *J. Gen. Microbiol.* **106**, 321-327.

Szende, K. (1971). *Plant Soil* Special Vol., pp. 81-84.

Szende, K., and Örödögh, F. (1960). *Naturwissenschaften* **47**, 404-405.

Szende, K., Sik, T., Ördögh, F., and Györffy, B. (1961). *Biochim. Biophys. Acta* **47**, 215-217.

Takahashi, J., and Quadling, C. (1961). *Can. J. Microbiol.* **7**, 455-465.

Tubb, R. S. (1974). *Nature (London)* **244**, 459-460.

Van Vliet, F., Silva, B., Van Montagu, M., and Schell, J. (1978). *Plasmid* **1**, 446-455.

Vincent, J. M. (1977). *In* "A Treatise on Dinitrogen Fixation" (R. W. F. Hardy and W. S. Silver, eds.), Section III, pp. 277-366. Wiley (Interscience), New York.

Vincent, J. M. (1980). *In* "Nitrogen Fixation" (W. E. Newton and W. H. Orme-Johnson, eds.), Vol. II, pp. 103-129. Univ. Park Press, Baltimore, Maryland.

Weidel, R. W. (1958). *Annu. Rev. Microbiol.* **12**, 27-48.

Wolpert, J. S., and Albersheim, P. (1976). *Biochem. Biophys. Res. Commun.* **70**, 729-737.

Wu, T.T. (1966). *Genetics* **54**, 405-410.

Yu, K. K.-Y., and Jordan, D. C. (1971). *Can. J. Microbiol.* **17**, 1283-1286.

Zelazna, I. (1964). *Acta Microbiol. Pol. Ser. A* **13**, 283-290.
Zelazna-Kowalska, I., and Kowalski, M. (1978). *Acta Microbiol. Pol.* **10**, 339-345.
Zelazna-Kowalska, I., and Lorkiewicz, Z. (1971). *Acta Microbiol. Pol. Ser. A* **3**, 11-20.
Zurkowski, W., and Lorkiewicz, Z. (1977). *Mol. Gen. Genet.* **156**, 215-219.
Zurkowski, W., and Lorkiewicz, Z. (1978). *Acta Microbiol. Pol.* **10**, 309-319.
Zurkowski, W., and Lorkiewicz, Z. (1979). *Arch. Microbiol.* **123**, 195-201.
Zurkowski, W., Hoffmann, M., and Lorkiewicz, Z. (1973). *Acta Microbiol. Pol. Ser. A* **5**, 55-60.

INTERNATIONAL REVIEW OF CYTOLOGY, SUPPLEMENT 13

Indigenous Plasmids of *Rhizobium*

J. Dénarié,* P. Boistard,* and Francine Casse-Delbart†

*Laboratoire de Biologie Moléculaire des Relations Plantes–Microorganismes, I.N.R.A.,
Castanet-Tolosan, and †Laboratoire de Biologie Cellulaire, I.N.R.A., Versailles, France*

A. G. Atherly, J. O. Berry, and P. Russell

Department of Genetics, Iowa State University, Ames, Iowa

I. Introduction: Early Genetic Evidence for Plasmid Control of Symbiotic Properties (1967–1976)	225
II. Physical Evidence for the Presence of Large Plasmids	227
A. Isolation Procedures	227
B. Characterization of *Rhizobium* Plasmids	231
C. The Presence of Large Plasmids in Fast-Growing Rhizobia	234
III. Genetic Methods for Plasmid Studies	235
A. Plasmid Genetic Markers	236
B. Transposon-Induced Mutagenesis	236
C. Plasmid Transfer and Curing	237
IV. Plasmid Control of Early Functions in Symbiosis	238
A. Host Range Specificity	238
B. Nodule Formation	239
V. Plasmid Control of Late Functions in Symbiosis	240
A. The *fix* Genes	241
B. The *nif* Genes	241
C. Transcription of Plasmid DNA in Bacteroids	243
D. The *hup* Genes	243
VI. Concluding Remarks	243
References	244
Note Added in Proof	246

I. Introduction: Early Genetic Evidence for Plasmid Control of Symbiotic Properties (1967–1976)

Symbiotic properties, such as the ability to induce nodule formation on the roots of a legume host (Nod$^+$) and the ability to fix nitrogen within a nodule (Fix$^+$), have long been known to be relatively unstable in certain laboratory cultures of *Rhizobium*. The first genetic evidence that plasmid genes in *Rhizobium* may be involved in the establishment of symbiosis and host range specificity was reported by Higashi (1967). He mixed together two *Rhizobium* species, *R. trifolii* K102 and *R. phaseoli* 110, in liquid culture for 30 minutes and then eliminated the *R. trifolii* cells from the mixed culture using a species-specific phage. The bacteria surviving this treatment were found to be infective on clover, leading to the conclusion that host specificity had been transferred from *R. trifolii* to *R. phaseoli*. Additional evidence reported in his paper supported the idea of conjugational transfer of host specificity mediated by an extrachromosomal element. Unfortunately, the possibility exists that a few phage-resistant *R. trifolii* cells survived this selection and were present when clover infectivity was tested. Although physiological and serological tests were made, it was not firmly established that all the *R. trifolii* cells had been eliminated and that conjugation had indeed occurred.

In the same study Higashi also found that loss of infectiveness (the ability to form nodules) occurred in both *R. trifolii* and *R. phaseoli* following treatment with acridine orange. Since acridine dyes are commonly used to eliminate plasmids from gram-negative bacteria, these results further suggested that plasmid genes may be involved in the control of nodule formation.

Several other early investigators also produced data suggesting a link between Fix and Nod functions with plasmids. Van Rensburg *et al.* (1968) described a loss of infectivity in *Rhizobium japonicum* following ultraviolet or acridine orange treatment. However, van Rensburg and Strijdom (1971) subsequently found that in fact these treatments had eliminated slow-growing infective bacteria and that a faster growing noninfective strain was less sensitive to acridine and ultraviolet light. Dunican and Cannon (1971) selected viomycin-resistant derivatives of *R. trifolii* strain T1 after acridine orange or ethidium bromide treatment. All the Vio^r variants examined were Fix^-, and the authors interpreted these results as giving support for plasmid control of Fix^+ functions in *R. trifolii*. Surprisingly the use of a potent mutagen, nitrosoguanidine, decreased the frequency of Vio^r variants as compared to their spontaneous appearance. Parijskaya (1973) found that sub-bacteriostatic doses of acridine orange in *Rhizobium meliloti* strain L-1 induced partial or total loss of infectivity at a high frequency. She suggested that this observation was consistent with the involvement of episomal elements in the determination of infectivity in nodular bacteria. Also, Zurkowski *et al.* (1973) observed irreversible elimination of infectivity after treatment of *R. trifolii* T37 with acriflavine or sodium dodecyl sulfate.

The loss of effectiveness or of nodule-forming ability associated with these episome-eliminating treatments provided early data for some plasmid-borne symbiotic properties in *Rhizobium*. However, no physical proof was obtained that plasmids had actually been eliminated from the nonnodulating strains. Actually, as pointed out in 1976 by Beringer, loss of such property after treatment with a "curing" agent could be due to a number of effects and in particular to the selection of a phenotype which is coincidentally defective in infectivity or effectiveness.

In a related study, Cole and Elkan (1973) found evidence suggesting plasmid-encoded antibiotic resistance genes in *R. japonicum*. They proposed that plasmids coding for resistance to chloramphenicol, penicillin G, and neomycin could be eliminated by treatment with acridine orange. In addition, these markers were transferable via conjugation with *Agrobacterium tumefaciens*. It would be interesting to determine if any relationship exists between these antibiotic resistance genes and nodulation ability in *R. japonicum*. Again, no physical evidence was presented for the actual presence of a plasmid or for plasmid removal in the treated strains.

II. Physical Evidence for the Presence of Large Plasmids

The physical presence of plasmids in *Rhizobium* was suggested by DNA density studies (Sutton, 1974) and demonstrated in some strains by dye-buoyant density ultracentrifugation (Klein *et al.*, 1975; Tshitenge *et al.*, 1975; Dunican *et al.*, 1976; Zurkowski and Lorkiewicz, 1976; Olivares *et al.*, 1977). The molecular weights reported were in the range $25-65 \times 10^6$. Simultaneously it was shown that another member of the Rhizobiaceae family, *A. tumefaciens* which induces tumors on dicotyledonous plants, harbored large plasmids with molecular weights ranging from about 100 to 160×10^6. These plasmids were shown to confer oncogenicity on the bacteria (Van Larebeke *et al.*, 1974; Watson *et al.*, 1975).

Agrobacterium and *Rhizobium* are related bacteria, and the induction of nodules has some analogies with tumor induction. With these facts in mind, Nuti *et al.* (1977) applied to *Rhizobium* the techniques developed for the extraction of large plasmids (molecular weights higher than 100×10^6) from *A. tumefaciens*. They were the first to demonstrate convincingly the presence of large plasmids in several *Rhizobium* species and to provide an estimate of their very high molecular weight (up to 400×10^6).

A. ISOLATION PROCEDURES

Since this key discovery by Nuti *et al.* (1977), several procedures have been developed for the extraction and characterization of large plasmids. In this section we shall discuss in some detail the principles of the procedures used for *Rhizobium* plasmid isolation (Table I).

The cleared lysate procedure (Clewell and Helinski, 1969; Guerry, *et al.*, 1973) for the isolation of plasmids is very effective with various gram-negative bacteria. Detergent cell lysis, followed by centrifugation, allows the separation of plasmids from the pelleted chromosome–membrane complex. This procedure is effective in the isolation of small plasmids in the molecular weight range $25-65 \times 10^6$ from *Rhizobium* (Dunican *et al.*, 1976; Zurkowski and Lorkiewicz, 1976; Olivares *et al.*, 1977; Julliot and Boistard, 1979); but difficulties have been encountered in applying these techniques to the isolation of large covalently

TABLE I

SOME ANALYTICAL AND PREPARATIVE PROCEDURES USED FOR *Rhizobium* LARGE-PLASMID ISOLATION

Reference	Lysis and dissociation from membrane–chromosome complex[a]	CCC DNA enrichment	DNA concentration[b]	Resolution step[c]	Molecular weight range ($\times 10^6$)
Analytical procedures					
Casse *et al.* (1979)	Alkaline SDS buffer	Alkaline denaturation extraction at high salt concentration	Ethanol	Agarose gel electrophoresis	Up to 300
Hirsch *et al.* (1980)	Pronase plus SDS	Alkaline denaturation, SDS-NaCl precipitation	PEG	Agarose gel electrophoresis	Up to 300
Eckhardt (1978), used in Casse-Delbart *et al.* (1981)	Lysozyme plus SDS RNase			Agarose gel electrophoresis	>300
Preparative procedures					
Hansen and Olsen (1978)	Lysozyme plus SDS heat pulses	Alkaline denaturation, SDS-NaCl precipitation	PEG	CsCl–Eth Br gradient	Up to 300
Schwinghamer (1980), used in Casse-Delbart *et al.* (1981)	Detergent washing (0.1% sarkosyl), sucrose lysozyme osmotic shock, sarkosyl shearing		—	CsCl–Eth Br	>300

[a] SDS, Sodium dodecyl sulfate.
[b] PEG, Polyethylene glycol.
[c] Eth Br, ethidium bromide.

closed circular (CCC) DNA molecules in *A. tumefaciens* (Zaenen *et al.*, 1974; Ledeboer *et al.*, 1976) and *Rhizobium* (Nuti *et al.*, 1977). These large plasmids are probably not dissociated from the chromosome–membrane complex during this treatment.

All the successful procedures for large-plasmid isolation from Rhizobiaceae are based on the use of a strongly polar detergent such as sarkosyl or sodium dodecyl sulfate (SDS) which gives a more complete lysis than neutral detergents. Large plasmids are difficult to isolate because of (1) their association with the chromosome–membrane complex, and (2) the fact that plasmid isolation procedures depend upon separation of the intact CCC form of the molecule: One nick in one strand results in loss of the molecule. Obviously the larger a plasmid is, the more sensitive it is both to enzymic (nucleases) or mechanical (shearing) degradation.

1. *Lysis and Dissociation from the Chromosome–Membrane Complex*

Usually cell lysis is performed by disrupting the cell wall by enzymes and a polar detergent. Enzymic degradation of the cell wall is generally achieved with lysozyme, but lysozyme has recently been replaced by pronase (Gross *et al.*, 1979; Hirsch *et al.*, 1980). Alternatively lysis can be purely chemical (Casse *et al.*, 1979a). The most effective polar detergent is SDS, except when the lysates are to be submitted directly to ultracentrifugation in cesium chloride gradients; in these cases SDS is replaced by sarkosyl (Nuti *et al.*, 1977; Spitzbarth *et al.*, 1979; Zurkowski and Lorkiewicz, 1979; Schwinghamer, 1980). Prewashing of the cells in a diluted solution of sarkosyl (0.1%), followed by an osmotic shock to facilitate access of lysozyme to the cell wall increased the efficiency of *Rhizobium* cell lysis (Schwinghamer, 1980), resulting in a more reproducible recovery of very large plasmids both for analytical and preparative procedures (Schwinghamer, 1980; Rosenberg *et al.*, 1981a). For some as yet unexplained reason, slow-growing rhizobia must be prewashed with at least 0.50 M sodium chloride before lysis, or no plasmid recovery is seen (A. G. Atherly and P. Russell, unpublished).

All isolation procedures for large plasmids involve treatments that contribute to separation of the plasmid from the folded chromosome–membrane complex. These include physical treatments such as shearing of the lysate (Nuti *et al.*, 1977; Zurkowski and Lorkiewicz, 1979; Gross *et al.*, 1979; Schwinghamer, 1980), heat pulses (Hansen and Olsen, 1978), chemical treatment such as alkaline lysis (Nuti *et al.*, 1977; Casse *et al.*, 1979a), and enzymic treatment with RNase (Eckhardt, 1978; Spitzbarth *et al.*, 1979), pronase (Gross *et al.*, 1979; Hirsch *et al.*, 1980), or both RNase and pronase (Bechet and Guillaume, 1978).

2. *Removal of the Chromosome and Cell Debris*

Several procedures have been described for plasmid DNA enrichment after cell lysis. The methods reported for CCC DNA enrichment make use of the irreversible denaturation of linear DNA at pH values above 12 (Currier and Nester, 1976; Casse *et al.*, 1979a; Gross *et al.*, 1979; Hirsch *et al.*, 1980). Denaturation of CCC molecules is reversible because of the intertwisting of both strands; however, chromosomal DNA is irreversibly denatured.

Removal of denatured DNA and cell debris can be achieved by phenol treatment in the presence of high salt concentrations (Currier and Nester, 1976). The remaining DNA is then precipitated by ethanol. These conditions allowed isolation of plasmids with molecular weights up to 160×10^6 from *A. tumefaciens* (Currier and Nester, 1976) and 118×10^6 from *R. japonicum* (Gross *et al.*, 1979). Decreasing the shearing at various steps of the method allowed isolation of plasmids of up to 300×10^6 molecular weight from *A. tumefaciens, Rhizobium meliloti*, and *R. leguminosarum* (Casse *et al.*, 1979b) and from *R. japonicum* (A. G. Atherly and P. Russell, unpublished results; C. Boucher and E. A. Schwinghamer, unpublished results).

Another plasmid isolation approach makes use of selective precipitation of the chromosome–membrane complex in the presence of SDS and a high concentration of sodium chloride (Hansen and Olsen, 1978). As opposed to the previous technique, DNA precipitation is performed with polyethylene glycol (PEG), not ethanol, since no deproteinization step is included. During ethanol precipitation proteins coprecipitate with the DNA, making the pellet very difficult to resuspend. However, the PEG procedure allows the isolation of plasmids of up to $250-300 \times 10^6$ molecular weight from *Pseudomonas aeruginosa* (Hansen and Olsen, 1978), *A. tumefaciens, R. leguminosarum* (Hirsch *et al.*, 1980), and *R. trifolii* (Beynon *et al.*, 1980).

Finally some procedures do not involve removal of the chromosomal debris; after lysis and dissociation of the plasmids from the complex, the lysates are directly submitted to separation either by ultracentrifugation in density gradients (Bechet and Guillame, 1978; Spitzbarth *et al.*, 1979; Zurkowski and Lorkiewicz, 1979; Schwinghamer, 1980) or by agarose gel electrophoresis (Eckhardt, 1978; Rosenberg *et al.*, 1981a).

The procedures which involve the least number of destructive operations (stirring, pipetting, precipitation, and resuspension) allow recovery of the largest plasmids. Plasmids of 460×10^6 molecular weight were found in *R. trifolii* by Zurkowski and Lorkiewicz (1979), and plasmids of more than 300×10^6 molecular weight were isolated from various *R. meliloti* strains (Rosenberg *et al.*, 1981b) using the procedures of Eckhardt (1978) and Schwinghamer (1980). Nevertheless a serious drawback of these procedures is that large-scale isolation requires ultracentrifugation of large volumes in cesium chloride-ethidium bromide gradients.

3. *Resolution of Plasmid DNA*

For preparative plasmid isolation ultracentrifugation in cesium chloride equilibrium gradients is required. Density differences between *Rhizobium* plasmid DNA and chromosomal DNA are generally not sufficient to allow their full resolution by equilibrium density gradients alone (Sutton, 1974; Bechet and Guillame, 1978; Schwinghamer and Dennis, 1979). Therefore an intercalating dye such as ethidium bromide, which discriminates between CCC DNA and linear or open circular molecules (Radloff *et al.*, 1967) on the basis of preferential binding of the dye to freely rotating strands, was used for *Rhizobium* large-scale plasmid preparation (Casse *et al.*, 1979a; Adachi and Iyer, 1980; Hirsch *et al.*, 1980; Rosenberg *et al.*, 1981b). In some instances ethidium bromide can be replaced by propidium diiodide, because propidium diiodide provides a better separation of CCC and linear DNA (Nuti *et al.*, 1977). Since dye-buoyant density ultracentrifugation is based upon structural differences between plasmid DNA and chromosomal DNA, it does not discriminate between different plasmids that can be present in the same strain.

For analytical purposes two methods have been used: sucrose gradients and agarose gel electrophoresis. An alkaline sucrose gradient was used in the final step after alkaline extraction by Nuti *et al.* (1977). The plasmids were recovered as denatured interwisted DNA. However, this does not prevent an estimation of plasmid molecular weight either by sedimentation coefficient or by a study of reassociation kinetics. In addition, neutral sucrose gradients were used with success for detection and estimation of the molecular weight of very large plasmids of about 460×10^6 (Zurkowski and Lorkiewicz, 1979).

The agarose gel elctrophoresis procedure developed by Meyers *et al.* (1976) is now commonly used for rapid characterization of *Rhizobium* plasmids (Casse *et al.*, 1979a,b; Gross *et al.*, 1979; Prakash *et al.*, 1980; Hirsch *et al.*, 1980; Brewin *et al.*, 1980a). Gel electrophoresis procedures present many advantages: (1) They avoid the spending of the time and money involved in cesium chloride centrifugation or DNA labeling with radioisotopes; (2) they allow the screening of a large number of strains; and (3) they can resolve CCC plasmid molecules of up to more than 300×10^6 molecular weight (Hansen and Olsen, 1978; Casse *et al.*, 1979a). One additional advantage in the case of large plasmids is that open circular forms cannot enter the gel, and linear forms migrate much more rapidly than CCC molecules; therefore resolution of plasmids present in a given strain is straightforward. Agarose electrophoresis of crude DNA extracts showed that the great majority of the *Rhizobium* strains investigated carried more than one large plasmid. Thus, there is a need to modify agarose gel electrophoresis procedures to make them useful for preparative isolation of the various plasmids contained in a strain.

B. Characterization of *Rhizobium* Plasmids

1. *Estimation of Molecular Weight*

The first evidence for the presence of plasmids of very high molecular weight ($> 300 \times 10^6$) in *Rhizobium* came from reassociation kinetics studies

(Nuti *et al.*, 1977; Prakash *et al.*, 1980). As previously mentioned, this technique does not require the isolation of intact CCC DNA and therefore provides a
procedure for molecular weight estimation for very large plasmids. Either alkaline or neutral sucrose gradients were used to calculate a sedimentation coefficient which could then be converted to a molecular weight (Dunican *et al.*, 1976;
Zurkowski and Lorkiewicz, 1979). *Rhizobium trifolii* strains T12 and 24 were
shown to carry plasmids of 190 and 460 $\times$ 10^6 molecular weight by neutral
sucrose gradients (Zurkowski and Lorkiewicz, 1979).

Contour length measurements obtained by electron microscopy provide more
accurate molecular weight estimations. Large plasmids with molecular weights
ranging from 90 to 270 $\times$ 10^6 were observed in *R. meliloti* (Casse *et al.*, 1979a;
Spitzbarth *et al.*, 1979), and up to 350 $\times$ 10^6 in *R. leguminosarum* (Schwinghamer and Dennis, 1979). Measurement of the relative mobility in agarose gel
electrophoresis under well-defined conditions allows an estimation of molecular
weight that compares favorably with contour length measurements within the
range 2-90 $\times$ 10^6 (Meyers *et al.*, 1976) and 90-140 $\times$ 10^6 (Casse *et al.*,
1979a). Within these ranges there is a linear relationship between the log of the
relative migration (RM) and the log of the molecular weight (determined by
electron microscopy). For plasmids of molecular weight higher than 150 $\times$ 10^6 it
is necessary to utilize the equation describing the nonlinear relationship between
RM in agarose; the molecular weights of plasmids pMG1 and pMG5 from *Pseudo-*

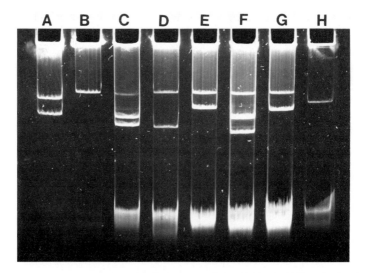

FIG. 1. Agarose gel electrophoresis of lysates according to Eckhardt (1978). (A) *Agrobacterium
tumefaciens* C58; (B) *R. meliloti* A145 (Netherlands); (C) *R. meliloti* RF22 (South Africa); (D) *R.
meliloti* L5-30 (Poland); (E) *R. meliloti* V7 (Canada); (F) *R. meliloti* 102F28 (United States); (G) *R.
meliloti* B294 (South America); (H) *P. putida* pMG1. From Hansen and Olsen (1978).

monas putida were estimated by electron microscopy to be 312 and 280 $\times$ 10^6, respectively (Hansen and Olsen, 1978), and that of plasmid pATC58 to be 273 $\times$ 10^6 (R. Villarroel-Mandiola and M. Van Montagu, personal communication). These reference plasmids were used to show that (Fig. 1) the megaplasmid present in all *R. meliloti* strains investigated had a molecular weight of 300 $\times$ 10^6 or more (Rosenberg *et al.*, 1981a).

2. Density of Plasmid DNA

Although buoyant density differences between chromosomal and plasmid DNA are not sufficient to allow preparative isolation of *Rhizobium* plasmids (Section II,A,3), the existence of a satellite DNA in a density gradient was used as an indication of the presence of plasmids (Sutton, 1974; Tshitenge *et al.*, 1975; Bechet and Guillaume, 1978). It has been demonstrated recently that *Rhizobium* plasmids are actually slightly less dense than chromosomal DNA (Schwinghamer and Dennis, 1979; Spitzbarth *et al.*, 1979). *Rhizobium meliloti* strain RCR 2011 has a plasmid density of 1.717 gm/cm^3, in contrast to the density of chromosomal DNA which is 1.722 gm/cm^3 (Spitzbarth *et al.*, 1979). The density of satellite DNAs of *Rhizobium* from *Lotus* nodules was 1.718 gm/cm^3, of a *R. trifolii* strain 1.718 gm/cm^3 (Sutton, 1974), and of *R. leguminosarum* 1.715 gm/cm^3 (Schwinghamer and Dennis 1979). Spitzbarth *et al.* (1979) suggested that these findings raised the possibility that indigenous *Rhizobium* plasmids may have a common nonchromosomal origin or carry conserved sequences.

3. DNA Sequence Homologies

Sequence homologies give an indication of the conservation of DNA fragments. Thus they suggest either a common origin or strong selection for the maintenance of DNA fragments having an essential role.

Prakash *et al.* (1981) found that *R. leguminosarum* plasmids shared a considerable degree of homology from strain to strain and with plasmids of *R. trifolii* RCR5, but much less homology with the smaller plasmid of *R. meliloti* strain L5-30. The smallest plasmid of *R. leguminosarum* strain LPR 1101 carries structural *nif* genes (Section V,B). This plasmid can be separated from the other plasmids of the same strain and has been found to possess considerable homology with total CCC DNA isolated from another strain of *R. leguminosarum*, but only slight homology with the smaller plasmid of L5-30 which is known not to carry the structural *nif* genes D and H. On the other hand, this plasmid carries one restriction fragment which hybridizes strongly with total DNA from two slow-growing rhizobia. This suggests that the hybridizing fragment carries structural *nif* genes known to be highly conserved in all nitrogen-fixing organisms (Ruvkun and Ausubel, 1980). Furthermore, hybridization studies revealed sequence homologies between *R. leguminosarum* and *R. trifolii* plasmids and plasmids from *A. tumefaciens* (Prakash *et al.*, 1981).

Similarly Jouanin *et al.* (1981) studied sequence homologies between plasmids isolated by the alkaline denaturation procedure from various strains of *R. meliloti*. Their data indicate that all the plasmids studied showed a significant extent of sequence homology. A similar short sequence ($< 5 \times 10^6$ in molecular weight) was present in all the plasmids studied.

C. The Presence of Large Plasmids in Fast-Growing Rhizobia

The different procedures described above (Section II,A and B) have been used to investigate the presence of large plasmids in various *Rhizobium* species. In the more than 60 fast-growing strains already investigated the presence of large plasmids ($< 90 \times 10^6$) was demonstrated without a single exception (Table II).

In most strains more than one large plasmid was detected. Most *R. leguminosarum* strains were found to carry at least three plasmids of more than 90×10^6 molecular weight (Casse *et al.*, 1979a,b; Hirsch *et al.*, 1980). In *R. meliloti* different isolation procedures revealed two classes of plasmids; one class (pRme) could be isolated by an alkaline denaturation procedure and had a molecular weight range of $90–200 \times 10^6$ (Casse *et al.*, 1979a), and a second class (pSym) of 300×10^6 molecular weight could not be detected by this procedure but only by the more sensitive Eckhardt's procedure. One such megaplasmid was found in 27 *R. meliloti* strains of various geographical orgins (Fig. 1) (Rosenberg *et al.*, 1981b). In 3 *R. meliloti* strains (RCR 2011, S26, and A145) no plasmids could be detected using the alkaline denaturation procedure (Casse *et al.*, 1979a), but these strains were found to carry one megaplasmid using the Eckhardt procedure (Rosenberg *et al.*, 1981b).

In *A. tumefaciens* it was found that in addition to the tumor-inducing (Ti) plasmid (MW $90–160 \times 10^6$) most strains contained a $> 250 \times 10^6$ molecular-weight plasmid (Casse *et al.*, 1979a).

It is worthy noting that among the *Rhizobium* strains studied the presence of plasmids of molecular weight lower than 85×10^6 is rarely reported (Casse *et al.*, 1979a,b), and no very small plasmids (MW $< 10 \times 10^6$) suitable as cloning vehicles for genetic engineering are found.

The generality of the presence of large plasmids is not yet firmly established for slow-growing rhizobia. With the exception of *R. japonicum* strains isolated from alkaline soils, for which an alkaline denaturation procedure is efficient (Gross *et al.*, 1979), difficulties have been encountered in reproducibly detecting and isolating plasmids from slow-growing strains of *R. japonicum* or of the cowpea group (C. Boucher, C. Schubert, and E. A. Schwinghamer, personal communication). An effective procedure remains to be devised for these bacteria. Nevertheless some positive results, even if not always positive for all the strains investigated (Nuti *et al.*, 1977; C. Boucher and E. A. Schwinghamer unpublished results; A. G. Atherly and P. Russell, unpublished results), suggest the presence of large plasmids in slow-growing bacteria.

TABLE II

Rhizobium STRAINS IN WHICH LARGE PLASMIDS HAVE BEEN STUDIED

Rhizobium species	Strains	References
Fast growing		
R. leguminosarum	RCR 1001, RCR 1016	Nuti et al. (1977)
	A 171, 300	Prakash et al. (1980)
	PRE	Krol et al. (1981)
	JB897, LPR 1705	Nuti et al. (1980, 1981)
	248, 300, 306, 309	Hirsch et al. (1981)
	300, TOM, 239, 248, 336	Brewin et al. (1980a)
	300, 128 C 53	Brewin et al. (1980b)
	L1, L4, L5, L18	Schwinghamer and Dennis (1979)
	1101	Prakash et al. (1981)
R. meliloti	M22 str, M9S, RCR 2011	Bechet and Guillaume (1978)
	L5-30, 102 F51, V7, 12,	Casse et al. (1979)
	1322, 41, 445, B 251,	
	U 54, Ls2a, 102 F28, V7,	
	3 DoA20a, S33, Balsac,	
	S14, 54032, I1, RF22,	
	B294, Ve8, Sa10, Lb1,	
	S26, A145, RCR2011, L530	
	RCR 2011, Sa10	Spitzbarth et al. (1979)
	RCR 2001	Prakash et al. (1980, 1981)
R. phaseoli	RCR 3605	Prakash et al. (1980, 1981)
	1233, 3622	Beynon et al. (1980)
R. trifolii	RCR 5, RCR 0402, 0403	Prakash et al. (1981)
	24, T12	Zurkowski and Lorkiewicz (1979)
	RT1	Ruvkun and Ausubel (1980)
	RCR 5	Hooykaas et al. (1981)
Slow growing		
R. japonium	RCR 3407	Nuti et al. (1977)
	3I1b117, 3I1b135, Rj 10B,	Gross et al. (1979)
	Rj 125, Rj19FY, Rj17W,	
	Rj23A, WA5099.1.1	
	61A76	Boucher and Schwinghamer (unpublished)
	61A76, 110, 3I1631	Atherly and Russell (unpublished)
	AA102, 74, 94, 138, 143	
Cowpea group	RD1	Nuti et al. (1977)

III. Genetic Methods for Plasmid Studies

Understanding the biological significance of *Rhizobium* plasmids requires genetic analysis. Genetic markers have been found to be located on plasmids, allowing their detection, curing, and transfer. The insertion of transposons has proved to be a useful method for the mutagenesis and tagging of plasmids.

A. Plasmid Genetic Markers

Hirsh (1979) identified three transmissible bacteriocinogenic plasmids of *R. leguminosarum,* each from a different field isolate. These plasmids were designated pRL1JI, pRL3JI, and pRL4JI and had molecular weights higher than 100 × 10⁶ (Hirsch *et al.,* 1980). The bacteriocins specified by these plasmids were of medium size. It was noted that the introduction of any of these three plasmids into *R. leguminosarum* strain 300, which normally produced a small bacteriocin, resulted not only in the expression of medium-sized bacteriocin production but also in the repression of small bacteriocin production.

A Nod⁻ derivative of *R. leguminosarum* strain A 171 cured of its smallest plasmid had a "rough" colony morphology which presumably reflected an alteration in the composition of its extracellular polysaccharides (Prakash *et al.,* 1980). This derivative was less sensitive to phage LPB 51 than the wild type. A colony morphology change was also associated with the deletion of a segment of a megaplasmid in *R. meliloti* strain 41 (A. Kondorosi, personal communication). Old colonies of *R. phaseoli* are able to produce a dark-brown pigment. This pigment production can be lost as a result of the deletion of a large plasmid (Beynon *et al.,* 1980).

B. Transposon-Induced Mutagenesis

A useful method for the mutagenesis of plasmids involves the insertion of transposons (Johnston *et al.,* 1978). This can be accomplished in several ways. First, since large plasmids represent as much as 10% of the total DNA, random samples of transposon-containing clones can be screened. To mutagenize the cells, an *Escherichia coli* strain was employed which possesses a plasmid carrying the transposon Tn5, which confers kanamycin resistance (Johnston *et al.,* 1978). It also has a phage Mu insertion to prevent establishment of the carrier plasmid in its new rhizobial host (Boucher *et al.,* 1977; Beringer *et al.,* 1978).

Alternatively, when a strain is known to carry a transmissible plasmid, it is possible to enrich specifically for insertions into this plasmid (Buchanan-Wollaston *et al.,* 1980).

Another procedure has been devised by Ruvkun and Ausubel (1981) for inducing localized mutations with a transposon even if the replicon is not transmissible. *Rhizobium meliloti* 2011 *nif* genes were first cloned into a small amplifiable plasmid in *E. coli.* Tn5 was inserted into *nif* in *E. coli,* and the *nif*::Tn5 segment was then recloned into a broad-host-range cloning vehicle derived from Inc.-P1 plasmid RK2. This *nif*::Tn5 plasmid was then conjugatively introduced into *R. meliloti* 2011 and chased by another Inc.-P1 plasmid. The Tn5 was maintained by "marker rescue" in the strain by growth on kanamycin, and Tn5 was inserted by recombination into the homologous *nif* gene of the resident

R. meliloti genome. This procedure, which was used to mutagenize *nif* genes specifically, could be employed to mutagenize any *Rhizobium* plasmid gene. A similar procedure based on *in vitro* cloning of *Rhizobium* plasmids into RP4 has been used to introduce transposons Tn5 and Tn7 into a *R. meliloti* 41 plasmid (Dénarié *et al.*, 1981).

C. PLASMID TRANSFER AND CURING

Most plasmids on which genetic markers have already been found have been shown to be self-transmissible. The bacteriocinogenic plasmids pRL1JI, pRL3JI, and pRL4JI are all transferred between strains of R. *leguminosarum* at high frequencies, i.e., 10^{-2} per recipient (Hirsch, 1979). Plasmid pJB5JI (pRL1JI::Tn5) is transmissible at high frequency to *R. trifolii* and *R. phaseoli* strains as well (Johnston *et al.*, 1978). Plasmid pRL5JI (MW 160 $\times$ 10^6), controlling host range specificity, can be transferred from *R. leguminosarum* strain TOM into *R. leguminosarum* strain 300 derivatives (Brewin *et al.*, 1980a). In *R. meliloti* 41 a plasmid of 140 $\times$ 10^6 molecular weight, pRme41a, tagged by a Tn5 transposon, has been transferred into *R. meliloti* strain L5-30 (Dénarié *et al.*, 1981). On the other hand, in the case of a plasmid of more than 200 $\times$ 10^6 molecular weight, as in *R. leguminosarum,* no transfer could be detected (Brewin *et al.*, 1980c). However, plasmid pRL6JI, which is apparently non-self-transmissible, could be mobilized by derivatives of the self-transmissible plasmids pRL3JI and pRL4JI (Brewin *et al.*, 1980b).

In vitro cloning of a segment of a *Rhizobium* plasmid into the broad-host-range vector RP4 provided a DNA homology between the two replicons and allowed intergeneric mobilization of a *R. meliloti* plasmid into *A. tumefaciens* (Dénarié *et al.*, 1981).

In some *A. tumefaciens* strains the large Ti plasmids (MW 90–150 $\times$ 10^6) can be eliminated by heat treatment. Similarly Zurkowski and Lorkiewicz (1978) incubated *R. trifolii* strains at 35°C for 7 days. Out of 10 strains exposed to this treatment 6 produced nonnodulating (Nod⁻) colonies at frequencies of 1–75%, depending upon the strain used. These strains showed no reversion to Nod⁺. When *R. trifolii* strains were treated with acridine dyes, no Nod⁻ clones could be found. Further studies on *R. trifolii* strains 24 and T12 and their Nod⁻ derivatives by sucrose gradient centrifugation showed that Nod⁻ clones had lost a plasmid of 190 $\times$ 10^6 molecular weight, but the largest plasmid, of about 560 $\times$ 10^6 molecular weight, was still present (Zurkowski and Lorkiewicz, 1979). Similarly Prakash *et al.* (1980) attempted to isolate cured strains of *R. leguminosarum* A171 by heat treatment (33°C for 2 weeks) and selecting for nonmucous colonies. All 10 such derivatives picked were nonnodulating. One Nod⁻ strain was characterized, and a single plasmid of 111 $\times$ 10^6 molecular weight was found to be missing as shown by dye–buoyant density centrifugation and gel electrophoresis. However, two larger plasmids were still present. Incubation of *R. meliloti* 41 at an elevated temperature (39°C) led to the appearance of colony morphology variants which were Nod⁻ (A. Kondorosi, personal communication). The Nod⁻ variants still possessed the smaller plasmid of 140×10^6 molecular weight but exhibited a deletion in the "megaplasmid."

As in *A. tumefaciens,* the class of large plasmids (MW $90-160 \times 10^6$) can be cured, at least in some strains. Thus we can conclude that plasmids do not carry genes essential for bacterial growth. Curing of a megaplasmid (MW $> 250 \times 10^6$) from *Agrobacterium* or *Rhizobium meliloti* has not been reported, suggesting that essential genes may be present on them.

IV. Plasmid Control of Early Functions in Symbiosis

The hypothesis of plasmid control of host range specificity and nodule formation, suggested by some early genetic evidence (Section I), has been validated recently thanks to the development of physical and genetic methods for studying *Rhizobium* large plasmids.

A. HOST RANGE SPECIFICITY

The first very convincing data presented for the presence of host specificity genes on a plasmid (or plasmids) came from the work of Johnston *et al.* (1978). *Rhizobium leguminosarum* strain T3, carrying a resident transmissible plasmid tagged by a Tn5 transposon (pJB5JI), was used as a donor in crosses with *R. trifolii* strains 6661 and 6710 and *R. phaseoli* strains 1233. Transconjugants arose at a frequency of about 10^{-2}, and all were capable of forming nodules on peas in addition to their normal hosts *R. trifolium* and *R. phaseolus,* respectively.

It is interesting that these authors reported that the transconjugant *Rhizobium* species which received the host specificity of *R. leguminosarum* were not as efficient in nodulating peas as the original parent strains. The *R. trifolii* and most of the *R. phaseoli* transconjugants nodulated peas a week later than normal, and the nodules were smaller than those produced by *R. leguminosarum*.

Rhizobium phaseoli strain 1233 contained two plasmids of about 200×10^6 molecular weight. Spontaneous deletions in the smaller plasmid abolished the ability to nodulate *Phaseolus* beans and brown pigment production (Beynon *et al.,* 1980). The majority of *R. phaseoli* (pJB5JI) transconjugants (97%) produced pigment and were not as efficient at nodulating peas and *Phaseolus* beans when compared with the parent strains; they contained the two plasmids (strain 1233) in addition to pJB5JI. The others (3%) did not produce pigment, nor did they form nitrogen-fixing nodules on *Phaseolus* beans, but they nodulated peas as well as the *R. leguminosarum* donor; they had received not only pJB5JI but also another plasmid from the donor and had lost their smaller plasmid. This confirms that the *R. phaseoli* strain 1233 smaller plasmid controls pigment production and the ability to nodulate *Phaseoli* beans. (Beynon *et al.*, 1980).

When these transconjugants were reisolated from pea nodules, most of them had lost the smaller resident plasmid and contained only the larger plasmid and pJB5JI. Thus, although certain plasmids controlling different host range specificities may coexist within a *Rhizobium* strain in culture, they may not be able to do so within the root of a given plant host. Beynon *et al.* (1980) suggested that there may be functional interference between genes specified by the pea nodulation plasmid pJB5JI and corresponding genes present on the *R. phaseoli* bean nodulation plasmid. This type of interaction could be described as "physiological incompatibility."

Even within the cross-inoculation groups some restrictions on host range are known. For instance, in the pea group, European *R. leguminosarum* field isolates cannot nodulate a pea cultivar (Afghanistan) and a primitive pea line JI241. However, the *R. leguminosarum* strain TOM isolated in Turkey is able to nodulate these cultivars as well as most of the commerical pea varieties. When crossed with strain 16015, a Nod⁻ derivative of a European strain, strain, TOM, is able to transfer the ability to nodulate to primitive cultivars. The transfer of the host range specificity is associated with the transfer of plasmid pRL5JI of molecular weight 160×10^6 (Brewin *et al.*, 1980a).

Therefore host range specificity genes in certain strains of the three species *R. leguminosarum, R. trifolii,* and *R. phaseoli* are borne on specific plasmids.

B. Nodule Formation

Both *R. trifolii* strains 12 and T24 carry two plasmids of molecular weight 460 and 190×10^6. After heat treatment Nod⁻ derivatives were obtained; they had lost the smaller plasmid of 190×10^6 molecular weight (Zurkowski and Lorkiewicz, 1979). Similar results were obtained by Prakash *et al.* (1980) with *R. leguminosarum* strain A 171. A Nod⁻ derivative obtained by culturing at an elevated temperature had lost its smallest plasmid pR1eA171a with a molecular weight of 110×10^6. The cured strain exhibited some modifications of surface properties, such as loss of agglutinability by a pea lectin (Prakash *et al.*, 1980). A Nod⁻ derivative of *R. leguminosarum* 300 obtained by ultraviolet treatment was shown to carry a deletion of a large plasmid of molecular weight higher than 200×10^6 (Hirsch *et al.*, 1980). This Nod⁻ mutant strain was restored to Nod⁺ Fix⁺ by the introduction of either the *R. leguminosarum* plasmid pJB5JI (Johnston *et al.*, 1978) or of the plasmid pRL5JI from the *R. leguminosarum* field isolate strain

TOM (Brewin *et al.*, 1980a). Similarly spontaneous Nod⁻ derivatives of *R. phaseoli* strain 1233 carried deletions of the smaller of its two plasmids (Beynon *et al.*, 1980). (See Note Added in Proof.)

Indirect evidence suggests plasmid control of nodule formation in *R. meliloti*. From strains L5-30, Lb1, and Sa10 spontaneous Nod⁻ derivatives were isolated; surprisingly all of them had lost the nitrogenase genes simultaneously, as revealed by DNA hybridization experiments with *Klebsiella pneumoniae* structural *nif* genes (Dénarié *et al.*, 1981). Similarly heat treatment at 39°C of *R. meliloti* 41 resulted in the appearance at high frequency of colony morphology variants exhibiting a Nod⁻ phenotype. These Nod⁻ variants had also lost the nitrogenase genes (A. Kondorosi, personal communication). In *R. meliloti* strains L5-30, Lb1, Sa10, and 41, the structural *nif* genes are located on a megaplasmid (MW > 300 × 10⁶). The fact that genes controlling nodule formation are lost together with *nif* genes suggests that the two classes of genes are located on the same large plasmid (Dénarié *et al.*, 1981). These spontaneous Nod⁻ derivatives of L5-30 have been inoculated on alfalfa seedlings to determine the symbiotic defect: They do not induce root hair curling, showing that the infection process is blocked at an early stage. We can only conclude that genes controlling an early step of root infection are plasmid-borne. We do not know whether these genes are different from those controlling host range specificity.

It is not known whether genes involved in the following steps of the nodule formation process—nodule meristem induction, nodule organogenesis, and differentiation—are plasmid-borne. One limitation is due to the lack of cytological, ultrastructural, immunological, and biochemical studies on the different Nod⁻ phenotypes. Another limitation could be due to the all-or-none response of most of the genetic studies reported above: even if the genes controlling all the steps of the process were plasmid-borne, plasmid curing or deletion of a large fragment of a plasmid would result in the phenotype corresponding to the block caused by the lack of the gene controlling the earliest step in the process. Further progress will require mutagenesis of the plasmids and precise studies on the symbiotically defective phenotypes.

V. Plasmid Control of Late Functions in Symbiosis

The development and functioning of nitrogen-fixing nodules consist of differentiation of the central tissue of the nodule (see Newcomb, this volume) and differentiation of bacteria into bacteroids (see Sutton *et al.*, this volume). It is likely that these complex processes are under the control of numerous *Rhizobium* genes. Various experimental approaches, genetic and physical, have led to the conclusion that, in fast-growing rhizobia at least, some of these genes are plasmid-borne.

A. THE *fix* GENES

Genetic studies on *R. leguminosarum* have shown that the genes that control the ability of rhizobia to fix nitrogen within a nodule (*fix* genes) are located on large plasmids. *Rhizobium leguminosarum* strain 6015, a Nod⁻ derivative of strain 300, carries a deletion on the largest of its four plasmids (Hirsch *et al.*, 1980). The pRL1JI plasmid is known to control the ability to nodulate peas (Johnston *et al.*, 1978) and restores the ability to form nitrogen-fixing nodules in strain 6015 (Buchanan-Wollaston, 1980). Selection was made for the insertion of transposon Tn5 into this plasmid. Then 160 independent mutant plasmids were introduced into the Nod⁻ strain 6015; there were 14 Fix⁻ transconjugants and 15 Nod⁻ transconjugants. These results indicate that both Nod and Fix functions are normally coded for by plasmid pRL1JI but also by the segment of DNA of the very large plasmid of strain 300 which is deleted in strain 6015 (Buchanan-Wollaston *et al.*, 1980). Similar experiments show that *fix* genes are located on the pRL5JI plasmid in *R. leguminosarum* strain TOM (Brewin *et al.*, 1980a). The precise symbiotic defect of the various Fix⁻ mutants has not yet been studied, and it is not known whether these *fix* mutations affect the development of effective nodules or the working of mature nodules.

B. THE *nif* GENES

Several recent experiments have provided evidence that the *nif* genes of *Rhizobium* species are located on extrachromosomal DNA.

Stanley and Dunican (1979), using *R. trifolii* as a donor, transferred via conjugation the ability to fix nitrogen to an avirulent strain of *A. tumefaciens* lacking a Ti plasmid and to a strain of *Klebsiella pneumoniae* with a deletion in the *nif* cluster. The matings were performed using the R plasmid RP1 to promote the transfers. Transconjugants of the two species expressed the RP1 resistance markers and could reduce acetylene under proper physiological conditions. These authors interpreted these results as indirect evidence that a plasmid containing the *Rhizobium nif* genes could be mobilized and transferred to another bacterial species.

A direct physical approach was made possible by the finding that the DNA sequences of the structural *nif* genes were highly conserved among various nitrogen-fixing bacteria (Nuti *et al.*, 1979; Ruvkun and Ausubel, 1980; Mazur *et al.*, 1980). The procedure utilized recombinant plasmids carrying *nif* genes isolated from *K. pneumoniae* which were constructed by Cannon *et al.* (1977, 1979). Plasmid pSA30 carries the nitrogenase structural genes *K, D, H,* and part of *nifE*, and pCM1 carries the remainder of *nif* genes except *nifJ* and part of *nifK*. These plasmids were radioactively labeled and used as probes with DNA blotted on a nitrocellulose filter using the hybridization methods of Southern

(1975). Ruvkun and Ausubel (1980) discovered that DNA isolated from 19 out of 19 nitrogen-fixing prokaryotic species hybridized to a ^{32}P-labeled pSA30 DNA fragment containing the *nif* structural genes *K*, *D*, and *H*. The other *nif* fragment (from pCMI) did not hybridize. It was concluded that the observed hybridization was due to the presence of *nif* genes sharing sequence homology with *K. pneumoniae nif* structural genes. Among the organisms examined were six *Rhizobium* species, all of which contained DNA which hybridized with the structural gene probe. It was not determined, however, whether this hybridization was with chromosomal or with plasmid DNA.

The same hybridization procedure was used by Nuti *et al.* (1979), employing *R. leguminosarum* purified plasmid DNA digested with *Eco*RI. They found that pSA30 hybridized with two plasmid DNA bands in *R. leguminosarum* strains A 171 and LPR180 and to three bands in strain JB897. Plasmid pCM1, which carries only part of the structural gene *nifK*, did not hybridize with *Rhizobium* plasmid DNA. These results strongly indicate that at least some of the nitrogenase structural genes in *R. leguminosarum* are located on the high-molecular-weight plasmids. LPR180 is a derivative of strain A 171 which has lost the pR1eA171a plasmid (MW 110 $\times$ 10^6) which seems to carry genes involved in nodule formation (Prakash *et al.*, 1980). Hybridization of pSA30 with CCC DNA from this strain shows that the nitrogenase genes are located on a very large plasmid (MW $>$ 200 $\times$ 10^6). Krol *et al.* (1981) showed that structural *nif* genes were also plasmid-borne in *R. leguminosarum* strain PRE.

Similarly, in two *R. meliloti* strains L530 and 41, nitrogenase genes were shown not to be on the smaller plasmids pRmeL530a (MW 90 $\times$ 10^6) and pRme41a (MW 140 $\times$ 10^6). They are located on the megaplasmids pRmeL530b and pRmeL541b (Rosenberg *et al.*, 1981b). The fact that structural *nif* genes are plasmid-borne in *R. meliloti* was confirmed with six other strains of various geographical origins: 1322 (Oceania), 102 F51 and Rm12 (North America), U45 (South America), and RF 22 (Africa) (Rosenberg *et al.*, 1981b). On the other hand in *R. leguminosarum* strain 1101 structural *nif* genes are located on the smallest of the large plasmids (Prakash *et al.*, 1981). It can be concluded that the nitrogenase genes are generally present on large plasmids of fast-growing rhizobia.

Ruvkun and Ausubel's studies (1980) on interspecies homology of nitrogenase genes included species of the Azotobacteraceae, Enterobacteraceae, Rhodospirillaceae, Bacillaceae, Rhizobiaceae, Actinomycetaceae, and Cyanobacteraceae. The conservation of *nif* genes discovered in different species of such distant taxonomic relationships is quite amazing. It was estimated that the sequence divergence between *K. pneumoniae* and *R. meliloti nif* structural genes was 8–20%, based on melting temperatures (T_m) of heteroduplex DNA. This is comparable to the sequence conservation of the eukaryotic genes coding for histone H4. The authors concluded that either there was extremely strong selective pressure for maintaining the integrity of these genes, or *nif* genes have been recently exchanged between nitrogen-fixing species through plasmid-mediated conjugation. The evidence that *nif* genes are located on extrachromosomal DNA could be interpreted as supporting the theory of recent radiation of the nitrogenase structural genes. However, it is also worth mentioning that in *K. pneumoniae* the *nif* gene cluster is located near the *his* operon on the bacterial chromosome (see Dixon *et al.*, 1980, for review), and in the nitrogen-fixing blue-green algae *Anabaena* Mazur *et al.* (1980) found that the *nif* structural genes were not located on the detectable plasmids. Further, in a nitrogen-fixing methanobacterium, no evidence could be found that *nif* genes were present on plasmids (A. Toukdarian and M. McConnell, personal communication).

Although the presence of structural *nif* genes on extrachromosomal replicons is now well documented, nothing is known about the location of other *nif* genes and of the genes which code for the various enzymes characteristic of the nitrogen-fixing bacteroid (see Sutton *et al.* and Imsande, this volume).

C. TRANSCRIPTION OF PLASMID DNA IN BACTEROIDS

A quite different approach was pursued by Krol *et al.* (1981) to investigate the role of plasmids in controlling the function of mature nodules. In *R. leguminosarum* strain PRE two large plasmids were detected. Plasmid DNA was isolated and hybridized with cellular RNA from broth-cultured bacteria and from endosymbiotic bacteroids. These hybridization experiments showed that, in addition to the structural *nif* genes, numerous plasmid genes were strongly transcribed in bacteroids and only weakly or not at all in bacteria. Therefore the large plasmids not only control nitrogen fixation itself but probably control various functions in the working of mature bacteroids.

D. THE *hup* GENES

Some strains of *Rhizobium* possess an active hydrogen uptake system (*Hup*) permitting hydrogen, a by-product of the nitrogenase reaction, to be recycled. Brewin *et al.* (1980b) has reported that determinants for hydrogenase activity (*hup*) in a field isolate *R. leguminosarum* strain 128 C53 are genetically linked to determinants for nodulation ability and are probably carried on a plasmid pRL6JI of $\sim 190 \times 10^6$ molecular weight. Although pRL6JI was not self-transmissible, the *nod* and *hup* determinants could be transferred to Nod⁻ *R. leguminosarum* strains after recombination with a derivative of a transmissible plasmid.

VI. Concluding Remarks

In fast-growing rhizobia (*R. leguminosarum, R. phaseoli, R. trifolii,* and *R. meliloti*) genetic and physical evidence has shown that some genes controlling the host range specificity, early steps of the infectious process, and the working

of mature bacteroids including nitrogen fixation itself are located on large plasmids. In all the strains investigated at least 10% of the genetic information was
extrachromosomal: these plasmids are likely to carry more than 300 genes. This
raises a question: are most *Rhizobium* genes *directly* involved in control of the
symbiotic process plasmid-borne? A systematic mutagenesis of *Rhizobium* with
transposons and genetic, cytological, and biochemical studies on the symbiotically defective mutants should provide an answer. Genetic tools have been devised both for chromosome (see Kondorosi and Johnston, this volume) and
plasmid analysis (Section III), which make this type of approach feasible in fast-
growing rhizobia. *Rhizobium* plasmid studies have benefited in some respects
from the studies performed earlier on *A. tumefaciens*. But one difficulty in
studying the *Rhizobium* system is that genes controlling symbiosis seem to be
located on plasmids with molecular weights much higher than those of *Agrobacterium* Ti plasmids. For instance, in *R. meliloti,* the *nod* and *nif* genes are
located on plasmids of more than 300×10^6 molecular weight.

   Another difficulty in studying the molecular biology of symbiotic nitrogen
fixation is due to the fact that methods for the study of chromosomes and plasmids
are available for fast-growing rhizobia which are not able to fix significant
amounts of nitrogen in the absence of the host plant (see Sutton *et al.,* this
volume). On the other hand, for slow-growing strains of *R. japonicum* or species
of the cowpea group, which are able to fix nitrogen in pure culture, very limited
information is available on the role of plasmids in the control of symbiotic
properties. The isolation of fast-growing strains able to fix nitrogen under free-
living conditions and the development of tools for the study of chromosomes and
plasmids of slow-growing rhizobia are required.

   An assessment of the role of plasmids in the control of symbiotic properties
should also be made possible following their transfer into different chromosomal
backgrounds, for example, within different *Rhizobium* species or into other members of the Rhizobiaceae family such as *Agrobacterium tumefaciens, A. rhizogenes,*
and *A. radiobacter.*

References

Adachi, T., and Iyer, V. N. (1980). *Analyt. Biochem.* **101,** 271–274.
Bechet, M., and Guillaume, J. B. (1978). *Can. J. Microbiol.* **24,** 960–966.
Beringer, J. E. (1976). *In* "Proc. 1st Int. Symp. on N$_2$ Fixation" (W. E. Newton and C. J. Nyman,
   eds.), pp. 358–366. Washington State Univ. Press.
Beringer, J. E., Beynon, J. L., Buchanan-Wollaston, A. V., and Johnston, A. W. B. (1978). *Nature
   (London)* **276,** 633–634.
Beynon, J. L., Beringer, J. E., and Johnston, A. W. B. (1980). *J. Gen. Microbiol.* **120,** 421–429.
Boucher, C., Bergeron, B., Barate De Bertalmio, M., and Dénarié, J. (1977). *J. Gen. Microbiol.*
   98, 253–263.
Brewin, N. J., Beringer, J. E., and Johnston, A. W. B. (1980a). *J. Gen. Microbiol.* **120,** 413–420.

Brewin, N. J., De Jong, T. M., Phillips, D. A., and Johnston, A. W. B. (1980b). *Nature (London)* **288**, 77-78.

Brewin, N. J., Beringer, J. E., Buchanan-Wollaston, A. V., Johnston, A. W. B., and Hirsch, P. R. (1980c). *J. Gen. Microbiol.* **116**, 261-270.

Buchanan-Wollaston, A. V., Beringer, J. E., Brewin, N. J., Hirsch, P. R., and Johnston, A. W. B. (1980). *Mol. Gen. Genet.* **178**, 185-190.

Cannon, F. C., Riedel, G. E., and Ausubel, F. M. (1977). *Proc. Natl. Acad. Sci. U.S.A.* **74**, 2963-2967.

Cannon, F. C., Riedel, G. E., and Ausubel, F. M. (1979). *Mol. Gen. Genet.* **174**, 59-66.

Casse, F., Boucher, C., Julliot, J. S., Michel, M., and Dénarié, J. (1979a). *J. Gen. Microbiol.* **113**, 229-242.

Casse, F., David, M., Boistard, P., Julliot, J. S., Boucher, C., Jouanin, L., Huguet, T., and Dénarié, J. (1979b). *In* "Plasmids of Medical, Environmental and Commercial Importance" (K. N. Timmis and A. Puhler, eds.), pp. 327-338. Elsevier, Amsterdam.

Clewell, D. B., and Helinski, D. R. (1969). *Proc. Natl. Acad. Sci. U.S.A.* **62**, 1159-1166.

Cole, M. A., and Elkan, G. H. (1973). *Antimicrob. Agents Chemother.* **4**, 248-253.

Currier, T. C., and Nester, E. W. (1976). *Anal. Biochem.* **76**, 431-441.

Dénarié, J., Rosenberg, C., Boistard, P., Truchet, G., and Casse-Delbart, F. (1981). *In* "Current Perspectives in Nitrogen Fixation" (A. H. Gibson and W. E. Newton, eds.), pp. 137-141. Griffin Press, Netley, S. Australia.

Dixon, R., Merrick, M., Filser, M., Kennedy, C., and Postgate, J. R. (1980). *In* "Nitrogen Fixation" (L. W. E. Newton and W. H. Orme-Johnson, eds.), Vol. I, pp. 71-84. Univ. Park Press, Baltimore, Maryland.

Dunican, L. K., and Cannon, F. (1971). *Plant Soil* Special Vol., pp. 73-79.

Dunican, L. K., and Tierney, A. B. (1974). *Biochem. Biophys. Res. Commun.* **57**, 62-72.

Dunican, L. K., O'Gara, R., and Tierney, A. B. (1976). *In* "Symbiotic Nitrogen Fixation in Plants" (P. S. Nutman, ed.), pp. 77-90. Cambridge Univ. Press, London and New York.

Eckhardt, T. (1978). *Plasmid* **1**, 584-588.

Gross, D. C., Vidaver, A. K., and Klucas, R. V. (1979). *J. Gen. Microbiol.* **114**, 257-266.

Guerry, P., Le Blanc, D. J., and Falkow, S. (1973). *J. Bacteriol.* **116**, 1064-1066.

Hansen, J. B., and Olsen, R. H. (1978). *J. Bacteriol.* **135**, 227-238.

Higashi, S. (1967). *J. Gen. Appl. Microbiol.* **13**, 391-403.

Hirsch, P. R. (1979). *J. Gen. Microbiol.* **113**, 219-228.

Hirsch, P. R., Van Montagu, M., Johnston, A. W. B., Brewin, N. J., and Schell, J. (1980). *J. Gen. Microbiol.* **120**, 403-412.

Hooykaas, P. J. J., van Brussel, A. A. N., den Dulk-Pas, H., van Slogteren, G. M. S., and Schilperoort, R. A. (1981). *Nature (London)* 351-353.

Johnston, A. W. B., Beynon, J. L., Buchanan-Wollaston, A. V., Setchell, S. M., Hirsch, P. R., and Beringer, J. E. (1978). *Nature (London)* **276**, 634-636.

Jouanin, L., De Lajudie, P., Bazetoux, S., and Huguet, T. (1981). (Submitted).

Julliot, J. S., and Boistard, P. (1979). *Mol. Gen. Genet.* **173**, 289-298.

Klein, G. E., Jemison, P., Haak, R. A., and Matthysse, A. G. (1975). *Experientia* **31**, 532-533.

Krol, A. J. M., Hontelez, J. G. J., Van Den Bos, R. C., and Van Kammen, A. (1981). *Nucleic Acids Res.* **8**, 4338-4347.

Ledeboer, A. M., Krol, A. J. M., Dons, J. J. M., Spier, F., Schilperoot, R. A., Zaenen, I., Van Larekeke, N., and Schell, J. (1976). *Nucleic Acids Res.* **3**, 419-463.

Mazur, B. J., Rice, D., and Haselkorn, B. (1980). *Proc. Natl. Acad. Sci. U.S.A.* **77**, 186-190.

Meyers, J. A., Sanchez, D., Elwell, L. P., and Falkow, S. (1976). *J. Bacteriol.* **127**, 1529-1537.

Nuti, M. P., Ledeboer, A. M., Lepidi, A. A., and Schilperoot, R. A. (1977). *J. Gen. Microbiol.* **100**, 241-248.

Nuti, M. P., Lepidi, A. A., Prakash, R. K., Schilperoot, R. A., and Cannon, F. C. (1979). *Nature* (*London*) **282**, 533–535.

Olivares, J., Montoya, E., and Palomares, A. (1977). *In* "Recent Developments in Nitrogen Fixation" (W. E. Newton, R. Postgate, and C. Rodriguez-Barrueco, eds.), pp. 375–385. Academic Press, New York.

Parijskaya, A. N. (1973). Mikrobiologiya **42**, 119–121.

Prakash, R. H., Hooykaas, P. J. J., Ledeboer, A. M., Kijne, J. W., Schilperoot, R. A., Nuti, M. P., Lepidi, A. A., Casse, F., Boucher, C., Julliot, J. S., and Dénarié, J. (1980). *In* "Nitrogen Fixation" (L. W. E. Newton and W. H. Orme-Johnson, eds.), Vol. II, pp. 139–163. Univ. Park Press, Baltimore, Maryland.

Prakash, R. H., Schilperoort, R. A., and Nati, M. P. (1981). *J. Bacteriol.* **145**, 1129–1136.

Radloff, R., Bauer, W., and Vinograd, J. (1967). *Proc. Natl. Acad. Sci. U.S.A.* **57**, 1514–1521.

Rosenberg, C., Casse-Delbart, F., Dusha, I., Michel, C., and Boucher, C. (1981a). Submitted.

Rosenberg, C., Boistard, P., Dénarié, J., and Casse-Delbart, F. (1981b). Submitted.

Ruvkun, G. B., and Ausubel, F. M. (1980). *Proc. Natl. Acad. Sci. U.S.A.* **11**, 191–195.

Ruvkun, G. B., and F. M. Ausubel (1981). *Nature* (*London*), **289**, 85–88.

Ruvkun, G. B., Long, S. R., Meade, H. M., and Ausubel, F. M. (1981). *Cold Spring Harbor Symp. Quant. Biol.* **45**, 492–499.

Schwinghamer, E. A. (1980). *FEMS Microbiol. Lett.* **7**, 157–162.

Schwinghamer, E. A., and Dennis, E. S. (1979). *Aust. J. Biol. Sci.* **32**, 651–662.

Sciaky, D., Montoya, A. C., and Chilton, M. D. (1978). *Plasmid* **1**, 238–253.

Southern, E. M. (1975). *J. Mol. Biol.* **98**, 503–517.

Spitzbarth, M., Pühler, A., and Heumann, W. (1979). *Arch. Microbiol.* **121**, 1–7.

Stanley, J., and Dunican, L. K. (1979). *Mol. Gen. Genet.* **174**, 211–220.

Sutton, W. D. (1974). *Biochim. Biophys. Acta* **366**, 1–10.

Thomashow, M. F., Panagopoulos, C. G., Cordon, M. P., and Nester, E. W. (1980). *Nature* (*London*) **283**, 794–796.

Tshitenge, G., Luyindula, N., Lurquin, P. F., and Ledoux, L. (1975). *Biochim. Biophys. Acta* **414**, 357–361.

Van Larebeke, N., Engler, G., Holsters, M., Van Der Elsacker, S., Zaenen, I., Schilperoort, R. A., and Schell, J. (1974). *Nature* (*London*) **252**, 169–170.

Van Rensburg, H. J., and Strijdom, B. W. (1971). *Phytophylactica* **3**, 125–130.

Van Rensburg, H. J., Strijdom, B. W., and Rabie, C. J. (1968). *S. Afr. J. Agric. Sci.* **11**, 623–626.

Watson, B., Currier, T. C., Gordon, M. P., Chilton, M. D., and Nester, E. W. (1975). *J. Bacteriol.* **123**, 255–264.

Zaenen, I., Van Larebeke, N., Teuchy, H., Van Montagu, M., and Schell, J. (1974). *J. Mol. Biol.* **86**, 109–127.

Zurkowski, W., and Lorkiewicz, Z. (1976). *J. Bacteriol.* **128**, 481–484.

Zurkowski, W., and Lorkiewicz, Z. (1978). *Genet. Res.* (*Cambridge*) **32**, 311–314.

Zurkowski, W., and Lorkiewicz, Z. (1979). *Arch. Microbiol.* **123**, 195–201.

Zurkowski, W., Hoffman, M., and Lorkiewicz, Z. (1973). *Acta Microbiol. Pol. Ser. A* **5**, 55–60.

NOTE ADDED IN PROOF. Very convincing data that Nod functions are present on a plasmid is given by Hooykaas *et al.* (1981). They were able to label a large plasmid (180×10^6 molecular weight) with Tn5 in *R. trifolii* (strain LPR 5020) and successfully eliminate this plasmid by heat treatment. Surviving bacteria lacked this large plasmid and were Nod⁻. Further, they transferred this plasmid by conjugation to *A. tumefaciens* which was then capable of nodule formation but not nitrogen fixation on clover. These findings strongly indicate that the plasmid (Sym) controls host specificity as well as most all of the steps leading to nodule formation.

INTERNATIONAL REVIEW OF CYTOLOGY, SUPPLEMENT 13

Nodule Morphogenesis and Differentiation

WILLIAM NEWCOMB

Biology Department, Queen's University,
Kingston, Ontario, Canada

I. General Introduction 247
II. Invasion of the Root Hair 248
   A. Deformation and Infection 249
   B. Ontogeny of the Infection Thread 249
   C. Infection Thread Structure and Growth 254
III. Role of the Root Cortex 256
   A. Mitotic Activity and Nodule Development 256
   B. Polyploid Cells and Nodule Development 266
   C. Release of Rhizobia from Infection Threads 270
   D. Release of Rhizobia in Nodules Lacking Infection Threads . 275
   E. Formation of Unwalled Regions of Infection Threads . . . 277
   F. Early Symbiotic Development of Infected Cells 277
   G. Late Symbiotic Development of Infected Cells 281
   H. Senescence of Infected Cells 281
   I. Uninfected Cells of the Central Tissue (or Zone) 282
IV. Ineffective Root Nodules 284
V. Cells Associated with Metabolite Transport 286
   A. Nodular Vascular Bundles 286
   B. Transfer Cells 286
VI. *Rhizobium* Nodules on Nonlegumes 289
VII. General Conclusions 293
   References 295
   Note Added in Proof 297

I. General Introduction

The establishment of a nodule and its subsequent development and morphogenesis involve complex interactions between two very different organisms, resulting in intricate structural and biochemical changes in each partner such that the resulting symbiotic association is capable of fixing (or reducing) atmospheric molecular nitrogen. The microsymbiont, a *Rhizobium* species or strain specific for a certain species or group of leguminous plants, proliferates in the rhizosphere and becomes attracted and attached to a root hair. Where rhizobia penetrate the host cell wall, a unique structure, the infection thread, forms and continues to grow toward the base of the root hair cell. Meanwhile, the cortical cells have been induced to undergo mitosis, and the resulting derivative cells are invaded by infection threads. Bacteria are released endocytotically from the

247

infection threads into the host cytoplasm, initiating differentiation of the bacteroid-containing cells which are the sites of nitrogen fixation and the synthesis of amino acids which are transported out of the nodule via nodular vascular bundles to the root stele to growing portions of the plant. While this scheme is widely known, many details of the development and ultrastructure remain poorly understood and controversial because of incomplete, inadequate, and/or contradictory reports.

This article seeks to detail the cytological features of developing nodules and provide the reader with the necessary anatomical and ultrastructural details so that she or he may relate biochemical and physiological aspects to the compartmentalization of developing nodules. No attempt has been made to provide a complete bibliography of the multitude of papers dealing with nodule structure; recent reviews on nodule structure include those by Dart (1974, 1975, 1977), Goodchild (1977), and Robertson and Farnden (1980). This article devotes special attention to controversial topics which are important developmentally or physiologically or have been confused by inadequate and/or conflicting observations. The intention is to clarify these problem areas and hopefully stimulate further investigations.

II. Invasion of the Root Hair

This event is crucial to the establishment and development of the legume–Rhizobium symbiosis and involves transition of the rhizobium from a free-living existence to a symbiotic one, as well as penetration of the host plant's cell wall by the future prokaryotic partner. In many leguminous species nodule development is presumed to commence with rhizobial invasion of a root hair, although actual evidence for this event is limited to a relatively few species (Dart, 1977). Invasion of epidermal cells occurs in the aquatic plant Neptunia (Schaede, 1940) and on rare occasions in soybean and clover (Bieberdorf, 1938; Nutman, 1959). Numerous environmental and microbiological factors have profound effects on the infection process and include temperature, pH, mineral nutrition, number of rhizobia, competition and specificity of rhizobial strains, antagonisms of other microbes, and delayed inoculation (Dart, 1974, 1975, 1977). These have been studied extensively. In contrast, structural studies of this event have been far fewer, perhaps partly because of difficulties imposed by the low percentage of infected root hairs. In a sample of 20,000 root hairs in Pisum sativum only 2.8% were observed to be infected (Haack, 1964). Thus perhaps it is not surprising that most structural observations of this event have utilized phase-contrast optics to examine living root hairs (Dart, 1974) rather than sections of fixed and embedded specimens.

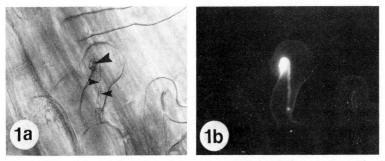

Fig. 1.   (a) Infected tip curled root hair of clover photographed with phase-contrast optics. Note the penetration site (large arrowhead) and infection thread (small arrowheads). ×570. (b) Same infection thread as in Fig. 1a photographed using broad-band blue light under which the root hair walls near the penetration site autofluoresce. ×570. Reprinted from Callaham (1979).

A. Deformation and Infection

Normally, uninfected root hairs are straight. Inoculation of legumes with either rhizobia or a culture filtrate produces a deformation, either a curling or branching, of the root hair (Dart, 1974). Curling is due to the displacement of cell wall growth from the tip of a growing root hair cell to the outside of the curl and results in curvatures up to or more than 360° (Figs. 1a and 2a) (Callaham and Torrey, 1980). Branching of root hairs (Fig. 3a) also occurs frequently and seems to involve a different mechanism including wall degradation (Callaham, 1979). Generally only deformed root hairs become infected. The first indication of infection is the presence of a "bright or hyaline" spot on the root hair cell wall. This area autofluoresces when illuminated with blue light (Fig. 1b) (Callaham and Torrey, 1980), and thus it does not appear to be comprised of callose as suggested by nonspecific aniline blue staining (Kumarasinghe and Nutman, 1977; Smith and McCully, 1979). The infection thread appears to originate from this hyaline region and to be continuous with it (Figs. 1a and b, 2, 3, and 4a).

B. Ontogeny of the Infection Thread

How does the infection thread originate? This question has aroused the interest of many and generated confusion and conflicting theories. Fortunately recent work has yielded information helpful to its resolution. Nutman (1956) has suggested that the rhizobia penetrate the root hair cell wall, make contact with the host plasma membrane, and bring about redirected wall growth. The redirected wall growth results in invagination of the root hair cell wall to form the tubular infection thread within which the rhizobia are surrounded by a polysaccharide thread matrix. The possibility that rhizobia interact directly with the host cell wall was first suggested by Fahraeus and Ljunggren (1959) who detected pectoly-

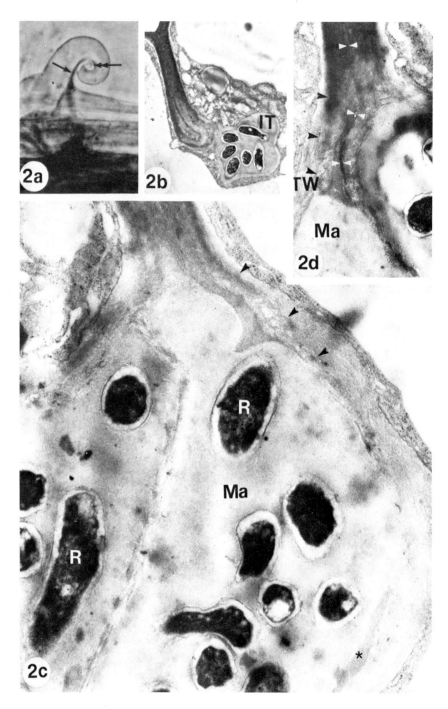

Fig. 2. See legend on p. 253.

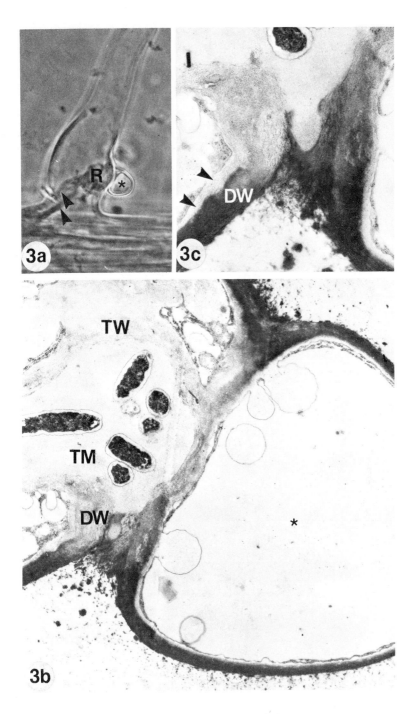

FIG. 3. See legend on p. 253.

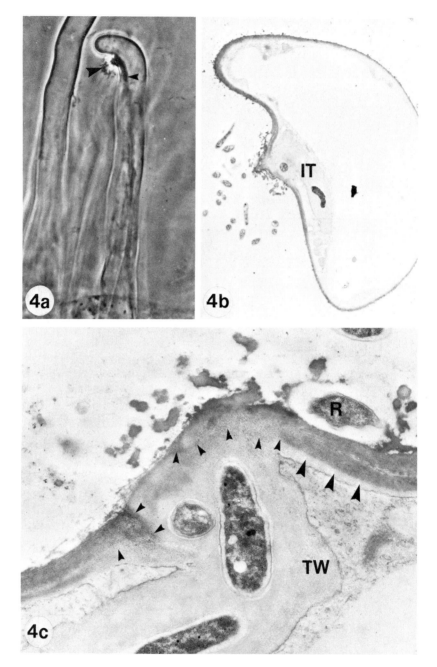

tic enzymes within infected leguminous roots. This idea received little sub-sequent support, because many other workers could not demonstrate such enzy-mic activities (Dart, 1974, 1975, 1977). However, recently pectolytic and cel-lulase activities have been demonstrated in *Rhizobium* cultures (Hubbell *et al.*, 1978; Martinez-Molina *et al.*, 1979), although direct evidence for their *in situ* involvement in the infection process is still lacking.

Ultrastructural observations of the infection process have been limited to four studies of which two (Sahlman and Fahraeus, 1963; Higashi, 1966) were carried out before adequate fixation techniques were established. Napoli and Hubbell (1975) presented evidence in favor of an invagination mechanism of infection thread formation based mainly on the apparent continuity of the thread wall and root hair cell wall and the absence of discontinuities (potential sites of bacteria-induced degradation) in the host cell wall. However, they illustrated that the wall at the junction of the infection thread wall and root hair stained only faintly; this could be cited as evidence of biochemical alterations in the cell wall.

FIG. 2. Light micrograph (a) and transmission electron micrographs (b–d) of a tip curled infected root hair of clover. (a) Phase-contrast micrograph of a deformed root hair showing an infection thread (single arrow) originating from the inside of the curl (double arrow). ×750. (b) Section through the adpressed regions of the root hair cell wall near the infection thread (IT). ×10,300. (c) Section through the site of infection showing where the original root hair cell wall has been degraded (*) and the newer less dense cell wall which is continuous with the thread wall. The demarcation between the new and old cell walls is marked by single arrowheads. Also shown are rhizobia (R) and the thread matrix (Ma) ×23,250. (d) Another section through the penetration site showing the degraded wall (white arrowheads) and the delimitation (black arrowheads) between the original root hair cell wall and the less electron-dense thread wall (TW) which surrounds the thread matrix (Ma) ×16,500. Reprinted from Callaham and Torrey (1981).

FIG. 3. An infected root hair of clover in which the infection thread arises from the axil of a lateral branch. (a) Phase-contrast micrograph of intact root hair with a lateral branch (*) showing an internal clump of rhizobia (R) and two infection threads (arrowheads). ×750. (b) Transmission electron micrograph of a section through the penetration site near the lateral branch (*). The electron-dense original root hair cell wall has been degraded (DW) near the thread matrix (TM). This degraded region is continuous with the less electron-dense thread wall (TW). ×9750. (c) Higher magnification of the penetration site shown in Fig. 3b. Note that the layer of less electron-dense cell wall (arrowheads) inside the original root hair cell wall appears less fibrillar than the degraded region (DW) of cell wall. ×15,000. (a,b) Reprinted from Callaham and Torrey (1981); (c) reprinted from Callaham (1979).

FIG. 4. Slightly curved infected root hair of clover. (a) Root hair embedded in epoxy resin. Rhizobial colony (large arrowhead) surrounds protruding region of root hair. An infection thread (small arrowhead) extends from the penetration site to the base of the root hair. ×400. (b) Section through the protruded region of a root hair showing rhizobia outside the root hair and within the infection thread (IT). ×3500. (c) Section median to the infection site showing an extracelluler rhizobium (R) firmly attached to the root hair cell wall. The inner wall layer (large arrowheads) continues into the infection thread wall (TW). The two ends of the original root hair cell wall at the infection site are marked by small arrowheads. ×29,500. Reprinted from Callaham and Torrey (1981).

Recently Callaham and Torrey (1981) examined the infection process in clover root hairs using correlated light and transmission electron microscopy such that their serial sections could be interpreted unambiguously. They confirmed that the pocket created by the root hair curl (Fig. 2a–d) or branching (Fig. 3a–c) was the site of infection thread formation. At this site in a curled root hair the outer electron-dense root hair cell wall is discontinuous and is replaced by a thread matrix, while the infection thread cell wall is continuous with a newly deposited inner layer of root hair cell wall which is less electron-dense (i.e., faintly staining) than the outer root hair cell wall.

Callaham and Torrey (1981) also examined the infection process in a slightly curled root hair (Fig. 4a–c) in which there was no pocket created by wall deformation. A rhizobial colony was present near a protruding wall region from which the infection thread originated. A single rhizobial cell was attached by its capsule to an area of root hair cell wall which appeared degraded, as it was thinner and less electron-dense than other regions of the outer host cell wall. They suggested that this positioning of the rhizobium may be crucial in preventing lysis of the root hair cell wall. As was the case in a curled root hair, the outer layer of the root hair cell wall ended abruptly and was in contact with the thread matrix, while the infection thread was continuous with the inner layer of the root hair cell wall. Thus these authors have provided strong evidence that the infection process involves a very localized degradation of the host cell wall and that, while the infection thread cell wall is continuous with the host cell wall, formation of the infection thread should not be regarded as consisting merely of invagination of the host cell wall. It is important to note that the continuity between the thread wall is with the newly deposited host cell wall and not with the original host wall with its localized invasion point which has been degraded presumably by the penetrating rhizobial cells.

C. Infection Thread Structure and Growth

The infection thread proceeds to grow toward the base of the root hair cell and may become branched while within the root hair cell. It is not unusual for the infection thread to abort within the infected root hair. In cross section (Figs. 10b and 14), the infection thread appears tubular, with the host plasma membrane separating the host cytoplasm from the thread wall. The thread wall encloses the matrix containing the invading rhizobia. Fine-structural examinations of infection threads (Figs. 5 and 15) reveal that the rhizobia are often surrounded by a thin electron-opaque area, considered either a bacterial capsule or an artifact, and a finely granular, moderately electron-dense matrix. On the basis of toluidine blue O, periodic acid–Schiff (PAS), and alcian blue staining, it has been suggested that the matrix is comprised of a mucopolysaccharide synthesized by the bacteria (Dart, 1974). It should be noted that the thread matrix does not stain

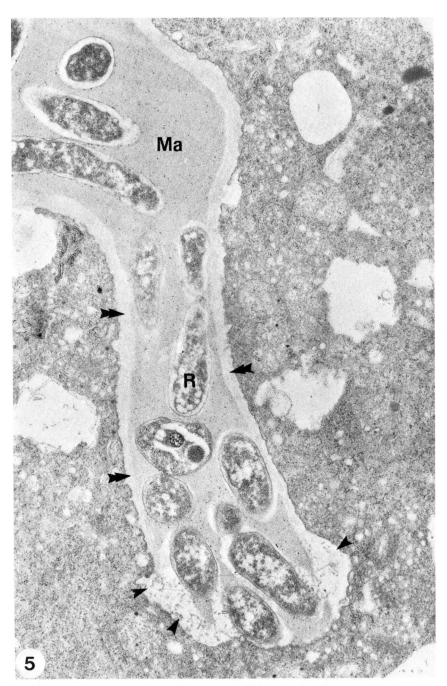

FIG. 5. Growing infection thread in an infected root of clover. The tip of the thread is unwalled, while the thread wall near the tip appears to be less tightly cross-lined (single arrowhead) in comparison to the other regions (double arrowheads) of the thread wall. The rhizobia are embedded in a finely granular matrix (Ma). ×16,200. Micrography courtesy of Dale Callaham, University of Massachusetts.

with the PAS, fluorescent PAS, or Thiery reaction and thus is chemically different from the thread wall which stains positively in all three reactions (Newcomb and McIntyre, 1980).

The growing tip of the infection thread within clover root hairs is unwalled, as revealed by the examination of serial sections (D. Callaham, unpublished results) (Fig. 5). Infection thread growth and wall deposition result from a tightly regulated control or balance between the two symbiotic partners; an unbalanced situation may result in thread abortion (Callaham and Torrey, 1981). Penetration of the root hair–cortical cell wall junction undoubtedly involves localized wall degradation followed by wall deposition such that the growth of the tubular infection thread continues into the cortical cell. Numerous workers (see Newcomb, 1976) have suggested that infection thread growth and penetration in cortical cells are due to invagination of the cortical cell walls (Fig. 11), although it seems unlikely that the host cell wall would be so plastic. Examination of the arrangement of microfibrils at the junction of thread walls and host cell walls suggests that invagination is not involved in thread penetration (Figs. 12 and 13). New wall components appear to originate from vesicles derived from Golgi bodies, and these fuse with the growing infection thread (Robertson *et al.*, 1978; Newcomb and McIntyre, 1981) and presumably play an important role in controlling the growth rate of the infection thread. Whether the growing tip of the infection thread in cortical cells is walled or unwalled is unknown, because an examination of serial sections of growing infection threads in the central infected tissues of nodules has apparently not been made. Within the root cortex, the infection thread may grow intracellularly and intercellularly (Fig. 12). (Although the infection thread is actually extracellular, it is convenient to use the term "intracellular" to distinguish these threads from the intercellular forms.) In the latter case the rhizobia and thread matrix occupy the intercellular spaces and utilize the preexisting cell walls as the thread wall; no new wall deposition is involved; thus this type differs ontogenetically from intracellular infection threads. In cross section, intercellular infection threads take the shape of the intercellular space and do not appear as circular tubes, as is the case with intracellular infection threads (Figs. 10a, 10b, and 14). Intercellular threads may give rise to intracellular infection threads (Fig. 11), presumably by localized degradation of cortical cell walls.

III. Role of the Root Cortex

A. Mitotic Activity and Nodule Development

Leguminous root nodules are categorized on the basis of shape into three groups: spherical or oval, such as mung bean (*Vigna radiata*), soybean (*Glycine max*), and peanut (*Arachnis hypogaea*); cylindrical or elongate club-shaped, such as pea (*P. sativum*), broad bean (*Vicia faba*), clover (*Trifolium*), and alfalfa (*Medicago*); and collar-shaped, such as lupine (*Lupinus*). Nodule shape is controlled by the host plant (Dart, 1977) and arises from the distribution and duration of mitotic activity in developing nodules (Dart, 1977; Syōno *et al.,* 1976; Newcomb, 1976; Newcomb *et al.,* 1979).

1. *Cylindrical Nodules*

The distribution of mitoses and the relationship of mitoses to nodule morphogenesis have been most extensively studied in pea nodules (Bond, 1948; Libbenga and Harkes, 1973; Newcomb, 1976; Syōno *et al.,* 1976; Truchet, 1978; Newcomb *et al.,* 1979). While these studies agree on many fundamental points, they disagree on some aspects, possibly because of different cultural conditions and cultivars. Pea nodules harvested 5 days after inoculation had only one infected cell—the root hair—which possessed a branched infection thread, but mitoses had already occurred in the inner and middle cortical cells of the root (Fig. 6). This suggests the involvement of growth substances produced by both the host and microsymbiont (Newcomb *et al.,* 1979). In contrast, Libbenga and Harkes (1973) did not observe mitotic divisions of the inner cortex until infection threads had penetrated to the middle cortex (see Fig. 2a and b in their paper). Two reports indicate that the early divisions of the inner and middle cortex are anticlinal (Truchet, 1978; Newcomb *et al.,* 1979), in contrast to a mixed orientation reported by Libbenga and Harkes (1973). There are large intercellular spaces between pea cortical cells, which facilitate determination of the patterns of cell division because parental and daughter cell walls are easily identified. The cortical cells which are stimulated to divide have prominent and often multiple nucleoli in large nuclei, while the nuclei and nucleoli of unstimulated cortical cells are much smaller (Fig. 6) (Truchet, 1978; Newcomb *et al.,* 1979). An increase in mitotic activity in the middle cortex coincides with penetration of the inner cortex by infection threads and the subsequent release of rhizobia within the invaded cells; these cells then cease mitotic activity after release of the rhizobia. Mitotic activity continues in the middle cortex and forms the nodule meristem. The proximal face of the meristem gives rise to derivative cells which for the most part become invaded by infection threads (Figs. 9, 10a, and 14) and ultimately develop into bacteria-containing, nitrogen-fixing cells. Other derivatives of the proximal face of the nodule meristem are never invaded by infection threads, hence remain infected but nevertheless are located in the central tissue of the nodule (Section III,I). The cells derived from the distal face of the nodule meristem differentiate to become the nodule cortex. Subsequent nodule growth is linked to the activity of the nodule meristem and the infection process. In pea nodules only cells penetrated by an infection thread become infected with "free" rhizobia and, in these cells, the bacteria population increases manyfold, as does

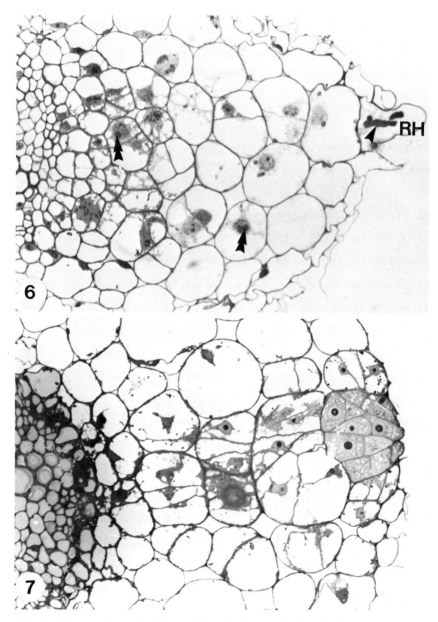

FIG. 6.   Five-day-old pea nodule containing rhizobia only in the infection thread (single arrowhead) present in the root hair (RH). Mitosis has already occurred in the inner and middle cortical
cells. Note the prominent nucleoli (double arrowheads) in the stimulated cortical cells. ×390.
Reprinted from Newcomb *et al.* (1979).

FIG. 7.   Preemergent soybean nodule possessing an outer layer of cytoplasmically rich cells and
several layers of vacuolate cortical cells. Note the predominance of anticlinal cell divisions which
have occurred in all tissue layers. ×275. Reprinted from Newcomb *et al.* (1979).

the volume of the host cell. Adjacent uninfected cells do not undergo this large increase in size (Section III,I). The growth of the infected cells pushes the nodule meristem outward and radially and results in the cylindrical shape of emergent pea nodules (Fig. 8) (Newcomb *et al.*, 1979). The mitotic activity of the nodule meristem of pea persists over a period of many weeks. In a study by Syōno *et al.*, (1976), mitotic activity occurred over a 4-week period, and at 5 weeks the

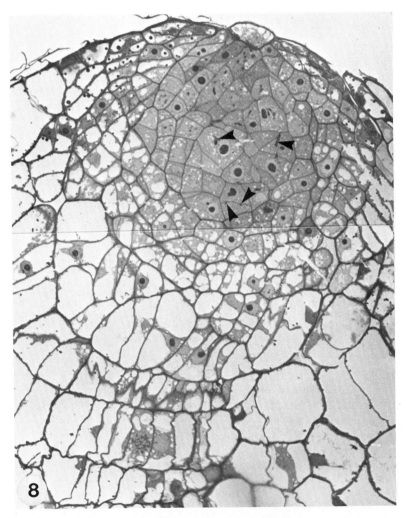

FIG. 8. Advanced preemergent soybean nodule containing a globular region of cytoplasmically rich cells, some of which possess infection threads (arrowheads). Note the regular orientation of the cell walls in the inner layers of the nodule. ×425. Reprinted from Newcomb *et al.* (1979).

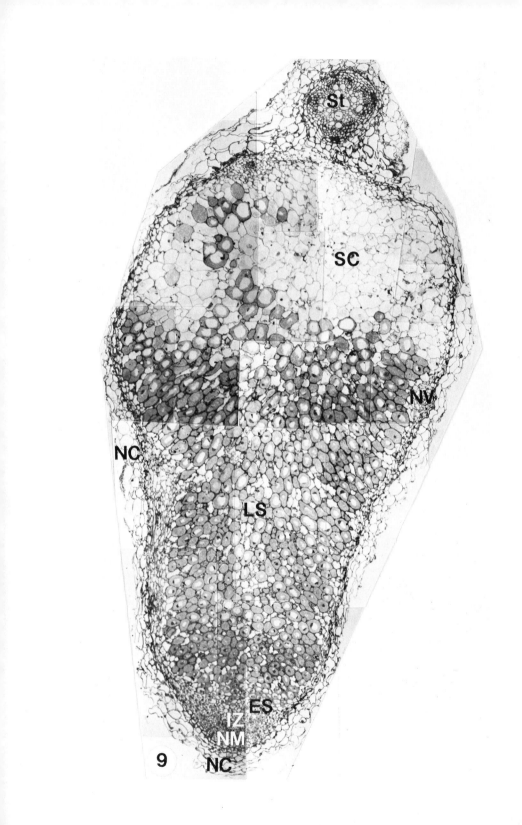

St

SC

NV

NC

LS

ES
IZ
NM
NC

9

mitotic index was 0 and no nodule meristem was present. Presumably a nodule meristem remains only as long as the rate of mitosis forming the derivative cells on the proximal face of the nodule meristem exceeds the rate of penetration by infection threads; when the latter rate exceeds the former, the nodule meristem cells become infected and cease mitosis after rhizobial release; shortly thereafter the nodule attains its maximum size. The rate of mitotic activity is closely correlated with cytokinin levels which are higher in the nodule meristem than in other zones of the nodule (Syōno et al., 1976). Pea nodules often possess several lobes which may arise by dichotomous division of the nodule meristem (Dart, 1975; Syōno et al., (1976).

Mature cylindrical nodules, as in the case of pea, consist of at least two distinctly colored zones: a distal white zone which includes the nodule meristem and invasion zone, and the outer layers of uninfected cells which are called the nodule cortex; and a pink zone which includes the large infected cells and the smaller uninfected cells of the central zone or tissue (Fig. 9). A proximal green zone may also be present, consisting of degenerating infected cells (Figs. 9, 25, and 26). Nodular vascular bundles are located in the nodule cortex and extend from the stele of the adjacent root toward the nodule meristem.

2. Spherical Nodules

Studies on the relationship of mitosis to the morphogenesis of spherical nodules have been less extensive than those dealing with cylindrical nodules and have dealt mainly with soybean nodules. In early preemergent soybean nodules, when the infection thread is confined to the deformed root hair or has just started to penetrate the outer cortex, numerous mitoses have already occurred in a thin sector of cortical cells between the infected root hair and the stele (Fig. 7) (Newcomb et al., 1979). Most of these divisions are anticlinal. The derivative cells of the outer cortex have large nuclei with prominent nucleoli and densely staining cytoplasm containing numerous small vacuoles. In marked contrast the derivative cells of the inner cortex possess large vacuoles, only small amounts of cytoplasm in addition to nuclei, and nucleoli which appear smaller than those in the derivative cells of the outer cortex. Subsequent mitotic activity leads to the formation of an outer spherical mass of cytoplasm-rich cells and an inner zone of vacuolate cells (Fig. 8). These nodules are still preemergent, and infection threads penetrate many of the cytoplasm-rich cells but not the vacuolate cells. Early emergent soybean nodules are spherical and contain a large central zone of cytoplasm-rich cells, many of which are invaded by infection threads. Mitotic figures are present in a few of the cells of the central zone. Some of the central

FIG. 9. Light microscope montage of a 4-week-old pea nodule illustrating the main zones of a cylindrical nodule. These zones are the nodule cortex (NC), nodule meristem (NM), invasion zone (IZ), early symbiotic stage (ES), and late symbiotic (LS) stage of developing infected cells, senescent cells (SC), nodular vascular bundles (NV), and the stele (St) of the adjacent root. ×60. Reprinted from Newcomb (1976).

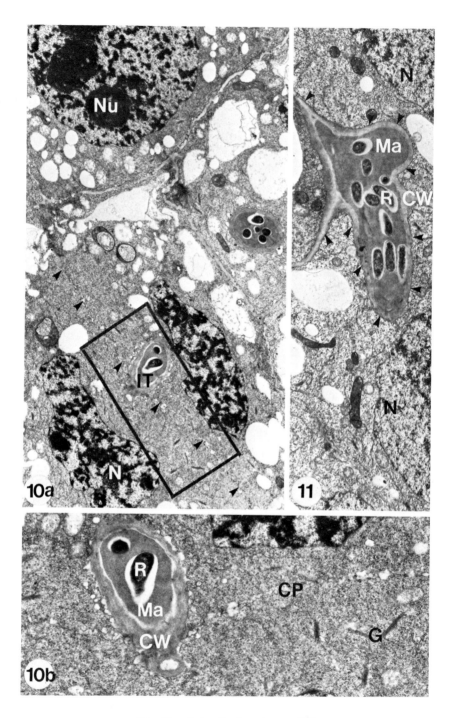

FIGS. 10 and 11.   See legends on p. 265.

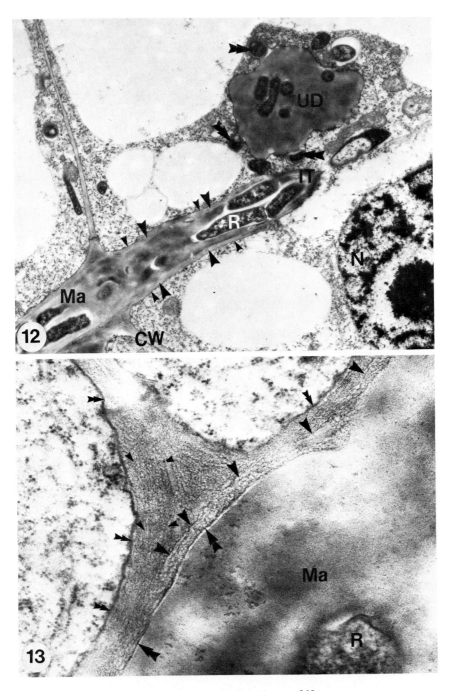

Figs. 12 and 13. See legends on p. 265.

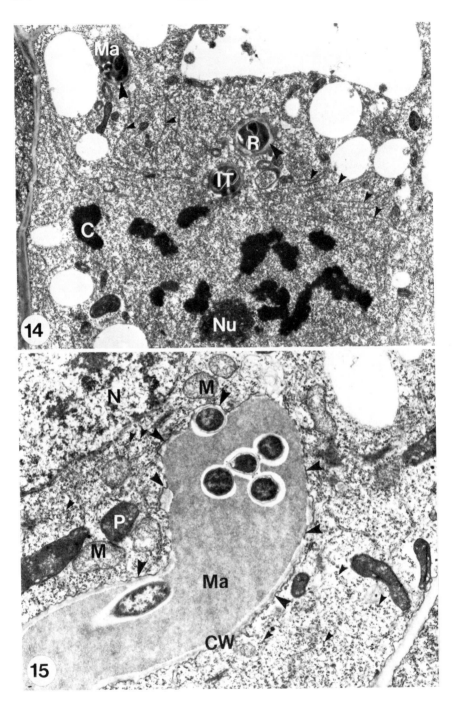

zone cells contain free rhizobia, and most of these cells also contain a large nucleus and prominent nucleolus; usually only small vacuoles occur in these cells at this stage. Surrounding the central zone are several layers of vacuolate cells which become differentiated into a nodule cortex, nodule vascular bundles, and schlerenchyma cells. After the rhizobia escape from the infection threads and proliferate in the host cytoplasm, the host cells cease mitosis and undergo a large increase in volume. The combined increase in volume of the numerous infected

---

FIGS. 10 and 11.   Infection threads in pea nodules. Reprinted from Newcomb (1976).

FIG. 10.   (a) Infection threads (IT) in meristematic cells containing numerous small vacuoles, many free ribosomes, prominent nucleoli (Nu), host nucleus (N), and a forming cell plate (arrowheads). ×4060. (b) Higher magnification of the outlined portion of Fig. 10a showing numerous Golgi bodies (G) and associated vesicles near the cell plate (CP) and an infection thread which is sectioned transversely and consists of the thread wall (CW), matrix (Ma), and rhizobia (R). The electron-translucent space between the rhizobia and matrix may be an artifact of preparation. ×9600.

FIG. 11.   Infection thread is in the process of growing into two adjacent cells. The contours of the cell wall give the (false?) impression that the process involves wall invagination. The plasma membrane (arrowheads) bounds the thread wall. Also shown are rhizobia (R) in the thread matrix (Ma) and portions of the host nuclei (N). ×5270.

FIG. 12.   Transmission electron micrograph of a *V. faba* nodule showing an intracellular infection thread (IT) which has penetrated the host cell wall (CW) and grown toward the host nucleus (N). The infection is bounded by a plasma membrane (single small arrowheads) and the cell wall (single large arrowheads) which are continuous with and similar in structure to the plasma membrane and cell wall of the host. Numerous rhizobia (R) are surrounded by the thread matrix (Ma) within the infection thread. Several rhizobia are in the process of escaping (double large arrowheads) from the unwalled droplet (UD) of the thread matrix, which is also surrounded by the host plasma membrane (single small arrowheads). ×7550.

FIG. 13.   Higher magnification of the junction of an intracellular infection thread cell wall and the host cell wall at the site of penetration in *V. faba*. The inner portion (single large arrowheads) of the thread cell wall adjacent to the matrix (Ma) contains densely stained cellulose microfibrils which are not continuous with those of the host cell wall. In contrast, the microfibrils of the host cell wall and the outer thread cell wall show some continuity in the area delimited by single small arrowheads. The dark line (double large arrowheads) at the interface between the thread cell wall and matrix does not possess the same thickness and substructure of the plasma membrane (double small arrowheads). ×35,500.

FIGS. 14 and 15.   Infection threads in pea nodules.

FIG. 14.   Transversely sectioned infection threads (IT) are present in a meristematic cell in late prophase as indicated by the condensed chromosomes (C) and still intact nucleolus (Nu). The infection thread wall (large arrowheads) surrounds the matrix (Ma) which contains rhizobia (R). Several profiles of RER (small arrowheads) are also shown. ×6270.

FIG. 15.   A rhizobium in the process of escaping from an unwalled region of an infection thread. Note that the host plasma membrane (large arrowheads) surrounds the thread wall (CW) and unwalled areas of the thread matrix (Ma). Also shown are numerous polyribosomes (small arrowheads), proplastids (P), mitochondria (M), and the host nucleus (N). ×13,900. Reprinted from Newcomb (1976).

cells of the central spherical zone causes the nodule to grow radially, resulting in a spherically shaped structure much larger than the early emergent nodule.

The mature emergent spherical nodule contains a pink central tissue comprised of many large infected cells and fewer uninfected cells and a white outer layer of nodule cortex which includes vacuolate parenchyma cells, several nodular vascular bundles, and in some species thick-walled schlerenchyma cells (Newcomb *et al.*, 1979).

3. *Collar-Shaped Nodules*

Less attention has been devoted to the morphogenesis of the collar-shaped nodules. It has been suggested that mitotic activity becomes localized in the peripheral regions of the nodule, causing it to grow around the root in a collar shape (Dart, 1975).

B. POLYPLOID CELLS AND NODULE DEVELOPMENT

With the possible exception of the intracellular location of leghemoglobin no topic associated with nodule development has generated more controversy and confusion than the relationship of polyploid cells and establishment of the intracellular rhizobial infection. The critical question whether preexisting polyploid cortical cells or diploid cortical cells (which subsequently become polyploid) give rise to the infected cells of a nodule has been the motivating theme of many studies for 40 years. Approximately four decades ago Wipf and Cooper (1940) made a number of observations leading to the formation of their famous hypothesis which stimulated much interest in nodule morphogenesis. They observed polyploid cells in the inner cortex in the region of root hairs in uninoculated roots. In inoculated pea roots they reported the occurrence of dividing nuclei, both diploid and polyploid, in the inner cortex near the tip of the growing infection thread. The interphase nuclei of other cortical cells were of various sizes, the largest of which were suggested to be polyploid (Wipf and Cooper, 1940). The same workers also prepared squash mounts of pea nodules and determined the chromosome numbers of mitotic figures, finding both diploid and tetraploid nuclei (Wipf and Cooper, 1938). On the basis of transverse sections of the apical tips of pea nodules, they stated that the outer cortical layers were diploid and the central portion of the nodule was tetraploid. Whether nuclear volume or cytoplasmic staining of sections coupled with chromosome counts of squash preparations were their criteria for determining ploidy in sectioned material is unclear. They further stated that the nuclei of invaded cells were larger than those of the nodule cortex. Wipf and Cooper (1940) suggested that hormonal substances secreted by bacteria within the infection thread stimulated the mitosis of both tetraploid and diploid cells. Subsequently the infection thread continued growing toward the meristematic cells whose tetraploid cells became infected (Wipf and Cooper, 1938; Wipf, 1939) and eventually differentiated into the nitrogen-fixing, bacteroid-containing cells.

Torrey and Barrios (1969), using squash preparations of pea nodules, set out to determine when the first polyploid mitosis occurred during nodule formation. The first poluploid mitoses were observed 6 days after inoculation; the polyploid nuclei were mainly tetraploid, with paired diplochromosomes, and were regularly associated with infection threads which often terminated in these cells. Infection threads were not associated with diploid nuclei. These workers concluded that the infection thread specifically stimulated preexisting endomitotic polyploid cortical cells to undergo mitosis. Since mitotic divisions occur in the cortex before the infection thread passes out of the infected root hair (Newcomb et al., 1979), it is curious that no polyploid mitoses occurred until 6 days after inoculation.

Libbenga and Harkes (1973) followed the ontogeny of pea nodules and observed that the earliest mitoses occurred in the inner cortex and that these cells also became the first cells infected with free rhizobia and subsequently did not divide. Therefore, any preexisting polyploid cells in the inner cortex could not contribute to the nodule meristem. Subsequent to the mitotic activity of the inner cortex, mitotic activity in the adjacent inner cortical cells gave rise to the derivative cells, in part comprising the nodule meristem which gives rise to most of the nodule cells. The inner cortical cells were able to incorporate [^{3}H]thymidine, but DNA levels were not determined. These authors concluded that the tetraploid cells in the nodule meristem arose from preexisting endoreduplicated cells or by Rhizobium-induced endoreduplication in diploid cortical cells.

Microspectrometric measurements of Feulgen-stained sections of pea nodules were carried out by Mitchell (1965) who reported the following DNA levels: nodule meristematic cells, 4c and 8c; nodule cortex, 2c and 4c; and bacteroid-containing cells, 8c and 16c. Mitchell concluded that his data were consistent with the work of Wipf and Cooper (1938) who, he claimed, stated that the meristem was comprised of tetraploid cells with a few diploid cells on the outside. While Wipf and Cooper (1938) may have intended this interpretation, their paper is not so specific in describing the meristem or in quantifying the numbers of diploid and tetraploid cells. Mitchell further states that the outer nodule cortex (with 2c nuclei) is derived by proliferation of the diploid nodule meristem, while the inner nodule cortex (with 4c nuclei) is derived from lateral proliferation of the tetraploid meristem. The central infected tissue originates from the inner surface of the tetraploid meristem, and following bacterial infection the DNA levels increase further.

Truchet (1978) has studied several parameters of nuclei and nucleoli in carefully defined zones and stages of developing pea nodules (Table I). Since the nuclei of stage-1 nodules and in zone I of stage-3 nodules had only 2c and 4c levels of DNA, he has concluded that the nodule meristem is diploid and thus

TABLE I
QUANTITATIVE MEASUREMENTS OF NUCLEI IN DEVELOPING PEA NODULES[a,b]

Parameter	Stage 1	Stage 2	Stage 3		
			Zone I	Zone II	Zone III
Nuclear diameter (nm)	9.0	9.8	8.0	11.1	14.8
Mitotic index (%)	—	—	8.7	0	0
Radioactive nuclei after exposure to [³H]thymidine (%)					
8-hour exposure	—	—	14.5	12.0	0
21-hour exposure	—	—	22.6	15.9	0
30-hour exposure	—	—	32.3	24.9	0
DNA levels	2c and 4c	2c → 8c, only a few 8c invaded cells	2c and 4c	Mostly 4c and 8c, some 2c and 16c, uninfected cells 2c and 4c	Mostly 8c and 16c, some 32c, uninfected cells mainly 4c but some 2c

[a] Data are from Truchet (1978).

[b] Stage 1: young preemergent nodules with mitosis and dedifferentiation of cortex. Stage 2: beginnings of the release of rhizobia from infection threads. Stage 3: three or four zones present. Zone I: meristematic cells without any sign of infection or invasion. Zone II: derivative cells with infection threads. Zone III: cells clearly invaded by bacteria. Zone IV: senescent cells.

does not arise from pre-existing tetraploid cells. He reasons that the 2c and 4c levels probably represent the G_1 and G_2 periods of the cell cycle of diploid cells, because if the 4c level were a G_1 value for a tetraploid cell, one would also expect G_2 values of 8c in the nodule meristem. However, 8c levels are not observed until rhizobia are released from infection threads in nodule stage 2 [ca. 6 days after inoculation, i.e., similar to the timing of nodule development observed by Torrey and Barrios (1969)] and in zone II of stage-3 nodules. Truchet states that the increase in DNA levels occurs in zone II cells which he has not observed dividing but which incorporate [³H]thymidine. However, the presence of only dividing diploid cells in his data contradicts the numerous observations of dividing polyploid cells in pea nodules by Wipf and Cooper (1938, 1940), Bond (1948), and Torrey and Barrios (1969). Since these authors illustrated polyploid mitotic figures, there seems to be little doubt that such polyploid divisions occur regularly in pea nodules. Quantitative measurements of DNA levels in mitotic figures of squash preparations of pea nodules would resolve this point. In addition, Truchet's observation of the increase in ploidy in zone III cells without accompanying [³H]thymidine incorporation is most unusual if thymidine in fact is transported into these cells. Thus these data could reflect the permeability of thymidine rather than DNA biosynthesis. I feel that these discrepancies cast doubt upon Truchet's conclusion that the release of rhizobia from the infection thread triggers the increase in ploidy rather than that the polyploid nucleus exists prior to the invasion of the growing thread. The attempt to quantitate DNA levels and biosynthesis in carefully defined zones and stages of developing nodules has much merit and clearly deserves further attention in elucidating the relationship between endoreduplication and nodule morphogenesis.

A recent developmental study on an ineffective alfalfa nodule infected by a leucine-requiring mutant of *R. meliloti* (Truchet *et al.*, 1980) attempted to provide further support and extend the earlier observations of Truchet (1978). These ineffective nodules contained numerous infection threads from which the rhizobia did not escape; the host cells remained diploid, as determined mainly from the nuclear measurements of the earlier study (Truchet, 1978). When urea was supplied as a source of combined nitrogen and when L-leucine or one of its precursors, γ-ketoisocaproate or γ-ketoisovalerate, was supplied, the resulting nodular tissue appeared similar to normal effective nodules (i.e., those infected with wild-type *R. meliloti*) with liberated rhizobia in infected cells which the authors claimed were polyploid. Since the nuclear measurements were based on the questionable data of the earlier study, their conclusion that the increase in ploidy of the host cell was associated with the escape of rhizobia from the infection thread is not well supported by undisputable data and must be confirmed by other criteria. Furthermore, the existence and proposed functions of a bacteria-synthesized "nodule organogenesis-inducing principle" and a "bacterial central tissue differentiation-inducing principle" proposed by these authors remain highly speculative.

It is known that rhizobia synthesize both auxins and cytokinins (Newcomb, 1980) and that the white distal portion, including the nodule meristem and invasion zone (Truchet's zones I and II) of pea nodules have high levels of cytokinins; in addition, cytokinin levels and mitotic indexes decline similarly in effective (Syōno *et al.*, 1976) and ineffective (Newcomb *et al.*, 1977) nodules. From studies on cultured pea root tissues it has been shown that auxins and cytokinins are involved in the induction of endoreduplication (Torrey, 1961) and in the division of these polyploid cells (Matthysse and Torrey, 1967). With cortical pea root explants it was demonstrated by Libbenga and Torrey (1973) that the DNA values of the cortical nuclei increased from 2c and 4c levels to 8c and 16c in the presence of auxins and cytokinins. These authors suggested that cytokinins in the presence of auxin stimulated two rounds of DNA synthesis prior to mitosis, the first round being associated with endoreduplication and the second linked to mitosis. Further data indicated that polyploid mitoses occurred at higher frequencies away from the root tip, while diploid mitoses occurred at lower frequencies away from the root tip. It should be noted that nodules do not form near the root tip. Thus it appeared that the pea cortical root explant system could be regarded as a model system for nodule initiation. It was thus suggested that rhizobia secrete auxins, cytokinins, and other metabolites necessary for initiating the DNA synthesis associated with endoreduplication and mitotic activity. Examination of Fig. 6, showing the mitotic activity of the inner and middle cortex in a 5-day-old nodule in which the bacteria are restricted to the infected root hair, reinforces this hypothesis.

C. Release of Rhizobia from Infection Threads

In nodules possessing infection threads the rhizobia escape from unwalled regions of the infection thread essentially by the process of endocytosis, a rare phenomenon in plants (Gunning and Steer, 1975). Subsequently the bacteria become free in the host cytoplasm. They are actually surrounded by a membrane called the peribacteroid membrane which is initially derived from the host plasma membrane. While this sequence has been clearly demonstrated in soybean (Goodchild and Bergersen, 1966), clover (Dixon, 1964, 1967), pea (Kijne, 1975a; Newcomb, 1976), lupine (Robertson et al., 1978), and mung bean (Newcomb and McIntyre, 1981), considerable confusion is caused by numerous reports which claim to demonstrate either de novo formation or other various origins for the peribacteroid membrane. A critical evaluation of these papers is appropriate not only because of the importance of this process in nodule development and physiology but also because of the serious deficiencies which I feel invalidate the conclusions of each of the papers purporting to demonstrate the origin of the peribacteroid membrane from sources other than the host plasma membrane.

1. Vesicles and Membrane Envelopes

The confusing usages of the terms "vesicle" and "membrane envelope" require clarification. The term "vesicle" has been used to indicate (1) unwalled drops of infection thread matrix (Nutman, 1963; Mosse, 1964; Dart and Mercer, 1964; Dixon, 1967; Dart, 1975); (2) the peribacteroid membrance (Burns and Hardy, 1975); (3) the club- or spherically shaped structures of the actinomycetous endophytes of Frankia-induced nodules (Becking, 1977; Torrey, 1978; Newcomb et al., 1978); (4) the terminal spherical oil-containing structures of vesicular arbuscular mycorrhizae (Gerdemann, 1975); and (5) saclike structures, bounded by a single membrane, derived from Golgi bodies or endoplasmic reticulum (ER) (Morre and Mollenhauer, 1974). In describing the Rhizobium–legume symbiosis the term "vesicle" seems most appropriate for Golgi- and ER-derived structures; but clearly the continued use of the term "vesicle" for unwalled droplets and the peribacteroid membrane is most confusing and unjustified. Use of the term "membrane envelope" for the host-derived membrane surrounding the free rhizobia is inappropriate, becuase microbiologists have previously used this term to refer to the outer membrane of gram-negative bacteria which include the genus *Rhizobium;* the term "peribacteroid membrane" has been proposed to eliminate this confusion (Robertson *et al.,* 1978). In this chapter the term "peribacteroid membrane" is used, and the reader is strongly encouraged also to adopt this usage.

2. Origin of the Peribacteroid Membrane from the Host Plasma Membrane

In pea (Newcomb, 1976; Newcomb *et al.,* 1977; Dixon, 1964; Libbenga and Bogers, 1974; Pankhurst and Schwinghamer, 1974) and broad bean (*Vicia faba*) (Newcomb *et al.,* 1981) nodules, rhizobia escape from large, unwalled droplets of thread matrix (Figs. 12 and 16a). These droplets may be either detached (as often seen in sections) or connected to the infection thread (Fig. 15) and are always surrounded by the host plasma membrane. The bacterium moves to the edge of the droplet, becoming closely associated with the surrounding host plasma membrane which forms a bulge around the rhizobium (Fig. 16a). The bulge becomes larger until the membrane pinches off so that the escaping rhizobium is surrounded by a peribacteroid membrane which is initially derived from the host plasma membrane (Fig. 16b). Usually only one rhizobium is enclosed by the peribacteroid membrane immediately following release from the infection thread or unwalled droplet. Similar observations of rhizobia escaping from unwalled droplets have also been noted in clover (Dixon, 1964, 1967), alfalfa (Fig. 17) (Vance *et al.,* 1980), and other leguminous nodules (Dart, 1974, 1975). In the nodules of soybean (Goodchild and Bergersen, 1966), lupine (Robertson *et al.,* 1978), and mung bean (Newcomb and McIntyre, 1981) rhizobia escape from unwalled regions of infection threads which contain much less thread matrix than pea, clover, and alfalfa nodules; the lesser amounts of thread matrix in soybean, mung bean, and lupine nodules presumably account for the lack of unwalled droplets in these nodules.

3. Other Proposed Origins of the Peribacteroid Membrane

Prasad and De (1971) reported that in mung bean nodules rhizobia were released without a surrounding membrane from infection threads into vacuoles lacking tonoplasts or into cytoplasm near vacuoles, and that later the bacteria became surrounded by a membrane derived from blebs of the outer membrane of the nuclear envelope. Newcomb and McIntyre (1981) could not confirm these observations; in their material the free rhizobia were always surrounded by a peribacteroid membrane, and no blebbing of the nuclear envelope was seen. It is possible that poor preservation of membranes accounts for the observations of Prasad and De (1971), although in addition their micrographs do not appear to support their conclusions. In *V. faba* nodules it was claimed that rhizobia were

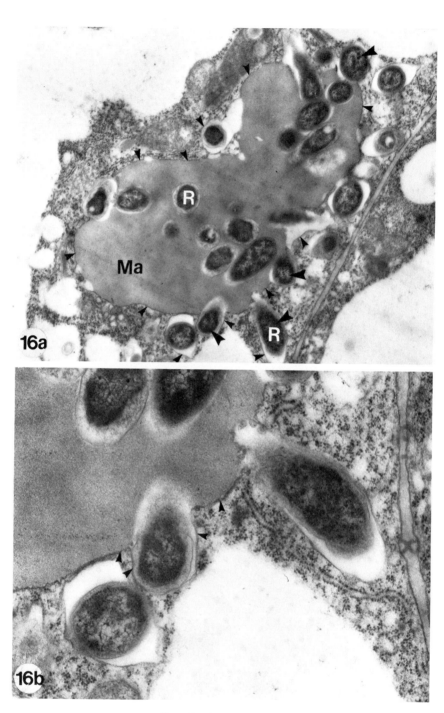

FIG. 16. See legend on p. 274.

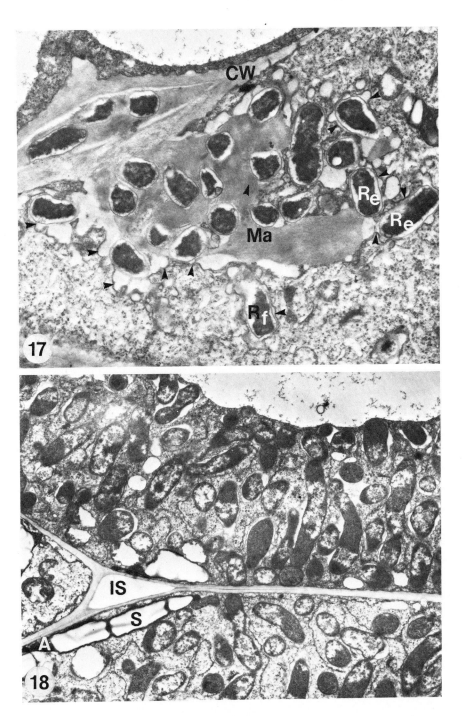

FIGS. 17 and 18.   See legends on p. 274.

released "naked" from infection threads (Generozova, 1979), but no micrographs were included to substantiate this point; lomasomes and endoplasmic reticulum were reported to be closely associated with the peribacteroid membrane, but the close proximity of two structures does not necessarily indicate a developmental interaction (O'Brien, 1972; O'Brien *et al.*, 1973). An earlier report (Generozova and Yagodin, 1972) suggested that the peribacteroid membrane in *V. faba* nodules was formed *de novo*, but again the micrographs are less than convincing, particularly since the cytology, including bacterial release, of pea and broad bean nodules is essentially identical (Figs. 12, 15 and 16a).

Mosse (1964) studied clover nodules fixed in either osmium tetroxide or potassium permanganate and suggested that the bacterial infection was localized between the paired membranes of ER; a single micrograph showed the continuity of a profile of ER with a peribacteroid membrane. This could merely represent the addition of membrane components (necessary for growth and turnover of the peribacteroid membrane) and not the initial formation. *De novo* formation of the peribacteroid membrane in clover nodules was also described (Dart and Mercer, 1963). In alfalfa nodules it was observed that rhizobia were released without a surrounding membrane and that the peribacteroid membrane arose *de novo* or from profiles of ER (Jordan *et al.*, 1963). Similarly in lupine, it was reported that the peribacteroid membrane had a *de novo* origin (Jordan and Grinyer, 1965); unfortunately, the micrographs do not substantiate this point. As pointed out above, other workers have convincingly demonstrated in clover, lupine, and alfalfa nodules that the free rhizobia are always surrounded by the peribacteroid membrane which is initially derived from the host plasma membrane.

Recently an interesting paper by Werner and Mörschel (1978) on soybean nodules included a micrograph illustrating an intercellular space containing several rhizobia surrounded by a membrane. These authors questioned whether the peribacteroid membrane always arose endocytotically from the host plasma

---

FIG. 16.   Release of rhizobia in *V. faba* nodules. Reprinted from Newcomb *et al.* (1980). (a) Rhizobia (R) in the process of escaping (large arrowheads) endocytotically from an unwalled droplet of thread matrix (Ma). The host plasma membrane (small arrowheads) surrounds the unwalled droplet and escaping bacteria. ×11,100. (b) Higher magnification of a portion of Fig. 16a showing the continuity of the host plasma membrane (arrowheads) surrounding the escaping rhizobia and unwalled droplet. ×33,500.

FIGS. 17 and 18.   Infected cells in alfalfa nodules.

FIG. 17.   An infection thread growing into the host cytoplasm from the host cell wall (CW) and rhizobia present within the matrix (Ma) in the process of escaping (R_e) and free (R_f) in the cytoplasm. Note that a membrane (arrowheads) surrounds the infection thread, escaping rhizobia (R_e), and free rhizobia (R_f). ×11,800.

FIG. 18.   An infected cell in the late phase of symbiotic growth containing many rhizobia in which the ribosomes are arranged such that a banded appearance results. Note the intercellular spaces (IS) and large, flattened starch (S) granules in the amyloplasts (A). ×6530.

membrane; they suggested that the intercellular membrane might be the result of degradation of the thread matrix and infection thread cell wall, leaving behind the host plasma membrane which formerly surrounded the thread with the enclosed bacteria. It is possible that these authors did not distinguish between the two types of infection threads and thus mistook an intercellular infection thread for an intracellular one. Unfortunately, no mention was made of the frequency of this structure which has not been reported in other studies on soybean nodule ultra-structure (Bergersen and Briggs, 1958; Goodchild and Bergerson, 1966; Bassett *et al.*, 1977). While Werner and Mörschel reported cutting and photographing serial sections, the quantity of sections was not indicated, although they stated that they took over 30 micrographs. This modest number of micrographs may indicate a similar number of serial sections, and this is simply insufficient to produce a complete series of sections through adjacent cells to detect ruptured or damaged cell walls. Furthermore, since they fixed small slices of nodules, it seems most likely that intact peribacteroid membranes were released from the cut surfaces of infected cells and that some became lodged in the intercellular spaces and subsequently were fixed, embedded, sectioned, and photographed. This observation must be confirmed before it can be regarded as being other than an artifact of preparation.

4. *Role of Vacuoles in Rhizobial Release*

Some authors have stated that rhizobia are released into vacuoles or infection vacuoles (Prasad and De, 1971; Werner and Mörschel, 1978; Werner *et al.*, 1980); this will remain unsubstantiated until it can be demonstrated that the surrounding membrane is a tonoplast (Robertson *et al.*, 1978). There is no evidence to support the existence of any connection, either physical or develop-mental, between the tonoplast and peribacteroid membrane.

D. Release of Rhizobia in Nodules Lacking Infection Threads

Some leguminous nodules do not have infection threads, and as a result rhizobia are distributed mainly by the mitosis of already infected cells (Dart, 1977). Initially there are no infected cells in the developing nodule, and it is not clear how rhizobia undergo the transition from being intercellular to becoming intracellular and surrounded by a peribacteroid membrane. The origin of the peribacteroid membrane in these nodules may also be the host plasma membrane, but this has not been convincingly documented in ultrastructural studies. This process has been studied in peanut nodules by Aufeuvre (1973) who claimed to illustrate infection threads and to demonstrate a plasma membrane origin for the peribacteroid membrane. However, no infection threads were found by Allen and Allen (1940) and Dart (1977). I find it impossible to draw any meaningful conclusions about the micrographs in Aufeuvre's study because of the poor quality of fixation.

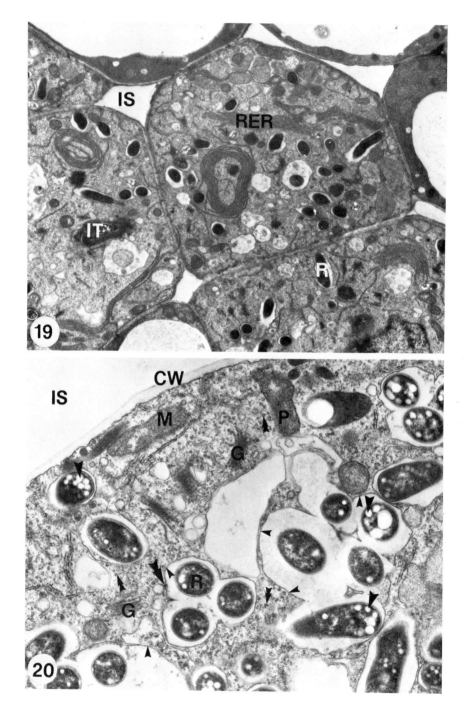

E. Formation of Unwalled Regions of Infection Threads

The mechanism by which wall-less sites if infection threads (Figs. 12 and 15) arise is unclear. Examination of serial sections of infected root hairs of clover has demonstrated that the growing tip of the infection thread is unwalled (Section I, C and Fig. 5), but unfortunately similar studies on infection threads in cortical cells have not been carried out. Such studies might reveal how rhizobia can escape from unwalled sites in certain cortical cells but cannot escape from the same sites (if they occur) in outer cortical or root hair cells. This is of fundamental importance in a cylindrically shaped nodule, because if rhizobia were released in the outer cortex instead of being liberated in the inner cortex during the early stages of nodule development, the nodule meristem would be surrounded by nondividing infected cells and as a result the nodule would remain very small.

In any case, it seems clear that rhizobial release occurs at unwalled sites in lateral positions of infection threads (Newcomb, 1976; Newcomb and McIntyre, 1981), although other unwalled sites may be situated at the tips of infection threads. In soybean nodules biochemical and histochemical studies have suggested that the plant cytoplasm synthesizes cellulase and that the rhizobia produce pectinase, resulting in localized degradation of the infection thread cell wall in a controlled and cooperative manner (Verma *et al.*, 1978a,b). Whether unwalled regions occur becuase of symbiotically regulated degradation or whether no cell wall is deposited at these sites in other leguminous nodules is unknown and certainly warrants further attention.

F. Early Symbiotic Development of Infected Cells

The release of rhizobia or bacteroids from infection threads initiates or at least coincides with a number of changes involved in the transition from a meristematic cell to an infected cell destined to become a site of symbiotic nitrogen fixation. [After rhizobia are released from infection threads, they may be referred to as bacteroids without distinguishing the morphological, cytological, or physiological condition (Bergersen, 1974; Goodchild, 1977; Robertson and Farnden, 1980).] In pea, broad bean, and mung bean nodules the cessation of mitotic activity of the host cell coincides with the release of rhizobia in it

---

Fig. 19. Infected cell in the early symbiotic phase of development in a soybean nodule. Illustrated are extensive profiles of RER, many rhizobia (R), an infection thread (IT), and several intercellular spaces (IS). ×6420.

Fig. 20. Infected cell in the early symbiotic stage of development in a *P. vulgaris* nodule, containing many bacteria with polyhydroxybutyrate deposits (single large arrowheads), vesicles fusing (double large arrowheads) with peribacteroid membranes (small single arrowheads), Golgi bodies (G), many polyribosomes (double small arrows), mitochondria (M), and proplastids (P). Also illustrated are the cell wall (CW) and an intercellular space (IS). ×16,850.

(Newcomb, 1976; Syōno et al., 1976; Truchet, 1978; Newcomb and McIntyre, 1981). Host cells containing free rhizobia *do undergo* mitosis in other leguminous nodules, including soybean and nodules which lack infection threads, such as peanut (Dart, 1977); this may be considered a method for distributing rhizobia when only a few or no infection threads are present, and thus increasing the number of infected cells. Eventually mitosis of infected cells also ceases in these nodules. The reason for the inhibition of host cell mitoses is uncertain but may involve supraoptimal levels of cytokinins which rhizobia synthesize (Syōno et al., 1976).

The number of rhizobia within the host cell increases markedly during the early symbiotic phase (Figs. 19 and 20). Initially the rhizobia are restricted to the outer regions of the host cytoplasm but eventually are fairly evenly distributed throughtout the host cytoplasm. As the bacteria increase in number by binary fission, the peribacteroid membranes must increase in size. Fusion of Golgi vesicles and possible ER vesicles with the peribacteroid membranes probably accounts for the additional membrane components (Robertson et al., 1978; Newcomb and McIntyre, 1981). In pea only 1 or 2 rhizobia are surrounded by a single peribacteroid membrane, while in soybean as many as 6 and in mung bean 16 are present (Figs. 21, 22, and 24). Fusion of peribacteroid membranes can also increase the number of rhizobia found within a single peribacteroid membrane. This is especially evident in mungbean (Newcomb and McIntyre, 1981), but in pea nodules apparently has little effect in the final number of rhizobia within a single peribacteroid membrane (Gunning, 1970). The host endomembrane system becomes much more extensive during the early symbiotic phase, with numerous Golgi bodies and profiles of rough endoplasmic reticulum (RER) becoming prominent. This is especially evident in soybean nodules whose young infected cells often contain extensive circular profiles of RER and numerous Golgi bodies with many associated vesicles of various sizes (Figs. 19 and 20). An abundance of free ribosomes and polyribosomes results in a densely staining cytoplasm in young infected cells in all legumes.

Increases in the numbers of proplastids, amyloplasts, and mitochondria occur. These organelles are usually located at the cell periphery near the cell wall, with the rhizobia situated more toward the center of the cell.

Interestingly, there are no published observations showing microbodies to be present in the infected cells of either *Rhizobium*- or *Frankia*-induced root nodules, although it has been suggested that microbodies are probably present in all higher plant cells (Gunning and Steer, 1975). It is known that catalase, which is usually localized in microbodies, is present in soybean and cowpea nodules (Atkins et al., 1980). Initially the starch granules of the amyloplasts are small and spherical. At the time when the rhizobia reach their final density, the starch granules become large, and in some species, such as pea and broad bean, are characteristically flattened (Fig. 21). The large central vacuole found during the early symbiotic

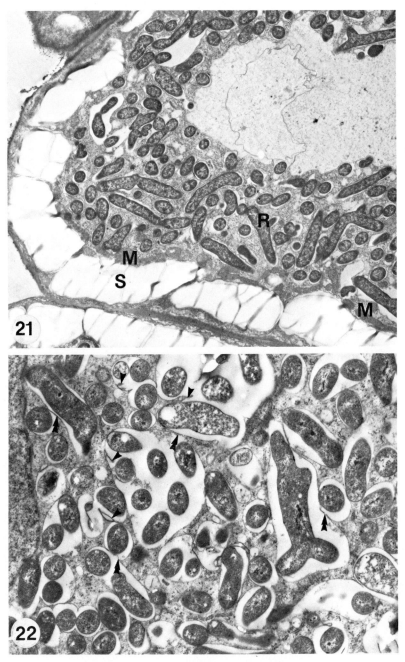

FIGS. 21 and 22.   See legends on p. 281.

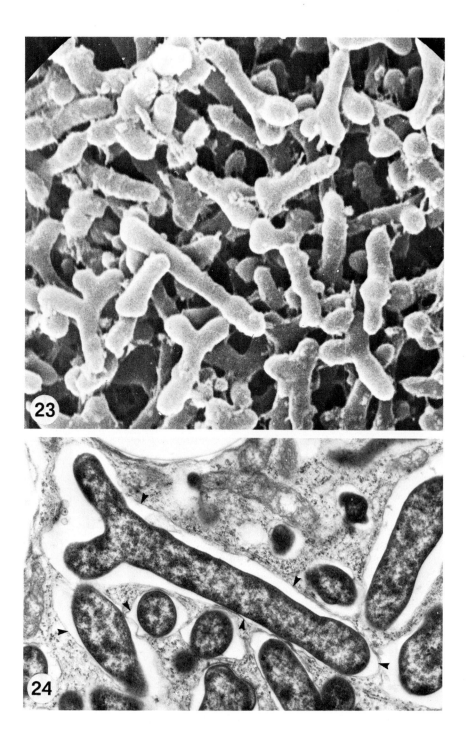

phase presumably forms in part by the coalescence of small vacuoles and by their subsequent growth. The nucleus increases in volume and DNA content (Mitchell, 1965; Truchet, 1978). In many nodules the host nucleus contains one, occasionally two, and rarely three prominent nucleoli.

G. Late Symbiotic Development of Infected Cells

This phase is characterized by changes in the size, morphology, and cytology of the rhizobia. Frequently the rhizobia become enlarged and assume various characteristic shapes (Figs. 18, 21, 22, and 23). In peanut the bacteroids become large and spherical (Aufeure, 1973; Dart, 1975, 1977), in pea and broad bean they are Y- and X-shaped (Figs. 23 and 24), and in mung bean only a few become branched (Fig. 22). The increase in size can be considerable; in peanut the bacteroids may be 20 μm in diameter (Aufeuvre, 1973), and in clover they have been estimated to occupy a volume 40 times that of the vegetative rhizobia (Gourret and Fernandez-Arias, 1974). In some other species the shape and size of rhizobia remain unchanged, but p-hydroxybutyrate deposits accumulate in the nitrogen-fixing stages of the bacteria (Goodchild and Bergersen, 1966; Ching et al., 1977) and may obscure the nucleoloid region (Vincent et al., 1962). A number of other inclusions including glycogen and polyphosphate granules occur in bacteroids (Craig and Williamson, 1972; Craig et al., 1973).

H. Senescence of Infected Cells

The functional lifespan of an infected cell is generally shorter than that of the nodule, but the senescence of nodules may be accelerated by harvest of the

---

Fig. 21. An infected cell in the late symbiotic phase of development in a pea nodule. Numerous rhizobia (R) are increasing in size and are also becoming branched. Mitochondria (M) and amyloplasts containing large, flattened starch (S) granules are located near the cell periphery. ×5100.

Fig. 22. Infected cell in the late symbiotic phase of development in a mung bean (V. radiata) nodule. One branched rhizobium is illustrated. Several peribacteroid membranes are fused (double arrowheads), and membrane remnants (single arrowheads) present within the areas bounded by these membranes may be derived from the fusions. ×10,800. Reprinted from Newcomb and McIntyre (1981).

Figs. 23 and 24. Infected cells in the late symbiotic phase of development in V. faba (Fig. 23) and pea (Fig. 24) nodules.
Fig. 23. Scanning electron micrograph showing Y-shaped bacteria in an infected cell of a V. faba nodule. ×5200. Micrograph courtesy of J. I. Sprent, University of Dundee.

Fig. 24. Y-Shaped bacteria (or bacteroid) in an infected cell from a pea nodule. The peribacteroid membrane (arrowheads) separates the host cytoplasm from the rhizobia. ×16,200. Micrograph by Susan Creighton and Deborah Jordan, Queen's University, Canada.

shoots, as in forage crops such as alfalfa (Vance *et al.*, 1980), or various environmental parameters which reduce the photosynthesis rate (Jordan, 1974). In effective pea nodules the lifespan of an infected cell was estimated as being between 7 and 10 days (Newcomb, 1976), but this time may be considerably longer in a spherical nodule which does not have a replenishment of infected cells from the nodule meristem but which fixes nitrogen for several weeks. In pea (Newcomb, 1976), alfalfa (Tu, 1974; Vance *et al.*, 1980), and soybean (Bergersen, 1974), the host cytoplasm senesces before degenerative changes are observed in the bacterial cells (Figs. 25 and 26). However, other authors have reported that the degenerative changes occur simultaneously in both host and bacterial cells (Dart, 1975; Kijne, 1975b) or that the rhizobia degenerate prior to the host cytoplasm (Grilli, 1963; Truchet and Coulomb, 1973). It is important to distinguish degenerative changes from damage induced by the preparative procedures. Freeze-fracture studies of senescing alfalfa nodules by Vance *et al.* (1980) confirmed that the host cytoplasm degenerated prior to the rhizobia; these workers observed multivesicular bodies which they suggested were similar to the lysosomes reported by Truchet and Coulomb (1973), but other authors (Newcomb, 1976; Robertson *et al.*, 1978) have not reported these structures.

I. Uninfected Cells of the Central Tissue (or Zone)

Numerous uninfected cells are present among the infected cells in the central zone or tissue of a legume nodule. The percentage of uninfected cells varies and is highest in ineffective nodules. In a nodule such as pea, in which every infected cell is invaded by an infection thread with the subsequent release of rhizobia, only very rarely is an infection thread observed passing through an uninfected cell (W. Newcomb, unpublished observations). According to Truchet (1978) the uninfected cells of the central zone have 2c and 4c levels of DNA. Until the question of whether polyploidy precedes infection, or vice versa, is resolved (Section III,B), it will not be known why infection threads do not regularly invade these uninfected cells. Typically these uninfected cells have a prominent nucleus and nucleolus, a large central vacuole, numerous amyloplasts containing large, spherical starch granules, few Golgi bodies and profiles of ER, and numerous mitochondria (Fig. 27).

Uninfected cells are much smaller than the adjacent infected cells (Figs. 9 and 27). In some nodules, such as soybean, the uninfected cells are arranged in radial

---

Figs. 25 and 26.   Sensescent infected cells in pea nodules.

Fig. 25.   The host cytoplasm has become degenerative, but the rhizobia (R) do not show signs of senescence. V, Vacuole; CW, cell wall. ×4100.

Fig. 26.   A more advanced stage of senescence than in Fig. 25. The host cytoplasm has degenerated except for remnants of the peribacteroid membrane (arrowheads). The rhizobia still appear to be healthy. ×7600.

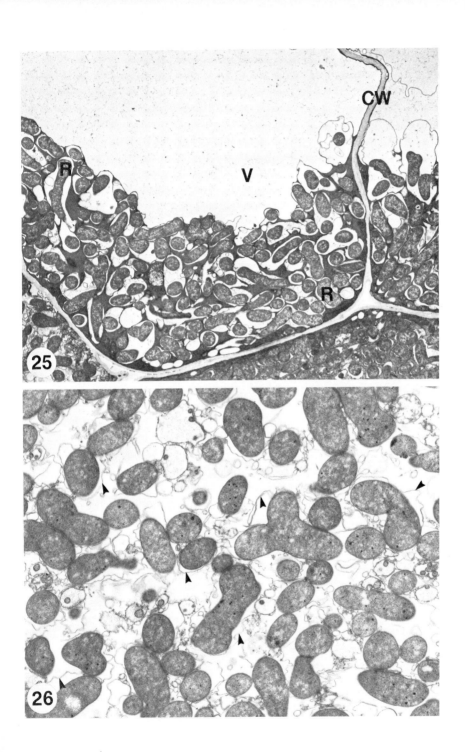

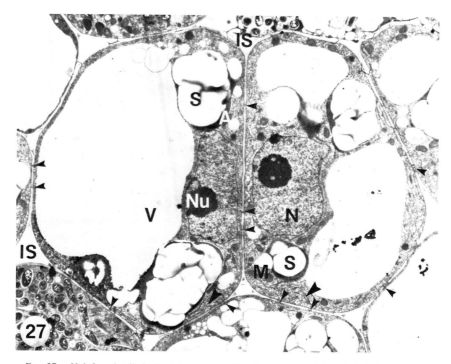

FIG. 27.   Uninfected cells in the central tissue of a *P. vulgaris* nodule. Illustrated are the large vacuoles (V), prominent nuclei (N) and nucleoli (Nu), amyloplasts (A) containing large starch (S) granules, mitochondria (M), RER (large arrowheads), and intercellular spaces (IS). Plasmodesmata (small arrowheads) interconnect the protoplasts of adjacent cells. ×3440.

files and may have a role in aerating the internal nodular tissues (Bergersen and Goodchild, 1973). Typically uninfected cells have a prominent nucleus and nucleolus, a large central vacuole, numerous amyloplasts containing large, spherical starch granules, many mitochondria, and few Golgi bodies or profiles of ER (Fig. 27). While free ribosomes and polyribosomes are numerous in uninfected cells, the number of ribosomes is markedly less than the numbers present in nearby infected cells.

IV. Ineffective Root Nodules

Ineffective nodules in which little or no molecular nitrogen is reduced or fixed may be caused by environmental conditions or genetic alterations, and genetic alterations of either (or both) the host or rhizobial genomes (Jordan, 1974; Dart, 1975, 1977). Jordan (1974) has characterized the associations arising from certain environmental factors as "transient ineffectiveness" because, when the stress condition is removed, an effective, i.e., nitrogen-fixing, association is reestablished. He provides examples of environmental stresses which are responsible for transient ineffectiveness, including boron deficiencies, low levels of carbohydrates (caused by light conditions, i.e., low rates of photosynthesis), high temperature, and oxygen deficiencies created during flooding.

Most studies on ineffective nodules have utilized associations in which the rhizobial partner was a natural or induced mutant. The growth and physiology of mutant rhizobia and methods for their production were reviewed by Jordan (1974). Ineffective nodules produced by alterations in the host genome have also received attention. More recently Holl and LaRue (1975) screened 2000 cultivars of the World Pea Collection and found 10 non-nitrogen-fixing cultivars; they estimated that at least 10 host genes controlled nodule development. In *P. sativum* 'Afghanistan', two genes are involved in suppression of nodulation: One of these genes controls the metabolism of nitrogen-fixation, while the other controls nodulation (Holl and LaRue, 1975). Grafting experiments demonstrated that the latter gene was associated with synthesis in the root of a nonnodulating factor. This factor controlled the proliferation of cortical cells but not the infection of root hairs (Degenhardt *et al.*, 1976). Another F_2 nodulating segregate of pea contained a functional nitrogenase but reduced acetylene only when nodule slices were incubated in succinate or pyruvate (Holl and LaRue, 1975); the cytology of nodules of this mutant appeared normal (Newcomb, 1980). Zobel (1975) used mutagenic treatments and in an examination of 96 M_2 seedlings found 75 plants with altered nodules and 5 plants with modified shoots; some nodules lacked leghemoglobin, while others contained leghemoglobin but were unable to reduce acetylene; other mutants were unable to become nodulated or varied in nodule size and number.

Commonly, ineffective root nodules are smaller than effective nodules of the same species, which may be caused by an alteration in levels of hormone or growth regulators. In ineffective pea nodules infected with a mutant strain of *Rhizobium leguminosarum,* cytokinin levels declined earlier than in effective nodules on the same pea cultivar (Syōno *et al.,* 1976; Newcomb *et al.,* 1977). These declines were correlated with declines in mitotic activity in the nodule meristem from which most nodule cells arise and thus could partially explain the smaller size of these ineffective nodules. In addition, an unidentified cytokinin was present in the ineffective nodules but was absent in the effective nodules. Ultrastructurally, the infection process including bacterial release was similar in both types of nodules; however, after a period of bacterial proliferation, some "free" bacteria had incomplete peribacteroid membranes or lacked these membranes completely; often the infected cells degenerated prematurely. This indicates that the stimulation of enhanced ER production in the infected cells occurring after bacterial release did not take place in these ineffective nodules; thus it appears that host membrane biosynthesis was also altered in these nodules. In addition, these nodules lacked leghemoglobin (Newcomb *et al.*, 1977). Ineffective soybean nodules produced by a mutant strain of *Rhizobium japonicum* also contained infected cells in which peribacteroid membrane growth was altered (Werner *et al.*, 1980).

V. Cells Associated with Metabolite Transport

A. NODULAR VASCULAR BUNDLES

Vascular bundles are located in the nodule cortex outside the central infected tissue (Fig. 9) in legume nodules. This is in contrast to the centrally located vascular bundle in actinomycete-induced nodules (Torrey, 1978) and in *Parasponia* (Section VI). The nodular vascular bundles join with the stele of the adjacent root tissue and are the routes of import for photosynthate and export for amino acids containing fixed nitrogen. Each nodular bundle contains xylem and phloem cells surrounded by a pericycle and an endodermis, respectively; thus it is necessary for compounds being transported into or from the central infected tissue to pass through the endodermis and pericycle of the nodular vascular bundle. Often these pericycle cells have wall ingrowths characteristic of transfer cells (Section V,B). The number of nodular vascular bundles varies among different species of legumes but seems to be constant for any given species. In cylindrical nodules the nodular vascular bundles end near the nodule meristem and form an "open" arrangement. In spherical nodules the nodular vascular bundles join or fuse at the distal region and form a closed pattern.

B. TRANSER CELLS

Certain highly specialized plant cells have a characteristic adaptation in which ingrowths of the secondary cell wall greatly increase the surface of the adjacent plasma membrane forming the so-called wall membrane apparatus (Figs. 28 and 29); cells possessing this wall membrane apparatus are called transfer cells and may be encountered in a variety of anatomical situations (Gunning and Pate, 1969, 1974; Pate and Gunning, 1972). Both the occurrence of transfer cells at sites of extensive short-distance transport and the marked increase in the area of the plasma membrane near the ingrowths suggest that these cells are uniquely

---

FIGS. 28 and 29.   Transfer cells located in the root tissue adjacent to effective pea nodules.

FIG. 28.   Pericycle transfer cells have numerous branched wall ingrowths (WI), many proplastids (P) and mitochondria (M), profiles of RER (arrowheads), and vacuoles. ×5670.

FIG. 29.   Xylem parenchyma transfer cells have unbranched wall ingrowths (WI) located near the pits in the xylem (X) cell walls. The middle lamella is designated by arrowheads. M, Mitochondrion. ×10,240.

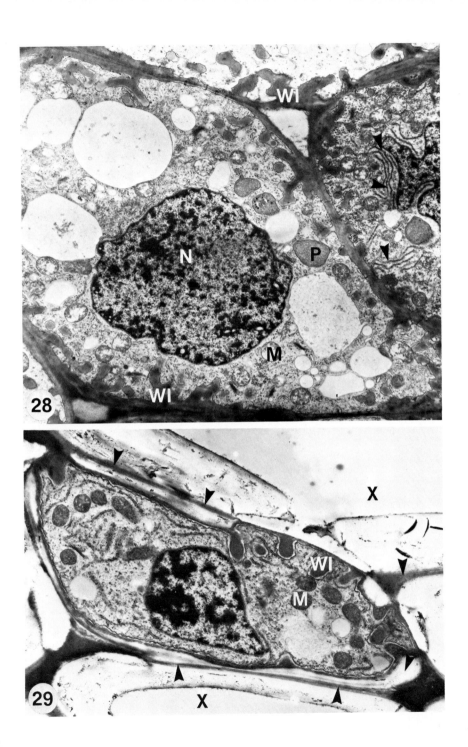

adapted for the short-distance movement of solutes. The occurrence of numerous mitochondria and profiles of ER also suggests a transport function for transfer cells. The arrangement of cellulose microfibrils within the wall ingrowths creates large channels, 2 nm or larger in diameter and having fixed anionic charges (Gunning and Pate, 1974), which may facilitate electroosmosis, a process that may be important for transport in higher plants (Tyree, 1970).

Pericycle transfer cells have been observed in the nodular vascular bundles of *Pisum, Medicago, Trifolium, Lupinus,* and *Lathyrus* (Pate *et al.,* 1969). This is especially interesting for two reasons: Transfer cells have been infrequently observed in roots (Gunning and Pate, 1974; Newcomb and Peterson, 1979), and it is an anatomical necessity for photosynthetic products or their derivatives and compounds containing nitrogen fixed by the bacteria to pass through these pericycle transfer cells (Pate *et al.,* 1969). Thus these pericycle transfer cells appear to have two roles: (1) to facilitate the movement of sugars from the seive elements toward the infected cells and (2) to facilitate the export of amides and amino acids to the xylem elements of the nodular vascular bundles. Analysis of sap exuding from detached nodules reveals concentrations of compounds containing fixed nitrogen up to 10 times those found in the central infected tissues of the nodule (Pate *et al.,* 1969). Interestingly the sap does not contain sugars (Pate *et al.,* 1969), and this may indicate the selective permeability or transport capabilities of the transfer cell plasma membrane.

Pericycle (Fig. 28), xylem, and phloem transfer cells are also present in the root tissue adjacent to the nodules of certain legumes. Xylem parenchyma cells occur in the root tissue adjacent to the nodules of pea (Fig. 29), soybean, broad bean, mung bean, and French bean (*Phaseolus vulgaris*) (Newcomb and Peterson, 1979), as well as in clover and alfalfa (W. Newcomb, unpublished observations). Wall ingrowths also occur in pericycle, xylem parenchyma, and phloem companion cells in root tissue near ineffective pea nodules (Newcomb *et al.,* 1977).

The numerous observations of the coincidence of wall ingrowth formation with the onset of solute movement (Gunning and Pate, 1969; Pate and Gunning, 1972) influenced formation of the hypothesis that the initiation of wall ingrowths is induced by the presence of the solute to be transported and that wall ingrowths do not form if the solute is absent (Gunning, 1977). However, the difficulties in studying the physiology of transfer cells and adjacent cells have prevented the obtaining of other than circumstantial evidence in support of this theory. In clover nodules the pericycle transfer cells of the nodular vascular bundle form after infection of the bacteroid-containing cells and before differentiation of the enlarged bacteroids (Pate *et al.,* 1969). Thus in clover nodules formation of the ingrowth occurs at about the initial period of nitrogen fixation or possibly at the initial flux of exported nitrogenous compounds. Whether compounds containing fixed nitrogen are in fact the inducers of wall growths in transfer cells near nodules has recently been questioned by observations that two kinds of ineffective pea nodules possess xylem and pericycle transfer cells (Newcomb and Peterson, 1979). In addition, the wall ingrowths near effective pea nodules form prior to the onset of nitrogen fixation (Newcomb and Peterson, 1979). Prior to the onset of nitrogen fixation, organic acids and amino acids are cycled to the nodules with the photosynthetic supply; in addition, organic acid production in nodules is not related to effectiveness (Newcomb and Peterson, 1979). Thus it is not clear what substances induce the formation of wall ingrowths associated with root nodules. The involvement of cytokinins and other growth regulators should not be dismissed (Gunning and Pate, 1969; Newcomb and Peterson, 1979). Interestingly transfer cells do not occur near or in actinomycete-induced root nodules.

VI. *Rhizobium* Nodules on Nonlegumes

Nodules form on certain species of *Parasponia,* a genus native to Java and Papua New Guinea (Trinick, 1973); (the host plants were at first misidentified as being members of the closely related genus *Trema:* Akkermanns *et al.,* 1978a,b). It has been demonstrated that cowpea-type rhizobia are the causal microorganisms responsible for the formation of these nodules in *Parasponia* (Trinick and Galbraith, 1980). The structure of these nodules is especially interesting, because they have features characteristic of both *Rhizobium-* and *Frankia*-induced root nodules.

On naturally growing *Parasponia* trees the nodules may be as large as 6 cm in diameter. They are often coralloid in appearance because of the presence of many branched nodule lobes, although laboratory-grown nodules exhibit limited branching (Trinick, 1979). Nodule roots do not occur on them (Trinick, 1979). The nodules contain a central vascular cylinder, a distally situated nodule meristem, a zone of infected cells, and a thick layer of nodule cortex (Figs. 30 and 31) (Trinick and Galbraith, 1976). Infection threads are involved in the infection of host cells but, unlike the situation in other effective legume nodules, the rhizobia do not always escape from the infection threads. In *Parasponia rugosa* (first identified as *Trema aspera* and then reidentified as *T. cannabina*) rhizobia are released from the infection threads in fewer than one-third of the infected cells (Figs. 32 and 33). Generally, if rhizobia are not released, the infection thread increases in size and the number of rhizobia within it also increases. Some cells contain both an enlarged infection thread and free rhizobia (Trinick and Galbraith, 1976). The infected cells do not divide after the release of rhizobia; thus the only method of cellular infection is through the invasion of an infection thread. Uninfected cells are scattered among the infected cells (Trinick and Galbraith, 1976). The infected cells may be as much as eight times as large as

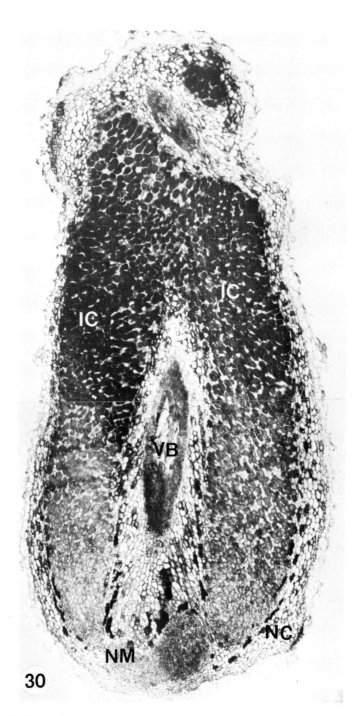

FIG. 30. Longitudinal section through a nodule of *P. rugosa* illustrating the distal nodule meristem (NM), the external layer of nodule cortex (NC), the zone of infected cells (IC), and the centrally located vascular bundle (VB). ×55. Reprinted from Trinick and Galbraith (1976).

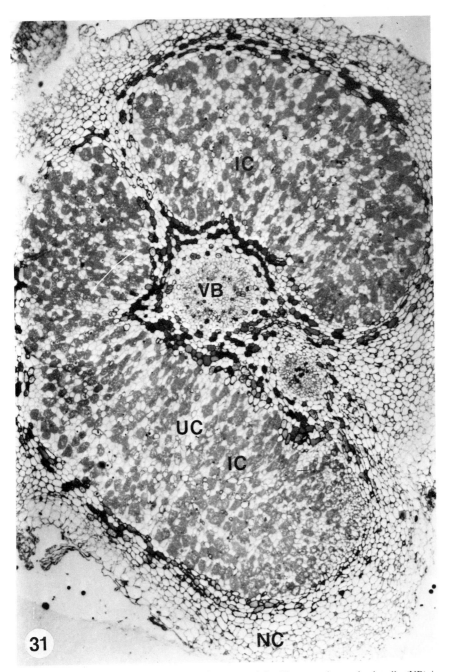

FIG. 31. Transverse section of a *P. andersonii* nodule. The central vascular bundle (VB) is surrounded by the zone containing numerous infected cells (IC) and fewer uninfected cells (UC), and an outer layer of nodule cortex (NC). ×67. Micrograph kindly provided by M. Trinick, CSIRO, Wembley, Western Australia.

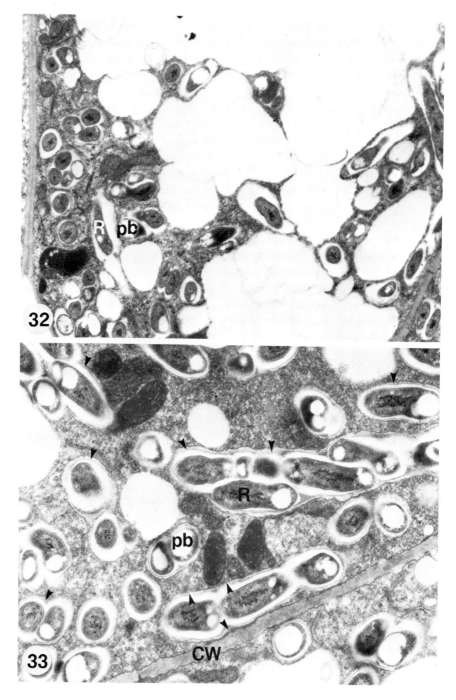

adjacent uninfected cells (Trinick and Galbraith, 1976). No information regarding DNA levels of the infected cells has been published.

Ultrastructural studies of *Parasponia andersonii* nodules (Trinick, 1979) have demonstrated that the infection thread cell wall and surrounding membrane are similar and continuous with the host cell wall and plasma membrane, respectively. The rhizobia are embedded in a thread matrix which usually lacks electron density and much contrast (Fig. 33). In some infection threads an electron-dense layer, identified as an inner membrane (Trinick, 1979), separates the thread matrix from the infection thread cell wall. Unfortunately, the published micrographs are of too low a magnification to resolve substructure characteristic of an unit membrane; however, if this structure is not an artifact of some nonmembranous boundary, one must raise the problem of its ontogeny. The infection thread cell wall may become thinner, perhaps not even forming, so that the rhizobia are only surrounded by a membrane, a situation similar to that observed in legume nodules.

Small inclusions, probably polyhydroxybutyrate, are present in rhizobia in early stages of infection and increase in size and number in more advanced stages. Young infected cells contain little host cytoplasm and large vacuoles (Fig. 32); in more advanced stages small vacuoles are dispersed throughout the cytoplasm. *Parasponia* nodules differ from other effective *Rhizobium* nodules in lacking leghemoglobin and in this respoect resemble actinomycete-induced nodules (Coventry *et al.*, 1976). Thus *Parasponia* nodules, in view of their resemblance to both *Rhizobium*- and *Frankia*-induced nodules, are in effect an intermediate form of these nitrogen-fixing associations.

VII. General Conclusions

Many studies have yielded a wealth of detailed information about nodule structure, but it is important that the reader be aware that our current knowledge represents only the tip of the iceberg. Significant questions regarding the relationship of infection thread growth and nuclear DNA levels, the control of bacterial release in nodules having extensive infection threads and in nodules lacking infection threads, and the compartmentalization of enzymes such as catalase and uricase, beg for answers. It is very likely that new structures will be

---

FIG. 32.   Infected cell of *P. andersonii* in the early symbiotic stage showing infection threads containing rhizobia (R) and the vacuolate host cytoplasm. Many of the rhizobia contain polyhydroxybutyrate deposits (pb). ×6000. Reprinted from Trinick (1979).

FIG. 33.   Infected cell of *P. andersonii* containing rhizobia (R) in walled (arrowheads) infection threads. Polyhydroxybutyrate granules (pb) are common in the bacteria. CW, Host cell wall. ×14,000. Micrograph courtesy of M. Trinick.

encountered in future fine-structural studies on nodules. For example, structures resembling cytoplasmic nucleoloids (Dickinson and Heslop-Harrison, 1970; Williams *et al.*, 1973) have been observed in certain spherical nodules (W. Newcomb, unpublished observations) and may represent an alternative to en-doreduplication for enhancing the number of copies of rRNA and subsequently the rate of protein synthesis. I have regularly seen other structures in leguminous nodules that did not resemble normal organelles or inclusions. While one must be sure one is not merely observing artifacts, the exciting prospects of unique structures and special control mechanisms in the cells of nodules should certainly not be dismissed.

The correlated utilization of various techniques has much merit, as exemplified by the careful study on the peribacteroid membrane by Robertson *et al.* (1978) who combined results from transmission electron microscope cytochemistry, freeze-fracturing, and biochemistry. The combined use of these methods strengthens the individual observations and reduces the possibility of inadvertently reporting an artifact, because one is unlikely to produce a new structure by very different procedures.

The goal of undertaking structural and developmental studies on nodules should be to increase our understanding of the morphogenesis of these symbioti-cally produced structures. The involvement of two very different organisms is in part responsible for the current limited knowledge of the control of nodule development. For example, the hormonal control of nodule development has received little attention. Hormone levels have been measured in several nodules (Wheeler *et al.*, 1979), but to date no study has analyzed all the major hormones in developing nodules, although suitable techniques exist (Wightman, 1977). Furthermore, the chemical identities of many of the hormones found in nodules have not been rigorously characterized (Greene, 1980). These are not unimpor-tant points for, although cytokinin levels are correlated with the mitotic activity of the nodule meristem (Syōno *et al.*, 1976), cytokinins usually act in conjuction with auxins and other growth substances (Wareing and Phillips, 1978). Such hormonol analyses must be correlated with the developmental stages of the nodule. This limited information becomes more important if investigations of the role of rhizobial plasmids in nodule morphogenesis are to proceed meaningfully, for it is difficult to understand genetic control of development if the develop-mental processes themselves are poorly understood.

One must ask interesting questions to obtain interesting and significant an-swers. It is hoped that this article will serve to stimulate some of these questions.

ACKNOWLEDGMENTS

The author is pleased to acknowledge the assistance of Faye Murrin and Susan Creighton who helped out in much of the research and provided a critical reading of the manuscript. I also thank

Professors J. G. Torrey and D. Davidson, along with Dale Callaham, for useful discussions of nodule development. I am also grateful for M. J. Trinick, J. I. Sprent, and D. Callaham for kindly supplying micrographs. The support of grants from Agriculture Canada, Natural Science and Engineering Research Council of Canada, and the Advisory Research Council of Queen's University is gratefully acknowledged.

References

Akkermans, A. D. L., Abdulkardir, S., and Trinick, M. J. (1978a). *Nature (London)* **274,** 190.

Akkermans, A. D. L., Abdulkardir, S., and Trinick, M. J. (1978b). *Plant Soil* **49,** 711.

Allen, O. N., and Allen, E. K. (1940). *Bot. Gaz.* **102,** 121.

Atkins, C. A., Rainbird, R., and Pate, J. S. (1980). *Z. Pflanzenphysiol* **97,** 2. 9.

Aufeuvre, M. A. (1973). *C. R. Hebd. Seances Acad. Sci. Ser. D* **277,** 921.

Basset, B., Goodman, R. N., and Novacky, A. (1977). *Can. J. Microbiol.* **23,** 573.

Becking, J. H. (1977). In "A Treatise on Dinitrogen Fixation" (R. W. F. Hardy and W.S. Silver, eds.), Section III, p. 185. Wiley, New York.

Bergersen, F. J. (1974). In "The Biology of Nitrogen Fixation" (A. Quispel, ed.), p. 473. North-Holland Publ., Amsterdam.

Bergersen, F. J., and Briggs, M. J. (1958). *J. Gen. Microbiol.* **12,** 482.

Bergersen, F. J., and Goodchild, D. D. J. (1973). *Aust. J. Biol. Sci.* **26,** 729.

Bieberdorf, F. W. (1938). *J. Am. Soc. Agron.* **30,** 375.

Bond, L. (1948). *Bot. Gaz.* **109,** 411.

Burns, R. C., and Hardy, R. W. F. (1975). "Nitrogen Fixation in Bacteria and Higher Plants." Springer-Verlag, Berlin and New York.

Callaham, D. (1979). M.Sc. thesis, University of Massachusetts, Amherst.

Callaham, D., and Torrey, J. G. (1981). *Can. J. Bot.* **59,** 1647.

Ching, T.M., Hedtke, S., and Newcomb, W. (1977). *Plant Physiol.* **60,** 771.

Coventry, D. R., Trinick, M. J., and Appleby, C.A. (1976). *Biochim. Biophys.* Acta **420,** 105.

Craig, A. S., and Williamson, K. I. (1972). *Arch. Mikrobiol.* **87,** 165.

Craig, A. S., Greenwood, R. M., and Williamson, K. I. (1973). *Arch Mikrobiol.* **89,** 23.

Dart, P. J. (1974). In "The Biology of Nitrogen Fixation" (A. Quispel, ed.), p. 381. North-Holland Publ., Amsterdam.

Dart, P. J. (1975). In "The Development and Function of Roots" (J. G. Torrey and D. T. Clarkson, eds.), p. 467. Academic Press, New York.

Dart, P. J. (1977). In "A Treatise on Dinitrogen Fixation" (R. W. F. Hardy and W. S. Silver, eds.), Section III, p. 367. Wiley, New York.

Dart, P. J., and Mercer, F. V. (1963). *Arch. Mikrobiol.* **46,** 382.

Dart, P. J., and Mercer, F. V. (1964). *Arch. Mikrobiol.* **49,** 209.

Degenhardt, T. L., LaRue, T. A., and Paul, E. A. (1976). *Can. J. Bot.* **54,** 1633.

Dickinson, H. G., and Helsop-Harrison, J. (1970). *Protoplasm* **69,** 187.

Dixon, R. O. D. (1964). *Arch. Mikrobiol.* **48,** 166.

Dixon, R. O. D. (1967). *Arch. Mikrobiol.* **56,** 156.

Fahraeus, G., and Ljunggren, H. (1959). *Physiol. Plant.* **12,** 145.

Generozova, I. P. (1979). *Fiziol. Rast.* **26,** 788.

Generozova, I. P., and Yagodin, B.A. (1972). *Fiziol. Rast.* **19,** 348.

Gerdemann, J. W. (1975). In "The Development and Function of Plant Roots" (J. G. Torrey and D. T. Clarkson, eds.), p. 575. Academic Press, New York.

Goodchild, D. J. (1977). *Int. Rev. Cytol. Suppl.* **6,** 235.

Goodchild, D. J. and Bergersen, F. J. (1966). *J. Bacteriol.* **92**, 204.

Gourret, J. P., and Fernandez-Arias, H. (1974). *Can. J. Microbiol.* **20**, 1169.

Greene, E. M. (1980). *Bot. Rev.* **46**, 25.

Grilli, M. (1963). *Caryologia* **16**, 561.

Gunning, B. E. S. (1970). *J. Cell Sci.* **7**, 307.

Gunning, B. E. S. (1977). *Sci. Prog. Oxford* **64**, 539.

Gunning, B. E. S., and Pate, J. S. (1969). *Protoplasma* **68**, 107.

Gunning, B. E. S., and Pate, J. S. (1974). *In* "Dynamic Aspects of Plant Ultrastructure" (A. W. Robards, ed.), p. 441. McGraw-Hill, New York.

Gunning, B. E. S., and Steer, M. W. (1975). "Ultrastructure and the Biology of Plant Cells." Arnold, London.

Haack, A. (1964). *Zentralbl. Bakteriol. Parasitenkd. Infektionskr. Hyg. Abt. 2*, **117**, 343.

Higashi, S. (1966). *J. Gen. Appl. Microbiol.* **12**, 147.

Holl, F. B., and LaRue, T. A. (1975). *In* "Proceedings of the 1st International Symposium on Nitrogen Fixation" (W. E. Newton and C. J. Nyman, eds.), Vol. II, p. 391. Washington State Univ. Press, Pullman.

Hubbell, D. H., Morales, V. M., and Umali-Garcia, M. (1978). *Appl. Environ. Microbiol.* **35**, 210.

Jordan, D. C. (1974). *Proc. Indian Natl. Sci. Acad. Part B*, **40**, 713.

Jordan, D. C., and Grinyer, I. (1965). *Can. J. Microbiol.* **11**, 721.

Jordan, D. C., Grinyer, I., and Coulter, W. H. (1963). *J. Bacteriol.* **86**, 125.

Kijne, J. W. (1975a). *Physiol. Plant Pathol.* **5**, 75.

Kijne, J. W. (1975b). *Physiol. Plant Pathol.* **7**, 17.

Kumarasinghe, R. M. K., and Nutman, P. S. (1977). *J. Exp. Bot.* **28**, 961.

Libbenga, K. R., and Bogers, R. J. (1974). *In* "The Biology of Nitrogen Fixation" (A. Quispel, ed.), p. 430. North-Holland Publ., Amsterdam.

Libbenga, K. R., and Harkes, P. A. A. (1973). *Planta* **114**, 17.

Libbenga, K. R., and Torrey, J. G. (1973). *Am. J. Bot.* **60**, 293.

Ljunggren, H. (1969). *Physiol. Plant. Suppl. V*.

Martinez-Molina, E., Morales, V. M., and Hubbell, D. H. (1979). *Appl. Environ. Microbiol.* **38**, 1186.

Matthysse, A. G., and Torrey, J. G. (1967). *Physiol. Plant* **20**, 661.

Mitchell, J. P. (1965). *Ann. Bot. N.S.* **29**, 371.

Morre, D. J., and Mollenhauer, H. H. (1974). *In* "Dynamic Aspects of Plant Ultrastructure" (A. W. Robards, ed.), p. 84. McGraw-Hill, New York.

Mosse, B. (1964). *J. Gen. Microbiol.* **36**, 49.

Napoli, C. A., and Hubbell, D. H. (1975). *Appl. Microbiol.* **30**, 1003.

Newcomb, W. (1976). *Can. J. Bot.* **54**, 2163.

Newcomb, W. (1980). *In* "Nitrogen Fixation" (W. E. Newton and W. H. Orme-Johnson, eds.), Vol. II, p. 87. Univ. Park Press, Baltimore, Maryland.

Newcomb, W., and McIntyre, L. (1981). *Can. J. Bot.* **59**, in press.

Newcomb, W., and Peterson, R. L. (1979). *Can J. Bot.* **57**, 2583.

Nwwcomb, W., Syōno, K., and Torrey, J. G. (1977). *Can J. Bot.* **55**, 1891.

Newcomb, W., Peterson, R. L., Callaham, D., and Torrey, J. G. (1978). *Can J. Bot.* **56**, 502.

Newcomb, W., Sippell, D., and Peterson, R. L. (1979). *Can. J. Bot.* **57**, 2603.

Newcomb, W., Creighton, S., and Latta, L. (1981). *Can J. Bot.* **59**, 1547.

Nutman, P. S. (1956). *Biol. Rev. Cambridge Philos. Soc.* **31**, 109.

Nutman, P. S. (1959). *J. Exp. Bot.* **10**, 250.

Nutman, P. S. (1963). *Symp. Soc. Gen. Microbiol.* **13**, 51.

O'Brien, T. P. (1972). *Bot. Rev.* **38**, 87.

O'Brien, T. P., Kuo, J., McCully, M. E., and Zee, S. Y. (1973). *Aust. J. Biol. Sci.* **26**, 1231.

Pankhurst, C. E., and Schwinghammer (1974). *Arch. Microbiol.* **100**, 219.

Pate, J. S., and Gunning, B. E. S. (1972). *Annu. Rev. Plant Physiol.* **23,** 173.

Pate, J. S., Gunning, B. E. S., and Briarty, L. G. (1969). *Planta* **85,** 11.

Prasad, D. N., and De, D. N. (1971). *Microbios* **4,** 13.

Robertson, J. G., and Farnden, K. J. F. (1980). *In* "The Biochemistry of Plants" (P. K. Stumpf and E. E. Conn, eds.), Vol. 5. Academic Press, New York.

Robertson, J. G., Lyttleton, P., Bullivant, S., and Grayston, G. F. (1978). *J. Cell Sci.* **30,** 129.

Sahlman, K., and Fahraeus, G. (1963). *J. Gen. Microbiol.* **33,** 425.

Schaede, R. (1940). *Planta* **31,** 1.

Smith, M. M., and McCully, M. E. (1979). *Protoplasma* **95,** 229.

Syōno, K., Newcomb, W., and Torrey, J. G. (1976). *Can. J. Bot.* **54,** 2155.

Torrey, J. G. (1961). *Exp. Cell Res.* **23,** 281.

Torrey, J. G. (1978). *Bioscience* **28,** 586.

Torrey, J. G., and Barrios, S. (1969). *Caryologia* **22,** 47.

Trinick, M. J. (1973). *Nature (London)* **244,** 459.

Trinick, M. J. (1979). *Can. J. Microbiol.* **25,** 565.

Trinick, M. J., and Galbraith, J. (1976). *Arch. Microbiol.* **108,** 159.

Trinick, M. J., and Galbraith, J. (1980). *New Phytol.* **85,** 37.

Truchet, G. (1978). *Ann. Sci. Nat. Bot. (Paris)* **19,** 3.

Truchet, G., and Coulomb, Ph. (1973). *J. Ultrastruct. Res.* **43,** 36.

Truchet, G., Michel, M., and Denarie, J. (1980). *Differentiation* **16,** 163.

Tu, J. C. (1974). *J. Bacteriol.* **119,** 986.

Tyree, M. T. (1970). *J. Theor. Biol.* **26,** 181.

Vance, C. P., Johnson, L. E. B., Halvorsen, A. M., Heichel, G. H., and Barnes, D. K. (1980). *Can. J. Bot.* **58,** 295.

Verma, D. P. S., Zogbi, V., and Bal, A. K. (1978a). *Plant Sci. Lett.* **13,** 137.

Verma, D. P. S., Kazazian, V., Zogbi, V., and Bal, A. K. (1978b). *J. Cell Biol.* **78,** 919.

Vincent, J. M., Humphrey, B., and North, R. J. (1962). *J. Gen. Microbiol.* **29,** 551.

Wareing, P. F., and Phillips, I. D. J. (1978). "The Control of Growth and Differentiation in Plants, 2nd ed. Pergamon, Oxford.

Werner, D., and Mörschel, E. (1978). *Planta* **141,** 169.

Werner, D., Mörschel, E., Stripf, R., and Winchenbach, B. (1980). *Planta* **147,** 320.

Wheeler, C. T., Henson, I. E., and McLaughlin (1979). *Bot. Gaz.* **140** (Suppl.), 552–557.

Wightman, F. (1977). *In* "Plant Growth Regulation" (P. E. Pilet, ed.), p. 77. Springer-Verlag, Berlin and New York.

Williams, E., Helsop-Harrison, J., and Dickinson, H. G. (1973). *Protoplasma* **77,** 79.

Wipf, L. (1939). *Bot. Gaz.* **101,** 51.

Wipf, L., and Cooper, D. C. (1938). *Am. J. Bot.* **24,** 87.

Wipf, L., and Cooper, D.C. (1940). *Am. J. Bot.* **27,** 821.

Zobel, R. W. (1975). *In* "The Development and Function of Plant Roots" (J. G. Torrey and D. T. Clarkson, eds.), p. 261. Academic Press, New York.

NOTE ADDED IN PROOF.   Recent observations of microbodies in the uninfected cells of the central tissues of soybean and *Phaseolus vulgaris* nodules have suggested a new physiological role for these cells (E. H. Newcomb and S. R. Tandon, *Science* **212,** 1394, 1981; W. Newcomb, unpublished observations) (Figs. 34a and b and 35). During nodule development the microbodies increase in size and the amount of smooth ER proliferates in the uninfected but not the infected cells (Newcomb and Tandon, 1981). These workers suggest that the final steps of the biosynthesis of ureides allantoin and allantoic acid, the principal forms of nitrogen exported from soybean nodules, involve the microbodies and smooth ER. The occurrence of microbodies in the uninfected cells of *P. vulgaris* nodules suggests that this hypothesis may apply to other legumes whose nodules synthesize ureides.

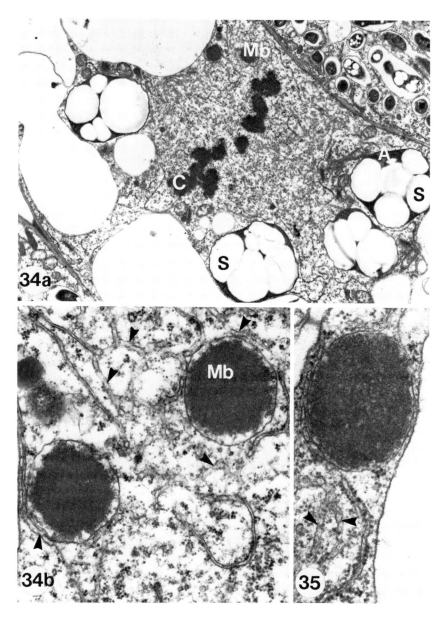

Fig. 34.   An uninfected cell and portions of adjacent infected cells in a soybean nodule. (a) Un-infected cell is undergoing mitosis and contains condensed chromosomes (C), microbodies (Mb), and amyloplasts (A) containing starch (S) granules. ×6950. (b). Higher magnification of a portion of Fig. 34a showing profiles of smooth ER (arrows) and the single bounding membrane of the micro-bodies (Mb) which have electron dense granular matrix. ×34,300.

Fig. 35.   Microbody and profiles of smooth ER (arrows) in an uninfected cell of the central tissue in a nodule of *Phaseolus vulgaris*. ×53,800.

INTERNATIONAL REVIEW OF CYTOLOGY, SUPPLEMENT 13

Mutants of *Rhizobium* That Are Altered in Legume Interaction and Nitrogen Fixation

L. D. KUYKENDALL

*USDA, ARS, Cell Culture and Nitrogen
Fixation Laboratory, Beltsville, Maryland*

I.	Introduction	299
II.	Spontaneous Derivatives	300
III.	Symbiotic Mutants Obtained by Direct Isolation	301
IV.	Contaminants Mistaken for Mutants	302
V.	Drug-Resistant Mutants	302
VI.	Auxotrophs	304
VII.	Glutamine Synthetase Mutants	306
VIII.	Polysaccharide-Deficient Mutants	307
IX.	Pleiotrophic Carbohydrate-Negative Mutants	307
X.	Interesting Mutants Retaining Symbiotic Nitrogen-Fixing Ability	307
XI.	Overview	308
	References	308

I. Introduction

Rhizobium bacteria characteristically form nodules on legume roots and reduce atmospheric nitrogen symbiotically for use by the host plant. Mutations are known which interfere directly or indirectly with the symbiotic capabilities of these special plant microsymbionts. Mutants of *Rhizobium* with altered symbiosis are important research tools for developing an understanding of the complex series of interactions involved in symbiotic nitrogen fixation. Mutations producing specific defects in symbiotic capacity are being used as genetic markers to map and identify genes required for infection, nodule formation, and nitrogen fixation.

The purpose of this chapter is to review the recent significant advances that have resulted from the study of *Rhizobium* mutants with altered symbiotic capabilities. For interpretative discussions of earlier work on mutations that affect symbiotic functions, the reader is referred to Schwinghamer (1975) and Dénarié *et al.* (1976). This chapter primarily focuses on developments that have occurred in the past 4 years. There are several instances in the literature of radically dissimiliar properties among distinct *Rhizobium* species and even among different strains of the same species. One should be cautious not to make generalizations for all rhizobia based on the results for one strain.

299

II. Spontaneous Derivatives

The isolation of certain spontaneous symbiotically defective derivatives of *Rhizobium japonicum* by Kuykendall and Elkan (1976) was the result of purifying a presumptive pure culture of *R. japonicum* strain 3I1b110 into component clones. These clones shared many properties and probably resulted from spontaneous divergence from a common ancestor as a result of maintenance in the laboratory on a selective medium containing D-mannitol as a carbon source. The resulting clonal derivatives (or substrains), strains I-110, S-110, L1-110, and L2-110, differed as much as 20-fold in symbiotic nitrogen-fixing ability. Differences in ability to utilize D-mannitol as a carbon source, as determined by the presence of an inducible D-mannitol dehydrogenase (EC 1.1.1.67) (Kuykendall and Elkan, 1977), allowed the different genetic types to be distinguished on the basis of colony size and morphology on a D-mannitol/L-arabinose medium. On this medium, which contains 5 gm/liter D-mannitol and only 0.5 gm/liter L-arabinose, the symbiotically deficient strains L1-110 and L2-110 form large colonies since they can utilize D-mannitol for growth (they possess an inducible D-mannitol dehydrogenase), whereas the symbiotically competent derivatives I-110 and S-110 form small colonies since they cannot utilize D-mannitol. No general correlation between D-mannitol utilization and symbiotic inefficiency in *R. japonicum* exist (except perhaps in the *R. japonicum* strain 110 genetic background), because many symbiotically competent *R. japonicum* strains, such as strains 3I1b71 and ATCC 10324, can utilize D-mannitol. Strain L1-110 fixes only 5% as much nitrogen as strain I-110 and does not utilize D-glucose as efficiently as strain I-110, since it preferentially uses the Entner–Doudoroff pathway rather than the higher-energy-yielding Embden–Meyerhof–Parnas pathway (Mulongoy and Elkan, 1977). We hypothesized that the greater nitrogen-fixing ability of strain I-110 might be due to its higher efficiency in glucose utilization. This does not appear to be the explanation, since these two strains differ by the same order of magnitude in their ability to express nitrogenase in microaerophilic culture *in vitro* with a carbon source such as D-gluconate, which they metabolize equally (Upchurch and Elkan, 1978), as in their ability to express nitrogenase in association with soybeans.

Working with colony-type derivatives of strain 61A76, as well as those from strain 110, Upchurch and Elkan (1978) demonstrated that the inefficient nitrogen-fixing strain derivatives had a greater capacity to assimilate ammonia, either in pure culture or as bacteroids, than the efficient derivatives. It is necessary to be cautious in interpreting studies on clonal substrains, since there is no clear causal relationship between loss of symbiotic ability and the other biochemical or physiological differences investigated.

Meyer and Pueppke (1980) studied the antibiotic sensitivity patterns of strain derivatives, including those from strain 110, and found differences in tetracycline, kanamycin, nalidixic acid, gentamycin, and trimethoprim resistance. These differences were not seen when the strains were first isolated (Kuykendall and Elkan, 1976) and, although they may have reflected divergence that had occurred since then, it seems clear that these strains, while they may be closely related, are more dissimilar than first believed. Meyer and Pueppke (1980) isolated yet another dissimilar strain from strain 110, namely, strain Y-110. It, unlike the other strain 110 derivatives, failed to bind radioactively labeled soybean lectin.

The occurrence and properties of clonal derivatives, or substrains, of *Rhizobium* need to be investigated further. The purity and authenticity of the *Rhizobium* cultures used in biochemical, genetic, and ecological investigations should not be assumed or taken for granted. Careful isolation of clones from colonies arising from single cells must be performed (Kuykendall and Elkan, 1976), and the symbiotic properties of pure clones ascertained, for the results obtained to be meaningful.

III. Symbiotic Mutants Obtained by Direct Isolation

Chemically induced mutants have an advantage over spontaneous derivatives that are symbiotically defective in that one can be more certain that the phenotypic changes observed have resulted from a single mutation. Hence a clear causal relationship between loss of symbiotic ability and other biochemical or physiological differences observed can be established. This argument is strengthened if there are revertants that coincidentally regain both correlated wild-type properties.

The most useful symbiotically defective mutants can be obtained only by using chemical mutagenesis followed by direct screening of survivors for symbiotic nitrogen fixation with the host legume. Maier and Brill (1976) mutagenized *R. japonicum* strain 61A76 with nitrosoguanidine (NTG) and found 5 mutant strains after screening about 2500 surviving clones for acetylene reduction with 2-week-old soybean plants. Two of the mutant strains failed to form nodules after 2 weeks, while the parent did. The other three mutant strains nodulated soybeans but did not symbiotically fix nitrogen. One of the virulent but nonfixing strains (SM5) formed nodules of wild-type appearance, whereas the other two formed only small nodules which were green rather than the normal pink. Strain SM5 specifically failed to produce active component II of nitrogenase, since (1) component II was shown immunologically to be present, and (2) acetylene reduction occurred when component II purified from *Azotobacter vinelandii* was added to a cell-free extract from nodules formed by strain SM5.

The nonnodulating mutants were later characterized by Maier and Brill (1978) as lacking a surface antigen characteristic of the parent strain 61A76. Antiserum prepared against the parent strain was still slightly active in agglutinating the parent strain after the nonnodulating mutant strain, SM1, absorbed most of the antibodies from the solution. Three components of the hydrolyzed lipopolysaccharide and O antigen were shown by paper chromatography to be missing in the nonnodulating mutant. These differences may not have a direct involvement in nodulation, since these mutants do nodulate after a delay and thus are not truly nonnodulating (R.J. Maier, personal communication).

Conditional ineffective mutants of *Rhizobium leguminosarum* were directly isolated by Beringer *et al.* (1977) following mutagenesis and screening of survivors for symbiotic performance with peas at two different temperatures. These workers screened 196 survivors from NTG mutagenesis and isolated 6 mutants that nodulated but did not fix nitrogen for peas at 26°C, and 3 of the 6 did fix nitrogen at 13°C. Two of the temperature-sensitive ineffective mutants were studied in temperature shift experiments. Acetylene-reducing nodules formed by strain 7155 changed from pink to green and stopped reducing acetylene 3 days after being shifted from 13°C to the nonpermissive temperature 26°C. At least 2 or 3 days of incubation at 26°C was required for disruption of symbiosis, since plants moved back to 13°C after 1 or 2 days at 26°C could, after 3 days, reduce acetylene. The other temperature-sensitive mutant (strain 7106) continued to reduce acetylene for more than 2 weeks after being shifted from 13° to 26°C. Such conditional symbiosis-defective mutants may prove to be very useful in determining the biochemical sequences of events in symbiotic nitrogen fixation.

IV. Contaminants Mistaken for Mutants

O'Gara and Shanmugan (1977) reported the isolation of slow-growing *Rhizobium trifolii* that reduced acetylene *in vitro*. In a subsequent communication, they (1978) reported that the presumptive *R. trifolii* mutants, unlike the parent, could nodulate soybean and mung bean plants. However, the presumptive *R. trifolii* mutants that grew slowly, induced nitrogenase *in vitro*, and nodulated soybeans and mung beans (and not clover) were shown to be culture contaminants, namely, *R. japonicum* strain 110 (Ludwig *et al.*, 1979; Mielenz *et al.*, 1979; Leps *et al.*, 1980). Fast- and slow-growing rhizobia are not interconvertible with a single mutation. Indeed, the differences between fast- and slow-growing *Rhizobium*, based on antigenic differences and flagellar arrangements alone, justify their assignment to distinct genera (Vincent *et al.* 1979).

V. Drug-Resistant Mutants

There have been several interesting reports relating drug resistance to symbiotic properties. Ram *et al.* (1978) reported the isolation of azide-resistant mutants of *R. leguminosarum* that fixed more nitrogen in symbiosis with peas than the wild-type parent. They reasoned that, since azide is an alternative substrate for nitrogenase, one way of overcoming toxicity due to azide might be increased synthesis of nitrogenase. Their report requires confirmation, since it seems that this procedure could work only if one performed the azide-resistant mutant selection with a species that induces nitrogenase *in vitro* and uses conditions under which nitrogenase is expressed.

Zelazna-Kowalskia (1979) and Zelazna-Kowalskia and Kowalski (1978) found that a mutant of *Rhizobium meliloti* deficient in symbiotic ability and possessing a low-level resistance to streptomycin could be restored to symbiotic proficiency by the selection of high-level streptomycin-resistant mutants or transduction of the *strA* gene. They hypothesized that changes in membrane permeability were responsible for failure in symbiosis and that the deficiency in symbiosis was corrected by the pleiotropic effect of the gene for high-level streptomycin resistance.

Spectinomycin-resistant mutants of fast-growing *R. trifolii* and *R. leguminosarum* which consistently express nitrogenase *in vitro* have been reported by Skotnicki *et al.* (1979). The spontaneous drug-resistant mutants were selected at 100 μg/ml spectinomycin, purified by at least two successive clonal isolations, and then confirmed as pure and authentic cultures by plant nodulation, phase susceptibility, and colony morphology and polysaccharide production on solid media. Also, a strain of *R. japonicum* (3I1b138) that did not reduce acetylene under the conditions tested was rendered inducible for nitrogenase *in vitro* after selection for spectinomycin resistance. This finding awaits confirmation. Perhaps the problem of inducing fast-growing *Rhizobium* strains to produce nitrogenase *in vitro* has been solved. This would permit *nif*$^-$ mutants to be isolated without laborious screening with plants.

As an approach in determining the role of oxidative phosphorylation in *Rhizobium* bacteriod metabolism and nitrogen fixation, Skotnicki and Rolfe (1979) used neomycin resistance selection followed by screening for inability to utilize succinate as a carbon source. Mutants which were uncoupled in oxidative phosphorylation were obtained, since the *neo*^{r} *suc*$^-$ mutants had only 8–20% of the ATPase activity of the wild type. Interestingly, the mutants regained a high level (about 70% relative to the parent) of ATPase activity in the bacteriod state where they competently fixed nitrogen. These investigators reasoned that the reacquired ATPase activity could be due to the plant supplying needed components, amplification of the defective bacterial ATPase, or production of a different bacteriodal ATPase.

The symbiotic properties of antibiotic-resistant mutants of *R. leguminosarum* were recently investigated by Pain (1979). None of the streptomycin, chloramphenicol, neomycin, viomycin, D-cycloserine, spectinomycin, or kanamycin mutants isolated from *R. leguminosarum* strain 3000 (a R751-carrying derivative of strain 300) were impaired in establishing normal nitrogen-fixing symbiosis with peas. This finding is difficult to reconcile with the findings of Schwinghamer (1967) who, working with *R. leguminosarum* and *R. trifolii,* reported that many antibiotic resistance mutations caused an impairment of symbiotic capabilities. None of the 56 independent viomycin-resistant mutants of *R. leguminosarum* isolated by Pain (1979) were ineffective, whereas all of the 17 viomycin mutants of *R. leguminosarum* examined by Schwinghamer (1967) were symbiotically defective. However, Schwinghamer did find that 3 out of 10 viomycin-resistant mutants of *R. trifolii* were symbiotically competent. In an earlier study, Schwinghamer (1964) had found that 29 out of 33 spontaneous viomycin-resistant clones derived from 11 different wild-type strains were ineffective. Levin and Montgomery (1974) isolated viomycin-, kanamycin-, and streptomycin-resistant mutants of 3 nitrogen-fixing strains of *R. japonicum* and found that they had retained their symbiotic nitrogen-fixing competence. In addition, many *R. japonicum* strains are naturally resistant to kanamycin, viomycin, and/or streptomycin (Cole and Elkan, 1979). The association of viomycin resistance with symbiotic defectiveness must be considered a phenomenon that is strain-specific. It may be conditional on the unique genetic background characteristic of a given strain of *Rhizobium.* All mutations with which ·symbiotic impairment is the indirect secondary consequence of a quite unrelated phenotypic change may ultimately be considered strain-specific.

The only drug-resistant mutants of *R. leguminosarum* that Pain (1979) found to be symbiotically defective were a small proportion of rifampicin-resistant (*rif^r*) mutants. Eleven out of 143 *rif^r* mutants examined were unable to fix nitrogen symbiotically. Since these mutations specified a high level of resistance to rifampicin (500 μg/ml) and were closely linked to *str* (known to be closely linked to an RNA polymerase gene in both *Bacillus subtilis* and *Streptomyces coelicolor*), the lesions were thought to be in the gene coding for RNA polymerase. A possible role for RNA polymerase in the development of competent nitrogen-fixing bacteriods is suggested. These observations may be analogous to certain *rif^r* mutants of *Bacillus* that are defective in both RNA polymerase and the ability to sporulate.

VI. Auxotrophs

With the exception of adenine auxotrophs, none of the other classes of auxotrophic mutants of *R. leguminosarum* studied by Pain (1979) were consistently symbiotically defective. Adenine auxotrophs were noninfective, as previously determined by Schwinghamer (1969) and Pankhurst and Schwinghamer (1974). However, a large proportion of the other classes of auxotrophs, such as methionine, tryptophan, and leucine auxotrophs, were described by Pain (1979) as giving rise to effective nodules containing prototrophs. These nodules were clearly formed by prototrophic revertants. Thus many amino acid-requiring mutants may be symbiotically defective as a secondary consequence of the auxotrophic marker, and prototrophic revertants have restored symbiotic competence.

When adenine was supplied in the plant growth medium, 1 of the 31 avirulent adenine auxotrophs studied by Pain (1979) formed nodules, albeit ineffective ones, on peas. Pankhurst and Schwinghamer (1974) had previously studied a similar nonnodulating auxotroph (L4-73) of *R. leguminosarum,* which formed ineffective nodules on peas when adenine was supplied in the plant growth medium. Schwinghamer (1970) studied an ineffective riboflavin-requiring mutant of *R. trifolii* that was rendered symbiotically competent by the addition of riboflavin to the plant growth substrate. The development of nitrogen-fixing bacteriods was shown to require either riboflavin or flavin mononucleotides (Pankhurst *et al.*, 1972).

From these studies it is now clear that adenine is required for the establishment of nitrogen-fixing symbiosis but, unlike riboflavin, exogenously supplied adenine probably does not enter the plant tissue in concentrations sufficient to restore the symbiotic fitness of adenine auxotrophs. While adenine auxotrophs of *R. leguminosarum* are unable to nodulate pea plants (Pain, 1979; Schwinghamer, 1969; Pankhurst and Schwinghamer, 1974), adenine auxotrophs of *R. meliloti* are able to form nodules, but they are ineffective (Scherrer and Dénarié, 1971; Fedorov and Zaretskaya, 1978). This difference may be species-specific; *R. meliloti* may be considered a species distinct from certain other fast-growing *Rhizobium* such as *R. leguminosarum, R. trifolii,* and *R. phaseoli* which may be justifiably consolidated into a single species (Vincent, 1979).

Leucine auxotrophs of *R. meliloti* were found to be ineffective (Dénarié *et al.*, 1975). Malek and Kowalski (1977) described an ineffective histidine auxotroph of *R. meliloti* which showed few host plant cells containing bacteria. This indicated that the mutant was defective in release from the infection thread and in intracellular proliferation in a manner similar to that previously observed in leucine auxotrophs of *R. meliloti* (Truchet and Dénarié, 1973). In both cases, supplementation of the plant growth medium with the required amino acid permitted release of the bacteria from the infection thread and normal proliferation, resulting in the development of a fully competent nitrogen-fixing symbiosis (Malek and Kowalski, 1977; Truchet and Dénarié, 1973).

Recently, Truchet *et al.* (1980) used a leucine-requiring mutant of *R. meliloti* with a leucine-dependent expression of effectiveness (as discussed above) to study in closer detail the sequence of events in nodule formation. Reasoning that the restoration of symbiotic competence by exogenous leucine might be due to a nonspecific improvement in the nitrogen nutrition of the host, they tested the effect of adding urea to the plant growth medium. Interestingly, the added urea brought about the development of nodules with the same external shape as wild-type nodules, and histological examination revealed a close similarity to those formed by the wild-type strain. Under these conditions, in the absence of specifically required leucine, *Rhizobium* remained in the infection threads. The differentiation of plant tissue to form nodules, or nodule organogenesis, thus appeared to be initiated by *Rhizobium* bacteria by a specific diffusible trigger or nodule organogenesis-inducing principle (NOIP) (Truchet *et al.* 1980). By complete analogy with crown gall tumorigenesis by *Agrobacterium tumefaciens* (Chilton *et al.*, 1977), the transfer of a segment of DNA from a large bacterial plasmid into the host plant genome may be involved in nodule organogenesis. The strain used by Truchet *et al.* (1980) carries a 91×10^6 dalton plasmid; these workers suggest that nodules formed by alfalfa inoculated with *R. meliloti leu* − mutants and supplied with urea would be suitable for testing for the incorporation of *Rhizobium* plasmid DNA into legume root nodule cells since the nodules would be free of rhizobia in the central tissue.

Organized nodules formed by alfalfa supplied with urea and infected by *leu* − mutants were reported by Truchet *et al.* (1980) to have central tissue that was entirely monosomatic. They discarded the notion that polyploid cells were preexistent and hypothesized that induction of polyploid followed from the release of *Rhizobium* into the host cytoplasm. Direct intracellular contact may be required for rhizobial induction of polyploidy in the infected host cells, and this hypothesis needs to be further tested.

Auxotrophs of *Rhizobium* that are symbiotically defective (as a consequence of their nutritional requirement) can be useful tools for studying the complex series of events in the formation of symbiotic nitrogen-fixing nodules.

VII. Glutamine Synthetase Mutants

Active glutamine synthetase (GS) appears to be required for symbiotic nitrogen fixation by *Rhizobium* (Kondorosi *et al.*, 1977; Ludwig and Signer, 1977). Kondorosi *et al.* (1977) isolated a glutamine-requiring mutant (Gln5) of a *his* − mutant of *R. meliloti* by selecting for D-histidine utilization. The Gln5 mutant isolate (a glutamine-requiring auxotroph) had only 5% of the GS activity of the parent strain and was symbiotically defective, forming small, white nodules on alfalfa that did not reduce acetylene. Revertants to glutamine independence were fully effective. A slow-growing *Rhizobium* cowpea strain 32H1 was used by Ludwig and Signer (1977) to isolate a GS mutant. Glutamine auxotrophs were found after screening survivors of NTG treatment and penicillin enrichment for glutamine-dependent mutants. A mutant with low GS activity was obtained, which formed nodules on *Macroptileum atropurpureum* that lacked detectable acetylene reduction ability. Revertants regained ability to fix nitrogen symbiotically. Both these reports postulate an active role for GS in the regulation of nitrogen fixation. This hypothesis, however, lacks any substantive evidence. The possibility that *Rhizobium* symbiotic cells may simply have a metabolic requirement for glutamine not satisfied by the host plant, as in the case of the adenine- and leucine-requiring mutants discussed above, has not been discounted.

VIII. Polysaccharide-Deficient Mutants

By selecting for physically small cells via repeated passages through 1.2-μm Millipore filters, Sanders *et al.* (1978) isolated a mutant of *R. leguminosarum* deficient in extracellular polysaccharide secretion and in nodulation ability. This mutant strain, Exo-1, produced small colonies as a result of diminished production of extracellular polysaccharide. The lipopolysaccharide of Exo-1 appeared to be unaltered. Revertants to production of extracellular polysaccharide regained the ability to nodule peas. Napoli and Albersheim (1980) further found that Exo-1 was unable to produce capsules. Two other mutants, Exo-22 and Exo-23, produced few nodules on peas and had fewer than 5% of the cells in culture encapsulated with polysaccharide. Exo-24 produced encapsulated cells in culture and formed about half as many nodules on peas as the parent strain. Exo-1 was found to be defective in a very early stage in the infection process, since it did not deform root hairs and did not induce infection threads. Useful information about the early infection process will probably be forthcoming from further study of these mutants.

IX. Pleiotrophic Carbohydrate-Negative Mutants

A mutant of *R. meliloti* isolated following mutagenesis and screening for inability to grow on a mixture of cellobiose and maltose formed few nodules on alfalfa and did not reduce acetylene (Ucker and Signer, 1978). This pleiotrophic carbohydrate-negative mutant, strain Rm 2620, was no longer inducible for β-galactosidase activity by lactose. Lactose-utilizing revertants did not occur in populations of 10^{10} cells. Revertants of Rm 2620 were apparently selected by the host plant, and these regained the capacity to nodulate normally and to fix nitrogen symbiotically.

X. Interesting Mutants Retaining Symbiotic Nitrogen-Fixing Ability

Ronson and Primrose (1979) have applied a mutant methodology approach to carbohydrate metabolism in *R. trifolii*. Interestingly, glucokinase (glk^-), fructose uptake (fup^-), and pyruvate carboxylase (pyc^-) mutants, as well as a double mutant $glk^- fup^-$ retained the ability to fix nitrogen symbiotically for clover. These results suggest that the bacteria are supplied tricarboxylic acid (TCA) cycle intermediates by the plant to provide the ATP and reductant for nitrogen fixation.

XI. Overview

The application of mutant methodology to symbiotic nitrogen fixation by *Rhizobium* has begun. Recent studies have pioneered new paths of research into the basic biology of nitrogen-fixing microsymbionts.

Spontaneous strain derivatives differing 20-fold in nitrogen-fixing ability have been shown for *R. japonicum*. A specific *nif⁻* mutant of *R. japonicum* that lacks active component II of nitrogenase has been isolated. Also, temperature-sensitive, symbiosis-defective mutants of *R. leguminosarum* have been isolated, and these may be useful in developing an understanding of the biochemical sequences of events occurring in symbiotic nitrogen fixation.

Recent work on leucine-requiring mutants of *R. meliloti* has demonstrated that the differentiation of plant tissue to form nodules, or nodule organogenesis, occurs even when the microsymbiont is retained in the infection thread. This finding suggests that nodule organogenesis is initiated by *Rhizobium* bacteria via a specific diffusible trigger. A capsular polysaccharide coat appears to be required for infection by *R. leguminosarum*. A role of catabolite repression-like phenomena in symbiosis is suggested by the occurrence of a pleiotrophic mutant of *R. meliloti* that is deficient in the ability to grow on several sugars and lacks the ability to fix nitrogen. A mutant methodology approach to carbohydrate metabolism in *R. trifolii* offers evidence that the bacteria are supplied TCA cycle intermediates by the plant to provide the ATP and reductant for nitrogen fixation.

The application of mutant methology promises to elucidate the biochemical pathways involved in symbiotic nitrogen fixation by *Rhizobium*. This knowledge is sought since methods of directly selecting *Rhizobium* mutants or recombinants with improved symbiotic nitrogen-fixing ability may then be developed.

REFERENCES

Beringer, J. E., Johnston, A. W. B., and Wells, B. (1977). *J. Gen. Microbiol.* **98**, 339–343.
Chilton, M. D., Drummond, M. H., Merlo, D. J., Sciaky, D., Montoya, A. L., Gordon, M. P., and Nester, E. W. (1977). *Cell* **11**, 263.
Cole, M. A., and Elkan, G. H. (1979). *Appl. Environ. Microbiol.* **37**, 867–870.
Dénarié, J., Truchet, G., and Bergeron, B. (1976). *In* "Symbiotic Nitrogen Fixation" (P. S. Nutman, ed.), pp. 47–61. Cambridge Univ. Press, London and New York.
Fedorov, S. N., and Zaretskaya, A. N. (1978). *Mikrobiologiya* **47**, 728–732.
Kuykendall, L. D., and Elkan, G. H. (1976). *Appl. Envriron. Microbiol.* **32**, 511–519.

Kuykendall, L. D., and Elkan, G. H. (1977). *J. Gen. Microbiol.* **98**, 291–295.
Kondorosi, A., Sváb, Z., Kiss, G. B., and Dixon, R. A. (1977). *Mol. Gen. Genet.* **151**, 221–226.
Leps, W. T., Roberts, G. P., and Brill, W. J. (1980). *Appl. Environ. Microbiol.* **39**, 460–462.
Levin, R. A., and Montgomery, M. P. (1974). *Plant Soil* **41**, 669–676.
Ludwig, R. A., and Signer, E. R. (1977). *Nature (London)* **267**, 245–247.
Ludwig, R. A., Raleigh, E. A., Duncan, M. J., Signer, E. R., Gibson, A. H., Dudman, W. F., Schwinghamer, E. A., Jordan, D. C., Schmidt, E. L., and Tran, D. T. (1979). *Proc. Natl. Acad. Sci. U.S.A.* **76**, 3942–3946.
Maier, R. J., and Brill, W. J. (1976). *J. Bacteriol.* **127**, 763–769.
Maier, R. J., and Brill, W. J. (1978). *J. Bacteriol.* **133**, 1295–1299.
Malek, W., and Kowalski, M. (1977). *Acta Microbiol. Pol.* **26**, 351–359.
Meyer, M. C., and Pueppke, S. G. (1980). *Can. J. Microbiol.* **26**, 606–612.
Mielenz, J. R., Jackson, L. E., O'Gara, F., and Shanmugam, K. T. (1979). *Can. J. Microbiol.* **25**, 803–807.
Mulongoy, K., and Elkan, G. H. (1977). *J. Bacteriol.* **131**, 179–187.
Napoli, C., and Albersheim, P. (1980). *J. Bacteriol.* **141**, 1454–1456.
O'Gara, F., and Shanmugan, K. T. (1977). *Biochim. Biophys. Acta* **500**, 277–290.
O'Gara, F., and Shanmugan, K. T. (1978). *Proc. Natl. Acad. Sci. U.S.A.* **75**, 2343–2347.
Pain, A. N. (1979). *J. Appl. Bacteriol.* **47**, 53–64.
Pankhurst, C. E., and Schwinghamer, E. A. (1974). *Arch. Microbiol.* **100**, 219–238.
Pankhurst, C. E., Schwinghamer, E. A., and Bergerson, F. J. (1972b). *J. Gen. Microbiol.* **70**, 161–177.
Ram, J., Grover, R. P. Riwari, R. B., and Kumar, S. (1978). *Indian J. Exp. Biol.* **16**, 1321–1322.
Ronson, C. W., and Primrose, S. B. (1979). *J. Gen. Microbiol.* **112**, 77–88.
Sanders, R. E., Carlson, R. W., and Albersheim, P. (1978). *Nature (London)* **271**, 240–242.
Scherrer, A., and Dénarié, J. (1971). *Plant Soil Special Vol.* pp. 39–45.
Schwinghamer, E. A. (1964). *Can. J. Microbiol.* **10**, 221–233.
Schwinghamer, E. A. (1967). *Antonie van Leeuwenhoek* **33**, 121–136.
Schwinghamer, E.A. (1969). *Can. J. Microbiol.* **15**, 611–622.
Schwinghamer, E. A. (1970). *Aust. J. Biol. Sci.* **23**, 1187–1196.
Schwinghamer, E. A. (1975). *In* "Dinitrogen Fixation" (R. W. F. Hardy, ed.). Wiley, New York.
Skotnicki, M. L., and Rolfe, B. G. (1979). *Aust. J. Biol. Sci.* **32**, 501–517.
Skotnicki, M. L., Rolfe, B. G., and Reporter, M. (1979). *Biochem. Biophys. Res. Commun.* **86**, 968–975.
Truchet, G., and Dénarié, J. (1973). *C. R. Acad. Sci. Paris Ser. D* **277**, 841–844.
Truchet, G., Michel, M., and Dénarié, J. (1980). *Differentiation* **16**, 163–172.
Ucker, D. S., and Signer, E. R. (1978). *J. Bacteriol.* **136**, 1197–1200.
Upchurch, R. G., and Elkan, G. H. (1978). *J. Gen. Microbiol.* **104**, 219–225.
Vincent, J. M., Nutman, P. S., and Skinner, F. A. (1979). *In* "Identification Methods for Microbiologists" (F. A. Skinner, ed.), 2nd Ed., Soc. for Appl. Bacteriol. Tech. Series, Vol. 14, pp. 49–69. Academic Press, New York.
Zelazna-Kowalska I. (1979). *Acta Microbiol. Pol.* **28**, 47–52.
Zelazna-Kowalska, I., and Kowalski, M. (1978). *Acta Microbiol. Pol.* **27**, 339–343.

INTERNATIONAL REVIEW OF CYTOLOGY, SUPPLEMENT 13

The Significance and Application of *Rhizobium* in Agriculture [1]

HAROLD L. PETERSON AND THOMAS E. LOYNACHAN*

*Department of Agronomy—Soils, Mississippi State University, Mississippi State, Mississippi, and *Department of Agronomy, Iowa State University, Ames, Iowa*

I.	Introduction	311
II.	Application of *Rhizobium* in Agriculture	312
	A. History	312
	B. Need for Inoculation with *Rhizobium*	313
	C. Selection of *Rhizobium*	315
	D. Techniques for the Inoculation of Leguminous Plants	319
	E. Inoculant Quality	323
III.	Significance of Inoculation in Agriculture	324
	A. Response in Soils Lacking Host-Compatible Rhizobia	325
	B. Response in Soils Containing Host-Compatible Rhizobia	326
IV.	Factors Limiting Inoculation in Agriculture	327
	References	328

I. Introduction

Agriculture faces enormous challenges during the 1980s. Many of these challenges are historically implicit to agricultural production. They interact and are complicated by global instability, making simple solutions highly unlikely. An example is the energy–food scenario in the United States. Faced with rising costs and uncertain supplies of fuel and related products, fermentation facilities are being constructed to convert "surplus" crop commodities (e.g., corn, wheat, and grain sorghum) to ethyl alcohol. If the United States and other food-exporting countries divert a significant portion of corn and cereal grain crops for energy synthesis, the immediate effect will be devastating in many countries that rely heavily on imported food.

It is in view of this and other conflicts that agriculture must somehow achieve what is seemingly impossible—the doubling or even tripling of food production for a projected world population of 6.2 billion people by the year 2000. Historically, much of the increased food production during the past 25 years has been directly related to technological and energy-related innovations. Fertilizers,

[1] Approved as journal article no. 4664 of the Mississippi Agricultural and Forestry Experiment Station, Mississippi State University, Mississippi State, Mississippi.

especially nitrogen, have been a key to increased production. Approximately 16 hl of natural gas is consumed in the synthesis of 1 kg of fertilizer nitrogen (anhydrous ammonia). Fertilizer production and use may not expand as rapidly during the next decade because of higher costs and diminishing supplies of fossil fuels.

Against this backdrop, biological nitrogen fixation (BNF) will be expected to assume increased responsibility for supplying more nitrogen in agricultural systems throughout the world. BNF is especially critical in many developing countries lacking abundant supplies of fossil fuels and industrial facilities for chemical synthesis of fertilizer nitrogen. Although there appears to be a potential for increased BNF through asymbiotic or associative symbioses, short-term to mid-term expansion of BNF will depend on leguminous plants and their symbiotic nitrogen-fixing partners, the *Rhizobium*.

In this article, we will highlight some of the background concerning the historical use of *Rhizobium* in agriculture. We will attempt to describe some of the recent, promising advances in the application of *Rhizobium* in agriculture and the implications of enhanced BNF by *Rhizobium*. Finally, we will examine several key problems and limitations of enhanced BNF by *Rhizobium* in agriculture and propose (in a general way) solutions to these problems and limitations.

II. Application of *Rhizobium* in Agriculture

A. HISTORY

The term "legume inoculation" is ordinarily used to describe the application of *Rhizobium* to seeds or the medium used for the growth of leguminous plants. It is not known if early agriculturists recognized the need to transfer a small amount of soil when introducing a leguminous crop into new soils. Salfeld (Fred *et al.*, 1932) is credited with examining "soil transfer" inoculation in field studies conducted during the late 1880s and subsequently describing the response of leguminous crops to inoculation. *Trifolium hybridum* L. (Alsike clover) inoculated with soil displayed vigorous vegetative development and nodulation, while uninoculated plants lacked nodules and were poorly developed.

Attempts were made to culture and prepare legume inocula within a decade of Beijerinck's isolation of bacteria (later designated *Rhizobium*) from nodules in 1888. In 1895, Nobbe and Hiltner (Fred *et al.*, 1932) applied for patents in England and the United States for a legume inoculant that was later marketed as Nitragin. This marked the beginning of the legume inoculant industry.

In its infancy, the legume inoculant industry suffered from overzealous and often exaggerated claims of benefit (Fred *et al.*, 1932). Despite these problems, the industry has survived and continues to provide a valuable and indispensable product throughtout much of the world.

B. Need for Inoculation with *Rhizobium*

"Inoculation" and "nodulation" are often used interchangeably (and incorrectly) to describe the process that culminates in the symbiotic association between various leguminous plants and *Rhizobium* spp. Inoculation is the process of introducing *Rhizobium* into the microhabitat of the leguminous plant. Nodulation is the production of a plant structure that contains the rhizobia and results from successful inoculation.

Inoculation is usually required when a leguminous plant is grown in a particular soil for the first time, unless reliable information indicates that compatible rhizobia are already present. Indeed, several years of inoculation may be necessary before a satisfactory population of rhizobia can become established in a soil.

The need for inoculation should not be left to uncertainty. Vincent (1970), Date (1976, 1977), Nutman (1976), and others have summarized the basic methodology for determining inoculation needs. Briefly, soils should be tested to determine the amounts and availability of plant nutrients. Fertilizer (excluding nitrogen) should be applied to alleviate nutrient limitations. An adapted legume cultivar or selection is grown in the soil with and without inoculation. Inoculation supplemented with sufficient fertilizer nitrogen (to ensure that nitrogen is not limiting) should be included (Date, 1976, 1977) as an additional control. Numbers of rhizobia in the soil should be determined immediately prior to planting using an acceptable procedure such as the most probable number (MPN) technique (Vincent, 1970). The need for inoculation can be determined in a growth chamber, greenhouse, or field. Ultimately, however, results obtained in controlled environments must be verified in the field.

The choice of the plant and strain of *Rhizobium* must be made carefully. The plant should be agronomically well suited to the location and should be resistant to plant pathogens known to be present. The strain of *Rhizobium* should be symbiotically competent, i.e., capable of nodulation and efficient fixation of nitrogen with the selected leguminous plant under the anticipated conditions of growth.

The results summarized in Table I illustrate the response of *Phaseolus vulgaris* L. to inoculation in soils lacking host-compatible rhizobia. Seed yields increased significantly in response to inoculation. Fertilizer nitrogen failed to increase yields significantly because of the low amount of nitrogen applied. Total nitrogen per plant indicates that a low rate of fertilizer nitrogen promoted nitrogen fixation by the inoculated beans. Similar results have been reported by Kang (1975) with *Glycine max* (L.) Merrill, and Sekhon *et al.* (1978) with *Lens esculenta* Moench. Other reports, however, indicate no stimulation or decreased nitrogen accumulation in response to nitrogen fertilization (Gibson, 1977).

If inoculation is needed, a major problem is to select the best plant–*Rhizobium* combinations for a given area. This selection process can be extremely expensive and time-consuming, depending on the heterogeneity of the soils and mangement

TABLE I

EFFECT OF INOCULATION ON NODULATION, NITROGEN FIXATION, AND SEED YIELD OF *Phaseolus vulgaris*[a]

Treatment	Nodulation (nodules/plant)		Total nitrogen (mg/plant)[b]	Seed yield (q/ha)
	6 weeks	11 weeks		
Uninoculated	0.7	0.6	240	7.93
Uninoculated plus nitrogen[c]	1.6	2.7	253	8.17
Inoculated	47.5	15.3	389	19.43
Inoculated plus nitrogen[c]	27.5	10.4	501	20.05
SE	3.3	2.5	68	1.89

[a] Data summarized from those reported by Habish and Ishag (1974), Nodulation of legumes in the Sudan. III. Response of Haricot bean to inoculation. *Exp. Agric.* **10,** 45–50 (Courtesy of Cambridge University Press).

[b] Total nitrogen was calculated from average shoot plus root weight per plant and the percent total nitrogen.

[c] Urea was applied at 43 kg nitrogen/ha at planting.

practices in an area. The selection process must deal realistically with the dominant soil series using accepted management practices. It should involve close interdisciplinary cooperation among plant breeders, plant pathologists, "rhizobiologists," entomologists, and agronomists. Unfortunately, this team approach is currently unattainable in many countries because of limited financial resources and personnel. In other countries it is complicated by a lack of emphasis on team research as a tool in problem solving.

Perhaps the constraints on greater team research can be overcome through increased international cooperation. A program designed to achieve this goal has recently been initiated through the Nitrogen Fixation by Tropical Agricultural Legumes (NifTAL) Program at the University of Hawaii. The International Network of Legume Inoculation Trials (Harris, 1979) represents an attempt to coordinate the testing of various leguminous plants and *Rhizobium* spp. in the developing countries of the tropics. It is a team approach in both the planning and experimental phases. It utilizes available personnel in each participating country, the International Centers (CIAT, ICRISAT, IITA, etc.), the NifTAL organization, universities, and industry. It recognizes the need for education and training as a prerequisite in these studies and has taken steps to organize and conduct regional workshops, conferences, and meetings. The objectives of the program are to determine the need for inoculation in participating countries and to develop the inoculation practices that will meet these needs.

The production of leguminous crops should proceed satisfactorily when the best legume–*Rhizobium* combinations are grown under intensive and correct management. However, several complications may occur. A serious problem is the presence of highly competitive, ineffective rhizobia in soil. Ineffective rhizobia may nodulate the host plant but fix little or no nitrogen. Hagedorn (1979) found that inoculation with two effective strains of *R. trifolii* significantly increased both dry weight and total nitrogen in *Trifolium subterraneum* (L.) 'Mt. Barker' in Oregon soils containing from 1.7×10^1 to 1.7×10^3 *R. trifolii* per gram. Uninoculated subclover plants were nodulated ineffectively. Furthermore, the increase in nitrogen fixation and dry matter production through inoculation was most pronounced and consistent in soils that received phosphorus, sulfur, and molybdenum fertilizer in accordance with soil test recommendations. Thus effective nitrogen fixation depends not only on the presence of competent strains of *Rhizobium* but also on adequate supplies of nutrients so that the plant can grow satisfactorily.

Another recurrent problem is obtaining a well-nodulated leguminous plant under adverse environmental conditions. The recent review by Gibson (1977) is an excellent and thorough examination of environmental and management factors that affect the symbiosis between leguminous plants and *Rhizobium*. Inadequate or excessive soil moisture, for example, interferes with establishment of the leguminous plant–*Rhizobium* symbiosis (Engin and Sprent, 1972). Satisfactory plant yields often require that planting be completed during a relatively short period of time. New approaches may be required for the management of soil water to meet planting deadlines. Technological and mechanical inovations such as the fluid drill (Hardaker and Hardwick, 1978) may help to overcome nodulation problems in dry soils. A current limitation to such activities in many countries is the lack of machinery and fuel.

Research results with *G. max* (L.) Merrill 'Lee-68' in Mississippi (H.L. Peterson, unpublished) suggest that inoculation with superior strains of *R. japonicum* may be important in limiting reductions in seed yields under conditions of drought and delayed planting. The potential for preventing yield reductions through inoculation needs to be examined in greater detail.

C. SELECTION OF *Rhizobium*

The selection of strains of *Rhizobium* to be used in legume inocula is expensive and laborious. The selection process must address several factors and unfortunately may involve compromising one factor for another.

Two critical characteristics must be considered and examined. First, a strain must be highly effective (efficient) in fixing nitrogen. This characteristic is usually determined in a controlled environment such as a growth chamber or greenhouse. For example, Wynne *et al.* (1980) evaluated 17 strains of *Rhizobium* from the "cowpea miscellany" for symbiotic competence with *Arachis hypogaea* L. subsp. *hypogaea* 'NC4' and *A. hypogaea* L. subsp. *fas-*

tigiata 'Argentine.' Strains varied from totally ineffective to highly effective (efficient) in fixing nitrogen. Comparison of the nitrogen-fixing efficiency of these strains to nonlimiting conditions of nitrogen availability is impossible because controls receiving nitrogen fertilizer were omitted. Rates of nitrogen fixation were estimated using acetylene reduction. Total plant nitrogen was as strongly correlated with plant color and weight as with acetylene reduction. Analysis of variance indicated that significant differences in total nitrogen fixed per plant were attributable to cultivars and strains. A significant cultivar–strain interaction occurred. This suggests that strains of peanut rhizobia vary in nitrogen fixation efficiency depending on plant genotype.

Cultivar–strain interactions have been receiving increased attention in several research laboratories (Holl, 1975; Diatloff and Brockwell, 1976; Devine and Weber, 1977; Mytton *et al.*, 1977; Zobel, 1980; Devine and Breithaupt, 1980). Although Seetin and Barnes (1977) and Duhigg *et al.* (1978) have noted extensive variation in acetylene reduction among selections made from several cultivars of *Medicago sativa* (L.), the use of mixed cultures of *Rhizobium meliloti* (commercial inocula) in these studies leads to difficulty in apportioning the components of the observed variation to the appropriate source (i.e., plant or *Rhizobium*). The argument against breeding a leguminous plant for compatibility with a specific, effective, highly efficient strain of *Rhizobium* because of difficulties in displacing indigenous strains is probably more of an excuse than a conclusion based on research. Likewise, legume-breeding programs that depend on indigenous strains of soil rhizobia may confound and limit the assessment of potentials for increasing BNF through complementary development of the plant and rhizobia. Preliminary results (G.R. Smith, unpublished) indicate that promising genetic lines of *Trifolium incarnatum* L. are being unknowingly eliminated from breeding programs because of highly specific requirements for strains of *Rhizobium* that are not present during selection.

Screening various cultivars of a leguminous plant against several strains of *Rhizobium* in the field is very expensive, time-consuming, and often prone to failure. However, screenings such as those reported by Wynne *et al.* (1980) can be performed under controlled environmental conditions. The decision to conduct such studies implies preliminary knowledge of existing variation in nitrogen fixation among plant genotypes and strains of *Rhizobium*. Inexpensive containers such as plastic bottles (Wacek and Alm, 1978), plastic pouches (Smith *et al.*, 1980), and serum bottles (Wacek and Brill, 1976) can be used to facilitate growth of the many plant–*Rhizobium* combinations in a limited space such as a growth chamber or growth room. Acetylene reduction can be performed on intact plants (Smith *et al.*, 1980) or nodulated roots (Wacek and Brill, 1976; Maier and Brill, 1978). Hydrogen uptake is being examined as an additional tool for screening certain leguminous plants and *Rhizobium* for symbiotic efficiency (Zablotowicz *et al.*, 1980). Perhaps when the biochemistry of the plant–*Rhizobium* interaction is more clearly understood, techniques such as two-dimensional polyacrylamide gel electrophoresis (Roberts *et al.*, 1980) and starch gel electrophoresis (Mytton *et al.*, 1978) may be used as part of a biochemical scheme for selecting superior combinations of leguminous plants and *Rhizobium*.

A second characteristic that must be determined is the competitiveness of highly effective nitrogen-fixing strains of *Rhizobium* with other soil microorganisms. Competitiveness among strains of *Rhizobium* can be determined through preliminary studies under environmentally controlled conditions (Marques Pinto *et al.*, 1974; Labandera and Vincent, 1975; Franco and Vincent, 1976; Materón and Vincent, 1980). A "competitive index" can be established through paired comparisons of nodulation by strains of *Rhizobium* (Marques Pinto *et al.*, 1974). Materón and Vincent noted little difference in the numbers of strain CB1809 (ineffective on cultivars of *G. max* with the Rj_2 genotype), strain 3Ilb 136/CRB (Beltsville Serogroup 122), and strain CC709 on roots of Lee, Hardee, and two (Lee × Hardee) hybrid lines. As expected, nodule morphology indicated that strain CB1809 was ineffective, forming fewer nodules on Hardee and (Lee × Hardee 31). The competitive index for CB1809 varied from 0.04 to 0.25, depending on the host. Single and double antibiotic markers were used to differentiate the strains of *Rhizobium* in the soybean nodules. Although no apparent problems were noted with the spontaneous mutants used by Materón and Vincent, Bromfield and Jones (1979) found that antibiotic-resistant isolates were less competitive than the parental strains when used to inoculate *Trifolium repens* L. 'S184.' The loss of competitiveness was postulated to have resulted from decreased heterogeneity of the antibiotic-resistant strains that interfered with infection of the relatively heterogeneous plants comprising the white clover population. Hagedorn (1979) found that antibiotic-resistant strains of *R. trifolii* were very competitive against soil rhizobia in nodulating *T. subterraneum* L. 'Mt. Barker.'

The competitiveness of strains of *Rhizobium* should be verified in field studies prior to the incorporation of a strain in commercial legume inocula. Caldwell (1969) found that strain 110 of *R. japonicum* from the Beltsville collection was more competitive in nodulating Kent soybeans than strain 38 or 76. While strain 76 formed more nodule mass than 110, the fresh weight of soybeans at midflowering and seed yields indicated a significant response to inoculation with strain 110. Rates of inoculation were comparable to those obtained using a high-quality commercial inoculant.

Gibson *et al.* (1976) found that strain WU95 of *R. trifolii* was more competitive and more persistent than four other strains of *R. trifolii* used as a mixed inoculant on *T. subterraneum* L. 'Woogenellup.' While the results are biased by a sampling technique involving removal of a single, prominent nodule from the crown root region of the plant, the high and increasing recovery of strain WU95 in each of three succeeding years after planting (78.9–92.5% of nodules sampled) suggested that WU95 was a very effective competitor against the other strains of rhizobia in the soils that were studied.

Diatloff and Brockwell (1976) noted that strain CB1809 of *R. japonicum* formed 53% of the nodules on Hardee soybeans but was absent when applied in competition with equal numbers of strain CB1795. The results obtained by Materón and Vincent (1980) are consistent with these findings.

Other factors such as temperature (Brockwell *et al.*, 1968; Hardarson and Jones, 1979), soil acidity (Bromfield and Ayanaba, 1980), and deficient or excessive supplies of nutrients (Franco, 1978; Keyser and Munns, 1979) are likely to affect competition among strains of *Rhizobium* in soil. The ability of strains to produce bacteriocins, release bacteriophage (Schwinghamer and Brockwell, 1978), or tolerate bacteriophage (Evans, *et al.*, 1979) may affect competitiveness in soil and culture media. Predation (Danso *et al.*, 1975; Habte and Alexander, 1978; Ramirez and Alexander, 1980) and antagonism by actinomycetes and other soil microorganisms (Damirgi and Johnson, 1966; Smith and Miller, 1974; Foo and Varma, 1976; Chowdhury, 1977) may affect the competitiveness of *Rhizobium* in soil, although this has not been evaluated under field conditions.

The mechanism of the leguminous plant affecting competition among strains of *Rhizobium* is poorly understood. Although studies (e.g., Mytton *et al.*, 1977; Stamford and Neptune, 1978; Jones and Hardarson, 1979) have shown host plant specificities in the symbiotic interaction with *Rhizobium*, the reasons for the selectivity are not well defined. Lectins, root exudates (see review by Schmidt, 1979), and seedcoat diffusates (Hale and Mathers, 1977) may be involved.

A "specific stimulation" hypothesis has been proposed to explain the role of the host plant in selecting specific strains of *Rhizobium* (Vincent, 1974). This hypothesis has been examined directly through recent work by Reyes and Schmidt (1979). In field experiments, the numbers of strain 123 of *R. japonicum* were determined by a quantitative immunofluorescence microscopic counting procedure (Schmidt, 1974). Although strain 123 increased numerically in the inner rhizosphere during a 16-day period following planting, numbers were essentially the same in both the inner and outer rhizosphere 30 days after planting. Numbers of strain 123 (3.12×10^6/gm) were greater in soil near the soybean taproot at harvest compared with 2.4×10^4/gm in the outer rhizosphere of *Zea mays* L. and 1.6×10^4/gm in an adjacent fallow. Unfortunately, no data are reported from 30 days after planting to soybean harvest at 120 days after planting. The author's suggestion that increased numbers of strain 123 at harvest occurred because rhizobia were released to the soil via nodule decay seems attractive but is weakly and only indirectly substantiated by greater numbers of strain 123 near the taproots as compared to lateral roots. Additional analyses conducted over the entire period of soybean growth would provide much needed information on the selective stimulation and population dynamics of *Rhizobium* in the rhizosphere.

An adequate determination of the reasons for host–rhizobia–environment interaction seem a mandatory prerequisite to increased protein production by leguminous plants through inoculation with superior nitrogen-fixing strains of *Rhizobium*.

D. Techniques for the Inoculation of Leguminous Plants

Brockwell (1977) has recently reviewed this subject in some detail. Inoculation of leguminous plants can occur naturally but often depends on the application of rhizobia to the seed or soil. The presence of a compatible strain of *Rhizobium* in the soil will result in nodulation. Rhizobia can be introduced naturally as seed-borne epiphytes, but nodulation failures have occurred frequently when leguminous plants were first introduced in soils void of compatible rhizobia. Inoculation with commercial preparations of *Rhizobium* has become fairly routine in agriculture.

1. *Inoculation Procedures*

a. *Seed.*   Inoculation of seed can be accomplished in several ways. Perhaps the easiest procedure is dry application of a peat-base inoculant to the seed in the hopper of a planter or in a container such as a pail. This procedure often results in poor nodulation, because the inoculant fails to adhere to the seed. The inoculant settles to the bottom of the hopper and is deposited in the first few meters of the row, or remains in the bottom of the pail as the seeds are planted in the soil.

Peat-base inocula can be mixed with water (with or without materials such as sugar, molasses, gum arabic, methyl ethyl cellulose) and applied to the seed. Care must be taken to use the correct amount of water when employing this approach, and seeds must be planted immediately after inoculation and mixing. Applying too much water to the seed can interfere with mechanical planting and increase seed damage. Delayed planting can result in extensive losses of rhizobia if environmental conditions cause excessive drying of the inoculant on the seed prior to planting (Burton, 1975).

Seed pelleting is another inoculation procedure. Legume inocula can be combined with a variety of materials (lime, gypsum, fertilizers, clay) and applied to the seed with an adhesive. This technique has been used extensively in Australia. It has the advantages of (1) permitting modification of the soil environment immediately adjacent to the seed, (2) allowing application of higher numbers of rhizobia to the seed, and (3) protecting rhizobia from inhibitory substances released by the seed or combined with the seed at planting (fertilizers, pesticides, and so forth).

Liquid suspensions of rhizobia can be used as seed inocula. For best results the inocula should be applied to the seed immediately prior to planting. Rhizobia applied to seeds in aqueous suspensions die more rapidly than those applied as aqueous suspensions containing peat (Burton and Curley, 1965). Davidson and Reuszer (1978) reported that the inclusion of peat fails to prevent approximately a 10-fold reduction in the number of rhizobia on the seed during 3 weeks' storage at 15°C. But the use of a stabilizer such as sucrose or charcoal may help promote rhizobial survival on the seed.

A special type of liquid inoculant that contains rhizobia in vegetable oil has been commercially available throughout soybean-growing areas of the United States since about 1977. The quality of this nonaqueous, liquid inocula may be limited by low numbers of viable rhizobia (Hiltbold *et al.*, 1980). Davidson and Reuszer (1978) noted slightly improved survival of *R. japonicum* when mineral oil and water were used to apply a moist, peat-base inoculant to soybean seeds.

Lyophilized rhizobia in talc or clay carriers have been used as seed inocula. The effectiveness of these products may also be limited by low or inadequate numbers of viable rhizobia (Skipper *et al.*, 1980; Hiltbold *et al.*, 1980).

b. *Soil.* Rhizobia can be added to the soil in proximity to the seed. Granular and liquid inocula have been used with some success, especially on leguminous plants such as peanuts that are susceptible to physical damage during the addition and incorporation of seed treatments.

The chemical composition of granular inocula varies considerably among products. Some are mixtures of calcium sulfate (Fraser, 1966, 1975), although others consist primarily of granulated peat (Burton, 1980). Still others involve clay aggregates or corncob residue.

Aqueous suspensions of rhizobia can be applied directly to soil. These may consist of peat-base inocula suspended in water (Schiffmann and Alper, 1968), or cell concentrates of rhizobia in aqueous suspension (Weaver and Frederick, 1974; Scudder, 1975; Dunigan *et al.*, 1980).

A principal advantage of granular or liquid inocula is that both afford an opportunity to inoculate at rates greatly exceeding the 10^5–10^6 rhizobia per seed obtained with high-quality peat-base inocula. Inoculation with aqueous suspensions of rhizobia at rates as high as 1.5×10^{10} viable rhizobia per centimeter of row (Kapusta and Rouwenhorst, 1973) did not increase plant yields in soils where soil nitrogen or the established strains of rhizobia were capable of supplying sufficient nitrogen for plant growth.

Another procedure involves inoculating a preceding crop of a nonlegume with rhizobia (Diatloff, 1969). This approach was combined with inoculation of the legume seed to improve nodulation in soils containing ineffective rhizobia. Gaur *et al.* (1980) reported that this strategy increased the percentage of nodulation by the applied strain of rhizobia, although nodule typing to determine the source (soil versus inocula) of the rhizobia was based on subjective characteristics such as nodule appearance. Nodule volume, shoot dry weight, pod numbers, and seed yield increased in response to inoculation of both the preceding and leguminous crop. No significant differences in nodule numbers were observed, and total nitrogen was not reported.

2. *Inoculant Carriers*

The carrier material used in the production of legume inocula may be very important in determining the success or failure of inoculation. The carrier is the material in which the rhizobia are mixed, packaged, distributed, and applied to the seed or seed environment. A satisfactory carrier must suuport and maintain a high number of rhizobia in a vigorous, viable condition. When applied to seeds or soil, the carrier should protect the rhizobia against both abiotic and biotic stresses, including chemical compounds released by the seed that are toxic to rhizobia.

The type of inoculant carrier can be important in promoting nodulation by an applied superior nitrogen-fixing strain of *Rhizobium*. The field results (H.L. Peterson, unpublished) summarized in Table II indicate that soybean seed yields responded significantly to inoculation at extremely high rates with a new oil-base carrier under development in our laboratory. The low numbers of rhizobia in the soil at planting probably contributed to the inoculation success. But the key is that no inoculant can perform successfully unless the rhizobia it contains are viable and can compete effectively with soil rhizobia to form an appreciable quantity of the nodules on the host plant.

Various carrier materials (Table III) have been used with legume inocula. Although finely ground peat is used extensively and is probably the preferred

TABLE II

EFFECT OF INOCULANT CARRIER ON NODULATION, NITROGEN FIXATION, AND SOYBEAN SEED YIELD (1978)[a]

Treatment[b]	Nodulation by the inoculant (%)	Nodule mass (mg/plant)[c]	Total nitrogen (mg/plant)[c]	Seed yield (q/ha)[c]
Uninoculated	0	543 (a)	462 (a)	18.91 (a)
Strain 8φ (water)	15	858 (b)	435 (a)	18.70 (a)
Strain 8φ (oil)	27	625 (a)	530 (a)	24.07 (b)

[a] Studies were conducted in Leeper silty clay containing 160 soybean rhizobia per gram at time of soybeam planting (27 June 1979). Whole plant samples were collected at the R2-R3 stage of soybean development (Fehr *et al.*, 1971).

[b] Rate of inoculation was 10^8 viable rhizobia per seed. Strain 8φ is a field isolate belonging to USDA Serogroup 110.

[c] Average values followed by the same letter are statistically equivalent at the 5% level of probability according to Duncan's new multiple range test.

TABLE III
INOCULANT CARRIER MATERIALS

Carrier material	Mixed	Pure	Seed	Soil
Agricultural lime	+		+	+
Bagasse	+	+	+	
Bone meal	+	+	+	+
Cellulose	+	+	+	
Charcoal	+	+	+	
City refuse	+	+	+	
Clay (various types)	+		+	+
Coal (various types)	+	+	+	+
Coir dust	+		+	
Compost, including				
City refuse	+		+	
Coffee husks	+		+	
Farmyard manure	+		+	
Peanut hulls	+		+	
Rice husks	+		+	
Straw	+		+	
Various other materials	+		+	
Farmyard manure plus tank silt	+	+	+	
Fertilizer (various types)	+			+
Granulated				
Clay	+			+
Peat	+			+
Pumice	+			+
Leguminous plant residues	+	+	+	
Maize				
Cobs	+		+	
Stalks	+		+	
Nodules	+		+	
Peat	+	+	+	+
Polyacrylamide gel		+		+
Seaweed	+			
Silica (sand)	+	+	+	+
Soil	+	+	+	+
Straw (various types)	+	+	+	
Tree leaves	+	+	+	
Vegetable oil	+	+	+	
Water	+	+	+	+

Type of culture[a]: Mixed, Pure. *Type of application*[b]: Seed, Soil.

[a] This indicates the microbiological status of the inoculant. A mixed culture usually consists of rhizobia and a wide range of other microorganisms such as bacteria, fungi, and actinomycetes. A pure culture should contain rhizobia, although in practical use the culture is usually contaminated during application to the seed or soil.

[b] General mode of inoculant application. Seed application is probably most common, but soil application is receiving considerable attention in situations that require increased numbers of rhizobia and where materials toxic to rhizobia are applied to the seed. See Brockwell et al. (1980) for a recent example of the comparative effectiveness of seed and soil inoculation.

carrier (Burton, 1967, 1976; Roughley, 1976; Strijdom and Deschodt, 1976; Date and Roughley, 1977; Brockwell, 1977), countries lacking high-quality peat have been forced to examine alternate carriers. Materials examined include bagasse (a residue from sugarcane) (Leiderman, 1971; Graham *et al.,* 1974), coal (Kandaswamy and Prasad, 1971; Dube *et al.,* 1973, 1975; Singh and Tilak, 1977; Paczkowski and Berryhill, 1979), composted tree leaves (Iswaran *et al.,* 1972; Rizk *et al.,* 1975), and a compost of coir dust (a coconut by-product) and soil (John 1966; Faizah *et al.,* 1980). The lack of physicochemical homogeneity in many of these carriers may result in widely variable inoculant quality. This, combined with strain heterogeneity and unpredicatable handling and storage during distribution and use, results in products that often fail to perform satisfactorily when used by the farmer.

E. INOCULANT QUALITY

Inoculant quality is a measure of the numbers, symbiotic effectiveness, and host compatibility of rhizobia in an inoculant. Providing farmers with high-quality inocula is a worldwide problem. Australia has developed the most comprehensive and effective procedures for monitoring and controlling inoculant quality (Date, 1969; Roughley, 1976; Date and Roughley, 1977; Vincent, 1977). Other countries such as Uraguay, Czechoslovakia, the Netherlands, Russia, South Africa, and Zimbabwa (Date and Roughley, 1977) have government agencies that control inoculant production. The United States allows the legume inoculant industry to determine and regulate inoculant quality. There has been increasing pressure from the public for legume quality assessment at the state level. States such as Indiana, Ohio, and Wisconsin have examined inoculant quality over the years, and other states are considering similar operations. Part of the increased pressure for quality monitoring of legume inocula is underscored by two recent investigations. Hiltbold *et al.* (1980) summarized evaluations of commercial soybean inocula obtained in Alabama during April and May of 1976, 1977, and 1978. Plate counts were performed at both Auburn, Alabama, and Clemson, South Carolina, while MPN analyses on 1978 samples were performed at Auburn. Greenhouse and field trials were employed. The results indicated that inoculant quality (the numbers of rhizobia and the ability to form nodules on the host plant) varied significantly among commercial products. Inocula containing less than 10^7 soybean rhizobia per gram by plate count were usually unable to promote satisfactory nodulation in greenhouse evaluations when applied to host legumes and grown in potting mixtures lacking compatible rhizobia. Field evaluations in 1978 indicated that only Nitragin Nitra Mo and Nitragin peat-base inocula increased soybean seed yield in comparison with the uninoculated soybeans. Seed yields were significantly and positively correlated ($p = 0.05$) with the number of rhizobia applied per seed. Greenhouse tests indicated that, to ensure adequate nodulation, inocula should supply at least 10^4 rhizobia per seed. This standard was affirmed in the 1977 field studies conducted under conditions of severe drought and high temperatures. A recommended standard of 10^4 rhizobia per seed is within the range of 10^3 per seed (Date, 1970) and 2×10^5 per seed (Burton and Curley, 1965). Skipper et al. (1980) obtained similar results with commercial inocula marketed in South Carolina and Georgia during 1975–1976.

The assumption that a legume inoculant will provide sufficient rhizobia for effective nodulation of the host plant under adverse field conditions may be incorrect and expensive. Inoculant quality is variable, and some inocula contain mixtures of strains that may be effective on one host but ineffective on another host for which the inoculant is recommended by the manufacturer. The presence of fungicides and micronutrients may reduce or even prevent rhizobia from surviving in the inoculant. People engaged in production of leguminous crops should be aware of the differences that exist among inocula and should use the best inoculant available (even though in many cases this may be hard to determine because of a lack of analytical data).

III. Significance of Inoculation in Agriculture

Legume inocula are needed in meeting two agricultural objectives. First, inocula are required during initial establishment of leguminous plants in soils lacking compatible rhizobia. The plant's yield response to inoculation is usually very dramatic in soils lacking compatible rhizobia if inoculation is successful and nitrogen is limiting. The second agricultural objective of inoculation is to replace an inefficient population of rhizobia with more efficient rhizobia. The success or failure of this operation may have a dramatic affect on crop yield, with establishment of the leguminous plant critically dependent on successful inoculation. Sometimes, however, inoculation is employed to increase the biological efficiency of the plant by promoting nodulation by strains of rhizobia that are more efficient in fixing nitrogen. The result of this "fine-tuning" operation may occasionally be successful, but reports indicating no significant response to inoculation with superior strains of Rhizobium are numerous (Kapusta and Rouwenhorst, 1973; Weaver and Frederick, 1974; Nelson et al., 1978; Boonkerd et al., 1978; Dunigan et al., 1980). It is very important to determine the reason for a lack of response to inoculation with superior strains of rhizobia. This information can be used to modify inoculation procedures and perhaps even the rhizobia to obtain the benefits from superior rhizobia and high-yielding varieties of leguminous plants.

A. Response in Soils Lacking Host-Compatable Rhizobia

Bezdicek *et al.* (1978) reported a substantial plant response to inoculation when soybeans were grown in soils void of host-compatible rhizobia. Granular and peat-base inocula (containing single or multiple strains of *R. japonicum*) were applied to soil or seed, respectively, during planting. Merit soybeans were planted in single-row plots, and treatments were replicated five times. Although the use of single-row plots usually leads to undesirable interactions among adjacent plots, the effects of inoculation on soybean seed yields were unquestionably spectacular (Table IV). Strains 110 and 138 are among the more effective nitrogen-fixing strains of *R. japonicum* (Johnson *et al.*, 1965; Caldwell and Vest, 1970). Inoculation with 110 and 138 was synergistic as compared to inoculation with each strain alone. The highest yield of 4489 kg/ha represented a 3454 kg/ha increase over the uninoculated control, a tremendous seed yield response to inoculation.

Bromfield and Ayanaba (1980) found in an acid soil from Nigeria that inoculation of TGm 294-4 cultivar of soybeans with two separate strains of *R. japonicum* resulted in yield increases of 1219 and 1185 kg seed/ha. The response to inoculation with cultivar TGm 80 was 1744 and 1601 kg additional seeds per hectare. In addition, both cultivars responded to lime. There also was evidence of an inoculation–liming interaction with cultivar TGm 294-4. The differences

TABLE IV

Effect of Inoculation on Nodulation and Seed Yield of Merit Soybeans[a]

Strain of *R. japonicum*	Inoculant carrier[b]	Calculated rate of inoculation (10^5/seed)[c]	Nodulation applied strain(s) (%)	Seed yield (kg/ha)[d]
Nitragin soil implant	G	16.0	—	4489 (a)
110	G	8.6	97	3683 (ab)
138	G	21.0	100	3824 (ab)
110 and 138	G	4.3 (110)	40 (110)	4368 (ab)
		10.0 (138)	60 (138)	
110	P	1.4	—	3837 (ab)
138	P	3.4	—	3756 (ab)
Uninoculated	—	—	—	1035 (c)

[a] Adapted from Bezdicek *et al.* (1978), by permission of the American Society of Agronomy.

[b] G, Granular; P, peat. The 110 and 138 treatment consisted of equal weights of granular inoculant with slightly greater numbers of strain 138 as indicated in the table.

[c] Calculated on the basis of the average MPNs of rhizobia applied per meter of row divided by the reported planting rate of 29 seeds/m.

[d] Average of five replications, 6.1 m harvested per plot. Values followed by the same letter are statistically equivalent at the 5% level of probability.

in seed yield between inoculated and uninoculated soybeans may have been even larger had the uninoculated soybeans not been contaminated by one of the strains of *R. japonicum* used in the study (CB1809 str[r]). Apparently the contamination occurred late in the growing season, because seed yields in the uninoculated plots still remained relatively low (397 and 331 kg/ha for the two cultivars).

In the same report, Bromfield and Ayanaba noted no significant increase in seed yield with inoculation in less acidic soils (pH 6.6). The strains applied as inocula produced 100% nodulation in both cultivars. Apparently indigenous rhizobia (which according to plant infection analyses were practically undetectable at time of planting) were highly prolific and as efficient in fixing nitrogen as the applied strains. An alternate explanation is that the two preceding crops of *Z. mays* L. failed to deplete available soil nitrogen significantly.

B. Response in Soil Containing Host-Compatible Rhizobia

Legume inoculation rarely results in a significant yield increase in soils with established populations of symbiotically competent rhizobia (Ham *et al.*, 1971; Kapusta and Rouwenhorst, 1973; Weaver and Frederick, 1974; Nelson *et al.*, 1978; Boonkerd *et al.*, 1978). Dunigan *et al.* (1980) reported results for 4 years of soybean inoculation trials at several locations in Louisiana. Commercial inocula were applied at rates ranging from those recommended by the inoculant manufacturer to massive rates (numerically undefined, but apparently >50 times the recommended rates). Seed yields increased in apparent response to inoculation in only one soil during one of 4 years. The authors speculate that this increased seed yield of soybeans with Nitragin Soil Implant and Hy-Rhize inocula occurred because of drought throughout the growing season that appeared to limit nodule numbers and nitrogen fixation (acetylene reduction) by the uninoculated soybeans.

Moisture and temperature are recognized as being very important in determining the numbers of rhizobia in soil. Osa-Afiana and Alexander (1979) found better survival of rhizobia at 8.0 bars moisture tension than at 1.5–0.02 bars. Van Rensburg and Strijdom (1980) found that strains of *R. japonicum* and cowpea *Rhizobium* (slow growers) withstood severe and rapid dessication better than strains of *R. meliloti* and *R. trifolii* (fast growers). The results were reversed under conditions of slow drying. These authors speculated that the lower moisture retention by the slow-growing rhizobia rendered them more sensitive to slow drying because of dysfunction of intracellular enzymes during drying. In any event, factors active during repeated wetting and drying of soil (Pena–Cabriales and Alexander, 1979) and flooding (Vandecaveye, 1927) can reduce the number of rhizobia in soil.

Flooding may be important in attempts to replace soil rhizobia through inoculation. Peterson (1979) found that the number of *R. japonicum* declined about

100-fold in a soil subjected to repeated and prolonged flooding. Inoculation with a superior nitrogen-fixing strain of *R. japonicum* at approximately 10^8 rhizobia per seed resulted in 21% nodulation by the inoculant strain. Nodule mass increased 38%, and total plant nitrogen at mid- to late flowering increased in response to inoculation with the superior strain of *Rhizobium*. Likewise, seed yield increased 2.3 q/ha in response to inoculation. The success of inoculation in the flooded soil seemed related to the low number of soybean rhizobia (160/gm) in the soil at the time of planting, since inoculation with the same strains in similar soils containing higher numbers of soybean rhizobia (10^4–10^5/gm) failed to increase soybean nodulation, nitrogen fixation, or seed yield.

Semu *et al.* (1979) examined the effects of inoculation and nitrogen fertilizer on soybean nodulation in soils that differed in their history of soybean production. In two soils that had been cropped previously to soybeans, inoculation with a commercial inoculant had no effect on total nodulation, indicating that the inoculant was incapable of displacing the dominant serotypes of *R. japonicum*. Numbers of soybean rhizobia (MPN) in these soils were 10^5–10^6/gm at planting. In another soil cropped to soybeans for the first time, strain 3Ilb 138 (present in the commercial inoculant) comprised 60–70% of the nodules. The study was repeated the following year at a site 3 km away but in the same soil series. The nodules contained strains of *R. japonicum* that were serologically different from those from the preceding year, which implies that either the strains in the inoculant changed from one year to the next or a native population present at undetectable levels at planting was very prolific and extremely competitive against the rhizobia used as inocula.

IV. Factors Limiting Inoculation in Agriculture

The successful application of legume inoculation in agriculture depends on recognizing and establishing the complementary conditions required by both partners in the symbiosis. It has been proposed that photosynthesis is a major limiting factor in the increased production of leguminous plants (Hardy and Havelka, 1976). Research is being directed toward increasing photosynthetic efficiency in leguminous plants. This research, however, should involve simultaneous improvement of nitrogen fixation efficiency in rhizobia. The existing variation in photosynthesis among leguminous plants needs to be exploited in coordination with efforts to harness variation among rhizobia for increased nitrogen fixation. The goal of this work should be to increase the performance of each partner in the symbiosis, while maintaining the proper balance needed for optimum plant production. It is encouraging that research along these lines is underway at several locations, with some promising results reported in other articles of this book.

We must expand our limited knowledge of the multifaceted, environment–soil–plant–rhizobia interaction. Suppose, for example, that a strain of *R. meliloti* were isolated and found to be 10 times more efficient than the best existing strains in fixing nitrogen with *Medicago sativa* L. Would the strain be compatible with a wide range of cultivars? How could the strain be applied to the host plant in the field to obtain preferential nodulation against high numbers of less efficient, host-compatible rhizobia in the soil? Would conventional management practices be capable of providing the additional nutrients required by plants with a much increased rate of nitrogen fixation? Would a program of increased investment and requiring new management techniques be accepted by the grower? The list of questions could go on almost indefinitely, but the point is clear. We need a firm research effort to address these and other questions at the systems level.

To many individuals this may seem a grandiose venture with little probability of success. The research investment is likely to be high, not only in monetary terms but also in terms of the time, effort, and the patience of many scientists and administrators. The public must be educated as to the extent and seriousness of the research problem and should be urged to become involved as much as possible. The public must understand the benefits and needs implicit in this type of research.

The complexity of research on increasing the contribution of leguminous plants to the global supplies of food, fiber, and forage dictates the need for team research under a well-managed, stable, long-term research commitment. The other contributions in this volume indicate a growing research base from which significant gains can be made. The problem that remains is to coordinate these efforts to unlock the benefits of increased food supplies to help feed a hungry world.

REFERENCES

Bezdicek, D. F., Evans, D. W., Abede, B., and Witters, R. E. (1978). *Agron J.* **70,** 865–868.
Boonkerd, N., Weber, D. F., and Bezdicek, D. F. (1978). *Agron J.* **70,** 547–549.
Brockwell, J. (1977). *In* "A Treatise on Dinitrogen Fixation" (R. W. F. Hardy and A. H. Gibson, eds.), Section II., pp. 277–309. Wiley, New York.
Brockwell, J., Dudman, W. F., Gibson, A. H., Hely, F. W., and Robinson, A. C. (1968). *In* "Transactions of the Ninth International Congress of Soil Science" (J. W. Holmes, ed.), Vol. 2, pp. 103–114. Angus-Robertson, Sydney.
Brockwell, J., Gault, R. R., Chase, D. L., Hely, F. W., Zorin, Margaret, and Corbin, E. J. (1980). *Aust. J. Agric. Res.* **31,** 47–60.
Bromfield, E. S. P., and Ayanaba, A. (1980). *Plant Soil* **54,** 95–106.
Bromfield, E. S. P., and Jones, D. G. (1979). *Ann. Appl. Biol.* **91,** 211–219.
Burton, J. C. (1967). *In* "Microbial Technology" (H. J. Peppler, ed.), pp. 1–33. Reinhold, New York.

Burton, J. C. (1975). *In* "Symbiotic Nitrogen Fixation in Plants" (P. S. Nutman, ed.), pp. 175–189. Cambridge Univ. Press, London and New York.

Burton, J. C. (1976). *Proc. Miss. Section Am. Soc. Agron.* pp. 51–68.

Burton, J. C. (1980). *In* "World Soybean Research Conference II: Proceedings" (F. T. Corbin, ed.), pp. 89–100. Westview Press, Boulder, Colorado.

Burton, J. C., and Curley, R. L. (1965). *Agron. J.* **57**, 379–381.

Caldwell, B. E. (1969). *Agron. J.* **61**, 813–815.

Caldwell, B. E., and Vest, G. (1970). *Crop Sci.* **10**, 19–21.

Chowdury, M. S. (1977). *In* "Exploiting the Legume-*Rhizobium* Symbiosis in Tropical Agriculture" (J. M. Vincent, A. S. Whitney, and J. Bose, eds.), pp. 385–411. Univ. of Hawaii.

Damirgi, S. M., and Johnson, H. W. (1966). *Agron. J.* **58**, 223–224.

Danso, S. K., Keya, S. O., and Alexander, M. (1975). *Can. J. Microbiol.* **21**, 884–895.

Date, R. A. (1969). *J. Aust. Inst. Agric. Sci.* **35**, 27–37.

Date, R. A. (1970). *Plant Soil* **32**, 703–725.

Date, R. A. (1976). *In* "Symbiotic Nitrogen Fixation in Plants" (P. S. Nutman, ed.), pp. 137–150. Cambridge Univ. Press, London and New York.

Date, R. A. (1977). *In* "Biological Nitrogen Fixation in Farming Systems of the Tropics" (A. Ayanaba and P. J. Dart, eds.), pp. 169–180. Wiley, New York.

Date, R. A., and Roughley, F. J. (1977). *In* "A Treatise on Dinitrogen Fixation (R. W. F. Hardy and A. H. Gibson, eds.), Section IV, pp. 243–275. Wiley, New York.

Davidson, F., and Reuszer, H. W. (1978). *Appl. Environ. Microbiol.* **35**, 94–96.

Devine, T. E., and Breithaupt, B. H. (1980). *Crop Sci.* **20**, 269–271.

Devine, T. E., and Weber, D. F. (1977). *Euphytica* **26**, 527–535.

Diatloff, A. (1969). *Aust. J. Exp. Agric. Anim. Husb.* **9**, 357–360.

Diatloff, A., and Brockwell, J. (1976). *Aust. J. Exp. Agric. Anim. Husb.* **16**, 514–521.

Dube, J. N., Namdeo, S. L., and Johar, M. S. (1973). *Res. Ind.* **18**, 94–95.

Dube, O. N., Namdeo, S. L., and Johar, M. S. (1975). *Curr. Sci.* **44**, 434.

Duhigg, P., Melton, B., and Baltensperger, A. (1978). *Crop Sci.* **18**, 813–816.

Dunigan, E. P., Sober, O. B., Rabb, J. L., and Boquet, D. J. (1980). *La. State Univ. Agric. Exp. Sta. Bull.* **726**, 1–16.

Engin, M., and Sprent, J. I. (1972). *New Phytol.* **72**, 117–126.

Evans, J., Barnet, Y. M., and Vincent, J. M. (1979). *Can. J. Microbiol.* **25**, 974–978.

Faizah, A. W., Broughton, W. J., and John, C. K. (1980). *Soil Biol. Biochem.* **12**, 211–218.

Fehr, W. R., Caviness, C. E., Burmood, D. T., and Pennington, S. S. (1971) *Crop Sci.* **11**, 929–931.

Foo, E. L., and Varma, A. K. (1976). *Folia Microbiol.* **21**, 315–319.

Franco, A. A. (1978). *In* "Limitations and Potential for Biological Nitrogen Fixation in the Tropics" (J. Döbereiner, R. H. Burris, and A. Hollaender, eds.), pp. 161–171. Plenum, New York.

Franco, A. A., and Vincent, J. M. (1976). *Plant Soil* **45**, 27–48.

Fraser, M. E. (1966). *J. Appl. Bacteriol.* **29**, 587.

Fraser, M. E. (1975). *J. Appl. Bacteriol.* **39**, 345–351.

Fred, E. B., Baldwin, I. L., and McCoy, E. (1932). "Root Nodule Bacteria and Leguminous Plants," pp. 229–256. Univ. of Wisconsin, Madison.

Gaur, Y. D., Sen, A. N., and Subba Rao, N. S. (1980). *Plant Soil* **54**, 313–316.

Gibson, A. H. (1977). *In* "A Treatise on Dinitrogen Fixation" (R. W. F. Hardy and A. H. Gibson, eds.), Section IV, pp. 393–450. Wiley, New York.

Gibson, A. H., Date, R. A., Ireland, J. A., and Brockwell, J. (1976). *Soil Biol. Biochem.* **8**, 395–402.

Graham, P. H., Morales, V. M., and Cavallo, R. (1974). *Turrialba* **24**, 47–50.

Habish, H. A., and Ishag, H. M. (1974). *Exp. Agric.* **10**, 45–50.

Habte, M., and Alexander, M. (1978). *Soil Biol. Biochem.* **10**, 1-6.

Hagedorn, C. (1979). *Soil Sci. Soc. Am. J.* **43**, 515-519.

Hale, C. N., and Mathers, D. J. (1977). *N. Z. J. Agric. Res.* **20**, 69-73.

Ham, G. E., Cardwell, B. V., and Johnson, H. W. (1971). *Agron. J.* **63**, 301-303.

Hardaker, J. M., and Hardwick, R. C. (1978). *Exp. Agric.* **14**, 17-21.

Hardarson, G., and Jones, D. G. (1979). *Ann. Appl. Biol.* **92**, 229-236.

Hardy, R. W. F., and Havelka, U. D. (1976). *In* "Symbiotic Nitrogen Fixation in Plants" (P. S. Nutman, ed.), pp. 421-439. Cambridge Univ. Press, London and New York.

Harris, S. C., ed. (1979). "Planning an International Network of Legume Inoculation Trials." NIFTAL Project, Univ. of Hawaii.

Hiltbold, A. E., Thurlow, D. L., and Skipper, H. D. (1980). *Agron. J.* **72**, 675-681.

Holl, F. B. (1975). *Euphytica* **24**, 767-770.

Iswaran, V., Sen, A., and Apte, R. (1972). *Curr. Sci.* **41**, 299.

John, K. P. (1966). *J. Rubber Res. Inst. Malaya* **19**, 173-175.

Johnson, H. W., Means, U. M., and Weber, C. R. (1965). *Agron. J.* **57**, 179-185.

Jones, D. G., and Hardarson, G. (1979). *Ann. Appl. Biol.* **92**, 221-228.

Kandaswamy, R., and Prasad, N. N. (1971). *Curr. Sci.* **40**, 496.

Kang, B. T. (1975). *Exp. Agric.* **11**, 23-31.

Kapusta, G., and Rouwenhorst, D. L. (1973). *Agron. J.* **65**, 916-919.

Keyser, H. H., and Munns, D. N. (1979). *Soil Sci. Soc. Am. J.* **43**, 500-503.

Labandera, C. A., and Vincent, J. M. (1975). *Plant Soil* **42**, 327-347.

Leiderman, J. (1971). *Rev. Ind. Agric. Tucumán* **48**, 51-58.

Maier, R. J., and Brill, W. J. (1978). *Science* **201**, 448-450.

Marques Pinto, C., Yao, P. Y., and Vincent, J. M. (1974). *Aust. J. Agric. Res.* **25**, 317-329.

Materón, L. A., and Vincent, J. M. (1980). *Field Crops Res.* **3**, 215-224.

Mytton, L. R., El-Sherbeeny, M. H., and Lawes, D. A. (1977). *Euphytica* **26**, 785-791.

Mytton, L. R., McAdam, N. J., and Portlock, P. (1978). *Soil Biol. Biochem.* **10**, 79-80.

Nelson, D. W., Swearingin, M. L., and Beckham, L. S. (1978). *Agron. J.* **70**, 517-518.

Nutman, P. S. (1976). *In* "Symbiotic Nitrogen Fixation in Plants" (P. S. Nutman, ed.), pp. 211-237. Cambridge Univ. Press, London and New York.

Osa-Afiana, L. O., and Alexander, M. (1979). *Soil Sci. Soc. Am. J.* **43**, 925-930.

Paczkowski, M. W., and Berryhill, D. L. (1979). *Appl. Environ. Microbiol.* **38**, 612-615.

Pena-Cariales, J. J., and Alexander, M. (1979). *Soil Sci. Soc. Am. J.* **43**, 962-966.

Peterson, H. L. (1979). *In* "World Soybean Research Conference. II:Abstracts" (F. T. Corbin, ed.), p. 14. North Carolina State University, Raleigh.

Ramirez, C., and Alexander, M. (1980). *Appl. Environ. Microbiol.* **40**, 492-499.

Rensburg, H. J., van, and Strijdom, B. W. (1980). *Soil Biol. Biochem.* **12**, 353-356.

Reyes, V. G., and Schmidt, E. L. (1979). *Appl. Environ. Microbiol.* **37**, 854-858.

Rizk, S. G., Farag, F. A., El Nady, M. A., and Lofti, M. (1975). *Agric. Res. Rev.* **53**, 173-179.

Roberts, G. P., Leps, W. T., Silver, L. E., and Brill, W. J. (1980). *Appl. Environ. Microbiol.* **39**, 414-422.

Roughley, R. J. (1976). *In* "Symbiotic Nitrogen Fixation in Plants" (P. S. Nutman, ed.), pp. 125-136. Cambridge Univ. Press, London and New York.

Schiffman, J., and Alper, Y. (1968). *Exp. Agic.* **4**, 219-226.

Schmidt, E. L. (1974). *Soil Sci.* **118**, 141-149.

Schmidt, E. L. (1979). *Annu. Rev. Microbiol.* **33**, 355-376.

Schwinghamer, E. A., and Brockwell, J. (1978). *Soil Biol. Biochem.* **10**, 383-387.

Scudder, W. T. (1975). *Proc. Soil Crop Sci. Soc. Fl.* **34**, 79-82.

Seetin, M. W., and Barnes, D. K. (1977). *Crop Sci.* **17**, 783-787.

Sekhon, H. S., Kaul, J. N., and Dahiya, B. S. (1978). *J. Agric. Sci.* **90**, 325-327.

Semu, E., Hume, D. J., and Corke, C. T. (1979). *Can J. Microbiol.* **25,** 739–745.

Singh, H. P., and Tilak, K. V. (1977). *Indian J. Agron.* **22,** 59–61.

Skipper, H. D., Palmer, J. H., Giddens, J. E., and Woodruff, J. M. (1980). *Agron. J.* **72,** 673–674.

Smith, G. R., Peterson, H. L., and Knight, W. E. (1980). *Proc. Forage Grassl. Conf.* pp. 55–64.

Smith, R. S., and Miller, R. H. (1974). *Agron. J.* **66,** 564–567.

Stamford, N. P., and Neptune, A. M. L. (1978). *In* "Limitations and Potentials for Biological Nitrogen Fixation in the Tropics" (J. Döbereiner, R. H. Burris, and A. Hollaender, eds.), pp. 337–338. Plenum, New York.

Strijdom, B. W., and Deschodt, C. C. (1976) *In* "Symbiotic Nitrogen Fixation in Plants" (P. S. Nutman, ed.), pp. 151–168. Cambridge Univ. Press, London and New York.

Vandecaveye, S. C. (1927). *Soil Sci.* **23,** 355–362.

Vincent, J. M. (1970). "A Manual for the Practical Study of the Root-Nodule Bacteria," pp. 105–112. Blackwell, Oxford,

Vincent, J. M. (1974). *In* "The Biology of Nitrogen Fixation" (A. Quispel, ed.), pp. 265–341. North-Holland Publ., Amsterdam.

Vincent, J. M. (1977). *In* "Exploiting the Legume-*Rhizobium* Symbiosis in Tropical Agriculture" (J. M. Vincent, A. S. Whitney, and J. Bose, eds.), pp. 447–456. University of Hawaii.

Wacek, T. J., and Alm, D. (1978). *Crop Sci.* **18,** 514–515.

Wacek, T. J., and Brill, W. J. (1976). *Crop Sci.* **16,** 519–523.

Weaver, R. W., and Frederick, L. R. (1974). *Agron. J.* **66,** 233–236.

Wynne, J. C., Elkan, G. H., Meisner, C. M., Schneeweis, T. J., and Ligon, J. M. (1980). *Agron. J.* **72,** 645–649.

Zablotowicz, R. M., Russell, S. A., and Evans, H. J. (1980). *Agron. J.* **72,** 645–649.

Zobel, R. W. (1980). *In* "World Soybean Research Conference II: Proceedings" (F. T. Corbin, ed.), pp. 73–87. Westview Press, Boulder, Colorado.

Index

A

Agriculture
 application of *Rhizobium* in, 312–324
 factors limiting inoculation in, 327–328
 significance of inoculation in, 324–327
Agrobacterium
 relationship to *Rhizobium*, 11–12
 taxonomy of, 2–5
Agrobacterium rhizogenes
 biology of
  host range, 111
  physiology of root infection, 107–110
  relationship to *A. tumefaciens*, 106–107
  taxonomy, 110–111
 future prospects
  biology of *A. rhizogenes*, 122
  molecular biology of *A. rhizogenes*, 122–123
Agrobacterium tumefaciens
 agricultural control of, 54
 chemical treatments and, 55–56
 control of crown gall disease
  biological methods, 35–38
  chemotherapy and soil fumigation, 38–40
  sanitation and cultural management, 40–41
 disease cycle
  dissemination and inoculation, 24–25
  infection, 26–32
  inoculum, 16–24
  multiplication of organisms, 33–35
  penetration and conditioning of host tissue, 25–26
 evaluation of present methods of control, 56
 host range, 50
 mode of action of agrosin on, 54–55
 relationship to *A. rhizogenes*, 106–107
 virulence and avirulence of, 52–54
Agrosin, mode of action on *A. tumefaciens*, 54–55
Ammonia, incorporation and translocation, 182–183
Arginine, catabolism, Ti plasmid and, 88

C

Carbohydrate metabolism, bacteroid state and, 169–170
Cell surface, of *Rhizobium*, 142–144
Cell wall-cell membrane complex, bacteroid state and, 162
Chromosome mapping, in *Rhizobium*, 208–212
Clover-*Rhizobium trifolii* system, recognition in, 136
 factors affecting expression of clover and *R. trifolii* receptors, 141–142
 root adsorption studies and localization of clover lectin, 139–141
 serological studies and *in vitro* lectin binding, 137–139
 summary, 142
Conjugation, in *Rhizobium*, 196–205
Crown gall disease
 around the world, 48–49
 control of
  biological methods, 35–38
  chemotherapy and soil fumigation, 38–40
  sanitation and cultural management, 40–41
 factors affecting development of, 50–52
 recovery from, 71–73
Crown gall teratoma
 control of tumor morphology and, 65–70
 definition of teratomas, 63–64
 description of teratomas, 64–65
 tumorigenesis, 62
Crown gall tumorigenesis
 general, 84–85
 Ti plasmid-determined traits
  arginine catabolism, 88
  conjugative transfer of plasmid, 89–90
  host range, 88–89
  oncogenicity and tumor morphology, 85–86
  opine synthesis in tumor cells and catabolism in agrobacteria, 86–88
  sensitivity to agrocin 84 and exclusion of phage AP1, 89

D

Dinitrogen fixation
  chemistry of, 182
  components of, 181–182
  cost of, 183–185
Disease cycle, of *Agrobacterium tumefaciens*
  dissemination and inoculation, 24–25
  infection, 26–32
  inoculum, 16–24
  multiplication of organisms, 33–35
  penetration and conditioning of host tissue,
    25–26
DNA
  plasmid, transcription in bacteroids, 243
  Ti plasmid, genetic analysis of, 90–95

E

Electron transport systems, bacteroid state and,
  168–169

G

Genes, arrangement in *Rhizobium,* 212–213

H

Hairy root plasmid, molecular biology of
  conjugal transfer of plasmid, 122
  isolation and restriction endonuclease diges-
    tion of plasmid, 115–118
  large plasmid role in crown gall, 111–114
  large plasmid role in hairy root, 114–115
  plasmid promoters, 118–120
Heme synthesis, bacteroid state and, 167–168
Hormone(s), regulation of legume-*Rhizobium*
  exchanges by, 180–181

L

Lectin
  clover
    root adsorption studies and localization of,
      139–141
    serological studies and *in vitro* binding,
      137–139
  soybean
    binding to *R. japonicum,* 129–132
    tissue distribution, 132–134
Legumes, infection and nodule development in,
  150–151

Legume-*Rhizobium* exchanges
  ammonia incorporation and translocation,
    182–183
  cost of dinitrogen fixation, 183–185
  dinitrogen fixation and
    chemistry of, 182
    components of, 181–182
  root-shoot interactions
    hormonal regulation of, 180–181
    translocation, 180
Lipid metabolism, bacteroid state and, 170–171

M

Metabolite transport, cells associated with
  nodular vascular bundles, 286
  transfer cells, 286–289

N

Neoplastic diseases, suppression and recovery in
  chemical modification of tumor phenotype *in
    vitro,* 74–76
  habituation, 73–74
  temperature-sensitive tumor growth, 76–77
  teratocarcinoma, 77–78
Nitrogenase, bacteroid state and, 165–166
Nitrogen metabolism, bacteroid state and, 166–
  167
Nodules, mass and composition of, 185–186
Nonlegumes, *Rhizobium* nodules on, 289–293
Nucleic acids, bacteroid state and, 163–164

O

Opine(s), synthesis in tumor cells and
  catabolism in agrobacteria, 86–88

P

Parasponia spp., nodule development in, 151–
  152
Plant, genotype, *Rhizobium* bacteroid state and,
  157–158
Plasmids, *Rhizobium,* bacteroid state and, 159
Protein synthesis, bacteroid state and, 164–165

R

Rhizobiaceae, taxonomy of
  genus *Agrobacterium,* 2–5

genus *Rhizobium,* 6–11
relationship between *Agrobacterium* and
    Rhizobium, 11–12
Rhizobiophages, genetics of
  maps of phage genome, 216–217
  phage-bacterium interaction, 214–216
  transfection, 217
Rhizobium
  application in agriculture
    history, 312
    inoculant quality, 323–324
    need for inoculation, 313–315
    selection of *Rhizobium,* 315–319
    techniques for inoculation, 319–323
  arrangement of genes in
    chromosomal, 212–213
    plasmid, 213
  cell surface of, 142–144
  chromosomal mapping
    construction of linkage maps, 208–210
    homology of chromosomes of different
      Rhizobium species, 210–211
    localization of nonselectable mutations,
      211
    replication mapping, 212
  evidence for plasmid control of symbiotic
    properties, 225–227
  evidence for presence of large plasmids
    characterization of plasmids, 231–234
    isolation procedures, 227–231
    presence in fast-growing rhizobia, 234–235
  factors limiting inoculation in agriculture,
    327–328
  gene transfer systems
    conjugation, 196–205
    transduction, 207–208
    transformation, 205–207
  genetic methods for plasmid studies, 235
    genetic markers, 236
    transfer and curing, 237–238
    transposon-induced mutagenesis, 236–237
  genotype, bacteroid state and, 158–159
  mutants
    auxotrophs, 304–306
    contaminants mistaken for mutants, 302
    drug-resistant, 302–304
    glutamine synthetase, 306–307
    interesting mutants retaining symbiotic
      nitrogen-fixing ability, 307–308
    isolation, 192, 194–196
    pleiotrophic carbohydrate-negative, 307
    polysaccharide-deficient, 307

    spontaneous derivatives, 300–301
    symbiotic mutants obtained by direct isola-
      tion, 301–302
  nodules on nonlegumes, 289–293
  plasmid control of early functions in sym-
    biosis
    host range specificity, 238–239
    nodule formation, 239–240
  plasmid control of late functions in symbiosis,
    240–241
    fix genes, 241
    hup genes, 243
    nif genes, 241–243
    transcription of plasmid DNA in bacteroids,
      243
  relationship to *Agrobacterium,* 11–12
  significance of inoculation in agriculture, 324
  response in soils containing host-
    compatible rhizobia, 326–327
  response in soils lacking host-compatible
    rhizobia, 325–326
  strain construction in, 217–218
  taxonomy of, 6–11
Rhizobium bacteroid state
  biochemistry and physiology
    bacteroid environment, 160–161
    cell wall-cell membrane complex, 162
    electron transport systems, 168–169
    heme synthesis, 167–168
    lipid metabolism, 170–171
    nitrogenase, 165–166
    nitrogen metabolism, 166–167
    nucleic acids, 163–164
    protein synthesis, 164–165
    respiration and carbohydrate metabolism,
      169–170
    senescence, 171
    stages of bacteroid development, 159–160
    viability, 161–162
  genetic aspects of
    concepts of infectiveness and effectiveness,
      156
    effects of plant genotype, 157–158
    effects of *Rhizobium* genotype, 158–
      159
    Rhizobium plasmids, 159
  structural aspects of
    infection and nodule development in
      legumes, 150–151
    mature bacteroid structure, 152–156
    nodule development in *Parasponia* spp.,
      151–152

Rhizobium japonicum, in vitro binding to soybean lectin, 129–132
Rhizobium trifolii, in vitro binding to clover lectin, 137–139
Root cortex
  role of
    early symbiotic development, 277–281
    formation of unwalled regions, 277
    late symbiotic development, 281
    mitotic activity and nodule development, 256–266
    polyploid cells and nodule development, 266–270
    release of rhizobia from infection threads, 270–275
    release of rhizobia in nodules lacking infection threads, 275
    senescence of infected cells, 281–282
    uninfected cells of central tissue, 282–284
Root hair, invasion of, 248
  deformation and infection, 249
  infection thread structure and growth, 254–256
  ontogeny of infection thread, 249–254
Root nodules, ineffective, 284–286

S

Soybean-*Rhizobium japonicum* system, recognition in

other aspects of, 134–135
summary, 135–136
tissue distribution of soybean lectin, 132–134
in vitro lectin binding to *R. japonicum*, 129–132

T

T-DNA
  evidence for DNA transfer, 95–96
  experimental approaches to study of T-DNA sequences, 96–97
    integration of T-DNA, 98–100
    reversion from neoplastic state, 100
    T-DNA organization, 97–98
Ti plasmid
  conjugative transfer of, 89–90
  DNA, genetic analysis of, 90–95
  traits determined by, 85–90
Transduction, in *Rhizobium,* 207–208
Transformation, in *Rhizobium,* 205–207
Translocation, legume-*Rhizobium* exchanges and, 180
Tumor, phenotype, chemical modification *in vitro,* 74–76
Tumor cell, potential of, 60–61
Tumorigenesis, crown gall teratoma and, 62